D1084619

to Frieda

Library of Congress Cataloging in Publication Data

Barna, Arpad.
   Integrated circuits in digital electronics.

   Includes bibliographical references. 1. Digital electronics. 2. Integrated circuits. I. Porat, Dan I., joint author. II. Title.
TK7868.D5B43   621.3819′58′35   73-6709

ISBN 0-471-05050-4

Printed in the United States of America

10 9 8 7 6 5 4 3 2 1

Preface

The rapid expansion in the availability of integrated circuits, particularly of large-scale integrated circuits, presents a complex problem to the logic designer who has to choose from a broad variety of device technologies and from circuits performing specific logic functions. To the novice it also presents an overwhelming problem, especially because logic design today is as much an art as it is a science: some of the most ingenious solutions are arrived at by a combination of insight and solid theory.

Several excellent texts are available covering the theoretical bases for the design of digital electronics, while at the practical end a plethora of devices is forthcoming from a dynamic and creative integrated circuits industry. This text aims to provide a link between elementary theory and practical applications. This book grew out of a course on Digital Circuits given at the University of Hawaii, to Electrical Engineering and Computer Science students. It also serves a broad spectrum of engineers and scientists who have a need to acquaint themselves with the latest techniques in digital equipment design utilizing integrated circuits, including large-scale integration.

Each chapter of the book is written to stand on its own with a minimum of cross-references added to guide the beginner. The reader with a background in digital techniques may single out the chapter(s) of his interest without concern for continuity. One chapter (Chapter 11) is devoted entirely to large-scale integrated circuits (LSI) starting with the description of various principles of device operation, proceeding to LSI arrays, and finally illustrating their application in digital systems.

Over 400 examples and problems with step-by-step procedures utilizing up-to-date components facilitate self-study and help the reader evaluate his success with which he has absorbed the new material. Fundamental principles are included in many of the examples to prepare the reader in using new and more complex devices as they become available; problems at the end of the chapters consolidate the learned material and broaden its scope. Answers to selected problems are given at the end of the book.

We wish to acknowledge the stimulating atmosphere provided by the Stanford Linear Accelerator Center of Stanford University; the quest for

CHAPTER 11
Large Scale Integrated Circuits   361

CHAPTER 1

Perspective

1.1 Digital Electronics

An electronic *system* is an ensemble of interacting electronic building blocks that comprise a functional unit. A *digital* electronic system performs the functions by dealing with discrete quantities, e.g., integer or fractional numbers. In contrast there are *analog* electronic systems which deal with continuously varying quantities such as temperature, pressure, acceleration. Many functions may be implemented by either a digital or an analog system. For example, a digital system may be utilized to measure continuous quantities provided that the least discrete step, or quantum, is sufficiently small to approach the final value with the desired accuracy. Conversely, manipulation of discrete quantities may be effected by use of analog electronic systems; the result is then obtained in analog form to the nearest least significant integer. The analog manipulation must have an accuracy and resolution that is consistent with the number of digits representing the discrete quantities.

Conceptually, we distinguish between continuous (analog) and discrete (digital) quantities in the macroscopic world. Modern physics has, however, discovered that microscopic phenomena may be entirely described "digitally" with help of a quantum theory: the emission or absorption of energy in the molecular, atomic, nuclear, or subnuclear regime is indeed effected in small discrete units, i.e., in quanta.

In digital electronic systems information is represented by binary digits, *bits*. A bit may assume either one of two values: 0 or 1. A mathematical system that uses two digits only is called a *binary system*; its foundations were laid by the British mathematician George Boole (1815–64) in his classic treatise "An Investigation of the Laws of Thought" (1854). This theory, based on the Aristotelian logic concepts of "true" (represented by a binary 1) and "false" (represented by a binary 0), found a practical application 84 years later when Shannon developed a *switching theory*

1

based on Boolean algebra.[1] In the late 1930's switching devices consisted mainly of relays since the vacuum tube was not deemed sufficiently reliable to serve as a component in large switching networks, such as telephone systems. The successful application of Boolean algebra to the analysis of switching networks may be viewed as one of the important breakthroughs in processing digital information.

Much emphasis was placed then on minimizing the number of components (relays) utilized in a switching network. A minimal network was defined as a circuit that required the least number of components to perform a specified function. Considering the cost, power consumption, and reliability of these components, the criteria of minimality were very appropriate.

The same criteria were applied in the synthesis of digital systems using vacuum tubes as building blocks. For example, one of the earliest computers, the University of Pennsylvania's ENIAC, used as many as 18,000 vacuum tubes. The statistical "mean time between failures" was not high, as a result of the limited lifetime of the vacuum tube.

The invention of the transistor by Bardeen, Brattain, and Shockley may be viewed as another major breakthrough in digital information processing. The volume occupied by a transistor is about one hundredth of that occupied by either a relay or a vacuum tube, and a similar ratio holds for the power dissipated by the devices under comparison. But, foremost, the transistor provided the technological base for the development of *integrated circuits*, an ensemble of many transistors which are interconnected to perform a specified function.

1.2   Integrated Circuits

The first integrated circuit was developed by J. S. Kilby in 1958, which may be marked as a start of a revolutionary change in the art of electronics design. It has been estimated that 50% of integrated circuits used today were not available only three years ago.

In the initial stages of transistor technology development, the devices were classified as either *discrete* or *integrated*. A digital integrated circuit consisted of several transistors and sometimes also of passive components (resistors, capacitors) that were interconnected to perform a logical function. Such logical functions implementing simple *logical operations* provide the building blocks for realization of switching networks. Further advances in technology enabled fabrication of many transistors on one semiconductor substrate. The level of complexity increased, broadening the scope of the circuits contained in one package. These new circuits could count, perform addition or subtraction of two binary digits, compare their binary values, and realize more complex logical operations.

Scientists and engineers engaged in the development of the transistor technology met the challenge of the next level of complexity. The individual components on a semiconductor substrate were further decreased in dimensions and similarly their power dissipation was reduced. Devices with some 30,000 transistors on one substrate became a reality, and practical limitations were now set primarily by the input and output terminals to the "outside" world rather than by the complexity and number of transistors on a substrate.

Thus today we differentiate between three levels of integrated circuit complexity. (i) small scale integration, *SSI*, which are circuits implementing a few simple logical functions. (ii) medium scale integration, *MSI*, which have the capacity to perform the equivalent of 12 to 100 logical functions; through suitable interconnection these functions are more complex and their potential variety is great. (iii) large scale integration, *LSI*, provides the highest level of complexity.

Digital circuits are being used in such diverse fields as information storage, medical instrumentation, process control, calculators and computers, air traffic and digital communications, and in voice and tone synthesis. Digital techniques are also penetrating into areas that were traditionally solved with analog techniques. An example will clarify the point. Assume it is desired to send analog information from point A to point B, where the value of an electric potential is made proportional to a measured analog quantity. The transmission medium may be a cable or a modulated radio-frequency carrier. In either case, electrical interference signals are present that add to the information signal modifying its content. Suppose that an accuracy of 1 part in 10^4 is required in the information signal. The interference must be kept below that value, or special techniques must be used to remove at the receiving end the unwanted signal component. On the other hand, consider that a signal is being transmitted as a pulse train which, via its sequence of present or absent pulses in the train, represents information in binary form. It is evident that this communication link can be more noise immune since no admixture of noise will change the information content unless its energy becomes comparable to the energy contained in the information pulse.

Higher power efficiencies of digital systems provide an additional reason for their applications in areas formerly dominated by analog techniques. Consider a transistor used as a switch in a digital circuit; the power dissipated in it is $P_D = I$ amperes $\times V$ volts. When the transistor switch is in the OFF state, I and hence P_D approach zero. Similarly, when the transistor switch is in the ON state, V and hence P_D also approach zero. Power is thus dissipated only during transitions between the ON and OFF states and in the load when the transistor is in the ON state. In contrast, in analog applications both I and V have nonzero values and the transistor dissipates power continuously.

Concurrent with the reduction in size and power consumption there has been a considerable increase in the operating speed of the new semiconductor devices. It is sometimes difficult to fully appreciate the revolutionary changes that were made possible through the invention of the transistor and the subsequent development of integrated circuits. For example, the largest early computers occupied a volume of hundreds of cubic meters and required many tens of kilowatts of electrical power and a sizeable air-conditioning installation to allow this amount of energy to be dissipated without raising the room temperature to unbearable values. The comparable computational capacity is obtained today in a desk-size computer dissipating one hundredth of the former's power.

To illustrate this point in a more quantitative manner let us establish a figure of merit, M, which expresses the speed × power product of a switching device. Short switching times will allow a greater number of switching operations to be performed by the device in a unit time, while a low power dissipation is also a measure of its efficiency. Thus a low value of M is desirable. For relays operating at a switching speed of about 10^{-2} second and a power dissipation of about 0.25 watt, $M = 10^{-2}$ second × 0.25 watt = 2.5×10^{-3} joules. The figure of merit decreases in vacuum tubes because of an appreciable decrease in switching time, although power consumption increases tenfold in comparison with relays. A reasonable average figure of merit for tubes is $M = 10^{-6}$ second × 2 watts = 2×10^{-6} joules. In integrated circuits, values of M range from 10^{-9} second × 5×10^{-2} watts = 5×10^{-11} joules down to 10^{-8} second × 2.5×10^{-4} watts = 2.5×10^{-12} joules. Thus the M value for tubes is about three orders of magnitude lower (better) than for relays, while integrated circuits outperform tubes by a factor of 10^5 to 10^6.

The preceding discussion was presented to impart to the reader a perspective of the enormous changes brought about by the new technology. Common to all technologies applied in digital systems is their utilization of switching devices. A switch has two states, either OFF or ON, which may be associated with the binary numbers 0 and 1; and a switching algebra may be applied for a systematic synthesis or analysis of switching circuits. The binary nature of digital circuits requires also an understanding of the binary number system, and for this we turn to Chapter 2.

CHAPTER 2

Number Systems

In this chapter we discuss number systems, and in particular we describe the various ways that numbers can be expressed in a binary representation, i.e., in a number system that uses two digits only: 0 and 1. The binary representation of information is generally utilized in computers and other digital systems because the basic elements from which such systems are built are of a binary nature. An understanding of the binary number representation is a prerequisite for most of the subsequent material and is specifically required as a background for Chapter 8 in which implementation of digital arithmetic circuits is described.

2.1 Positional Notation

Representation of numbers in positional notation has received such universal acceptance that we tend to overlook its importance and the long historical evolution that brought us to this point. When we are given a sequence of decimal digits, e.g., 40795, we attach to each digit a multiplying factor, or *weight* that is some power of 10 and which depends on the position of the digit in the number; for example, $40795 = 4 \times 10^4 + 0 \times 10^3 + 7 \times 10^2 + 9 \times 10^1 + 5 \times 10^0$.

Thus every decimal number of n digits is a sum of the weighted coefficients

$$N_{10} = a_{n-1}(10)^{n-1} + a_{n-2}(10)^{n-2} + \cdots + a_1(10)^1 + a_0(10)^0 = \sum_{i=0}^{n-1} a_i(10)^i,$$

(2.1)

where a_i is the coefficient of the weight 10^i.

The positional notation is a short representation of eq. (2.1) in which the plus signs and the weights have been omitted. Thus

$$N_{10} = a_{n-1}a_{n-2} \cdots a_1 a_0.$$

(2.2)

5

Representation of numbers as shown in eq. (2.2) is barely ten centuries old. Early man used tally marks for counting. The Romans utilized a decimal system in which the position of the digit had some limited significance. The symbols used were I(1), V(5), X(10), L(50), C(100), CIↃ(1000), CCIↃↃ(10^4), and CCCIↃↃↃ(10)[5]. Though an improvement over the tally system, arithmetic with Roman numbers was extremely cumbersome. Notice that the symbol for 0 is absent in this system. It appears that the Mesopotamians understood the concept of 0 and had a positional system using 60 symbols. This system was abandoned around 1700 B.C. although its influence is still witnessed today: 60 seconds in a minute, 60 minutes in one hour, 360 degrees in a circle. The decimal system with a positional notation was developed by the Hindus around the 5th century and was introduced to western civilization by the Arabs who also added the symbol 0.

Referring back to eq. (2.2) we notice the general form of this positional representation. The base, or *radix*, 10 is only implied in that representation. We could have indeed used any other radix r. Equation (2.1) in its general form may thus be written

$$N_r = a_{n-1}(r)^{n-1} + a_{n-2}(r)^{n-2} + \cdots + a_1(r)^1 + a_0(r)^0 = \sum_{i=0}^{n-1} a_i(r)^i, \quad (2.3)$$

where r is the base in which the number is represented. r different symbols are required in a number system of radix r having a range of integers from 0 to $(r-1)$ as shown in Figure 2.1 for $r = 2, 3, 8, 10, 12$, and 16. Notice that for $r > 10$ we had to introduce additional symbols.

Radix r	Number system	r Digits used in number system
2	Binary	0, 1
3	Ternary	0, 1, 2
8	Octal	0, 1, 2, 3, 4, 5, 6, 7
10	Decimal	0, 1, 2, 3, 4, 5, 6, 7, 8, 9
12	Duodecimal	0, 1, 2, 3, 4, 5, 6, 7, 8, 9, α, β
16	Hexadecimal	0, 1, 2, 3, 4, 5, 6, 7, 8, 9, A, B, C, D, E, F

FIGURE 2.1.   Digits (symbols) used in several number systems.

The above discussion can be easily extended to express fractional numbers. In the decimal system we represent a fractional number by the digits that are placed to the right of the *decimal point*. This point of division between the integer and fractional parts of a number in any base is called the *radix point*. For example, the octal number (base 8) .1472 has the value

$1 \times 8^{-1} + 4 \times 8^{-2} + 7 \times 8^{-3} + 2 \times 8^{-4}$. In general, the value of any fractional number n_r to the base r having m digits after the radix point is

$$n_r = a_{-1}(r)^{-1} + a_{-2}(r)^{-2} + \cdots + a_{-(m-1)}(r)^{-m+1} + a_{-m}(r)^{-m} = \sum_{i=-m}^{-1} a_i(r)^i.$$

$$(2.4)$$

Combining eqs. (2.3) and (2.4) we obtain a general expression for any integer, fractional, or mixed number of $(n + m)$ digits in base r

$$N = N_r + n_r = [a_{n-1}(r)^{n-1} + \cdots + a_0(r)^0 . a_{-1}(r)^{-1} + \cdots + a_{-m}(r)^{-m}]_r$$

$$= \sum_{i=-m}^{n} a_i(r)^i. \qquad (2.5)$$

Note the radix point between the r^0-th and the r^{-1}-th terms above.

A subscript after a number will be used to indicate the radix of that number. For example 143_{10} is a number in base 10, while 143_{16} and 143_8 are numbers in base 16 and 8, respectively. All three numbers have different numerical values, each given by eq. (2.5).

2.2 The Binary Number System

In the binary number system $r = 2$, thus the two digits used to represent any number are 0 and 1. The digit 2 does not exist in this system. However, the numerical equivalent of 2 represented in the binary system can be obtained by application of simple arithmetic rules, remembering that a number greater than 1 generates a *carry*. Thus $1 + 1$ equals 0, carry 1, i.e., $1 + 1 = 10_2$; $10_2 + 10_2 = 100_2 = 4_{10}$; etc. Figure 2.2 shows decimal numbers 0 through 20_{10} in column 1 and their corresponding binary, octal, and hexadecimal numbers in column 2, 3, and 4, followed by selected numbers up to 1000_{10}.

The common use of binary and decimal numbers in digital systems requires procedures for converting a number given in one base to an equivalent number in another base. These will be discussed in detail in Section 2.4. In what follows here, a few simple binary-to-decimal and decimal-to-binary conversion methods are described.

Binary-to-Decimal Conversion. Two binary-to-decimal conversion methods will be given; the first one of these is based on eqs. (2.3)–(2.5).

Decimal	Binary	Octal	Hexadecimal
0	0	0	0
1	1	1	1
2	1 0	2	2
3	1 1	3	3
4	1 0 0	4	4
5	1 0 1	5	5
6	1 1 0	6	6
7	1 1 1	7	7
8	1 0 0 0	10	8
9	1 0 0 1	11	9
10	1 0 1 0	12	A
11	1 0 1 1	13	B
12	1 1 0 0	14	C
13	1 1 0 1	15	D
14	1 1 1 0	16	E
15	1 1 1 1	17	F
16	1 0 0 0 0	20	10
17	1 0 0 0 1	21	11
18	1 0 0 1 0	22	12
19	1 0 0 1 1	23	13
20	1 0 1 0 0	24	14
32	1 0 0 0 0 0	40	20
50	1 1 0 0 1 0	62	32
60	1 1 1 1 0 0	74	3C
64	1 0 0 0 0 0 0	100	40
100	1 1 0 0 1 0 0	144	64
255	1 1 1 1 1 1 1 1	377	FF
1000	1 1 1 1 1 0 1 0 0 0	1750	3E8

FIGURE 2.2.   Representation of selected numbers in various number systems.

EXAMPLE 2.1.   Convert to base ten the binary integer number $N_2 = 110101_2$.

From eq. (2.3), this number represents the sum

$$N_2 = 1 \times (2)^5 + 1 \times (2)^4 + 0 \times (2)^3 + 1 \times (2)^2 + 0 \times (2)^1 + 1 \times (2)^0$$
$$= (32 + 16 + 4 + 1)_{10} = 53_{10}.$$

EXAMPLE 2.2.   Convert to base ten the binary fraction $n_2 = .1101_2$.

From eq. (2.4),

$$n_2 = 1 \times (2)^{-1} + 1 \times (2)^{-2} + 0 \times (2)^{-3} + 1 \times (2)^{-4}$$
$$= (0.5 + 0.25 + 0.0625)_{10} = (.8125)_{10}.$$

Another binary-to-decimal conversion technique is the so called *double-and-add* method: starting with the most significant binary digit (bit*), double the 1 present in that bit and proceed to the next bit. If the next bit is 0, double the number obtained in the previous step; if it is 1, add a 1 after the doubling. Continue until the least significant bit has been processed.

EXAMPLE 2.3. Convert to base 10 the binary number 101101_2. The process is shown in Figure 2.3, yielding 45_{10}.

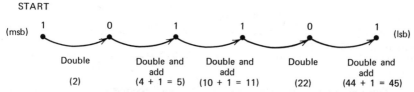

FIGURE 2.3. Binary-to-decimal conversion using the *double-and-add* method.

Decimal-to-Binary Conversion. One method for decimal-to-binary conversion of small numbers is based on recognition of the sum of powers of 2 contained in the decimal number. The largest power of 2 is recognized first and subtracted from the decimal number. The same method is repeated on the difference obtained until the least significant bit has been processed.

EXAMPLE 2.4. Convert to binary the decimal number 81_{10}.
The highest power of 2 contained in 81_{10} is $2^6 = 64_{10}$. Thus $(81 - 64)_{10} = 17_{10}$. The highest power of 2 contained in 17_{10} is $2^4 = 16_{10}$. Thus $(17 - 16)_{10} = 1$. The remainder represents the least significant bit, $1 = 2^0$. Thus $81_{10} = 1 \times 2^6 + 1 \times 2^4 + 1 \times 2^0$, and in positional notation $81_{10} = 1010001_2$.

The method of Example 2.4 is cumbersone for large numbers. Another algorithm, that will be developed in Section 2.4, is given here: The decimal number is divided by 2; the first remainder, 0 or 1, represents the least significant bit (lsb); the result is further divided by 2, the remainder representing the next bit (weight = 2^1), etc.

---

* The word *bit* is a contraction of *binary digit*. *Most significant bit* is commonly abbreviated *msb*; *least significant bit* is abbreviated *lsb*.

EXAMPLE 2.5.   Convert to binary the decimal number 100_{10}.

	Remainder	Binary Weight
$100 \div 2 = 50$	0	2^0 (lsb)
$50 \div 2 = 25$	0	2^1
$25 \div 2 = 12$	1	2^2
$12 \div 2 = 6$	0	2^3
$6 \div 2 = 3$	0	2^4
$3 \div 2 = 1$	1	2^5
$1 \div 2 = 0$	1	2^6 (msb)

The answer is obtained by reading the remainder column from *bottom to top*: $100_{10} = 1100100_2$.

2.3   Ternary, Octal, and Hexadecimal Number Systems

The *ternary* number system is of interest because theoretically it is the most efficient radix for representation of numbers. Work in ternary logic has also been stimulated since this represents a first step toward a many-valued logic. One computer at least has been built in which several subsystems operate with ternary components. However, the reliability and low cost of binary devices at present greatly outweigh the advantages that may be derived by use of any other radix in digital systems.

One inconvenience associated with the use of binary representation of information is the large number of bits required to express a given quantity. This disadvantage can be easily removed by converting the binary number to another base that is also a power of 2. The *quaternary* system, $r = 2^2$, is one such possibility; but it would still require a large number of digits for representation of information.

In the *octal* number system $r = 2^3$; thus eight symbols, 0 through 7, are required to express all possible numbers in base 8. An octal number, N_8, of n integer and m fractional digits has the numerical value

$$N_8 = a_{n-1}(8)^{n-1} + a_{n-2}(8)^{n-2} + \cdots + a_1(8)^1 + a_0(8)^0$$
$$+ a_{-1}(8)^{-1} + \cdots + a_{-m}(8)^{-m}$$

$$= \sum_{i=-m}^{n} a_i(8)^i. \tag{2.6}$$

Selected octal numbers are shown in the third column of Figure 2.2.

Eight combinations may be represented by exactly three bits, 000 through 111. Thus conversion from binary to octal is obtained from the following rule: starting at the radix point divide the integer part and the fractional part of the binary number into groups of three bits. Each group represents an octal digit.

EXAMPLE 2.6. Convert the binary number $N_2 = 111101000.0111$ into an octal number.

$$\begin{array}{lccccc}
\text{Binary} & 111 & 101 & 000 & 011 & 1 \\
\text{Octal} & 7 & 5 & 0 & 3 & 4
\end{array}$$

We have assumed two additional 0's after the lsb, yielding $N_8 = 750.34_8$.

In the *hexadecimal* number system $r = 2^4$; thus sixteen different symbols are required to express all possible numbers in base 16. A hexadecimal number, N_{16}, of n integer and m fractional digits has the numerical value

$$N_{16} = a_{n-1}(16)^{n-1} + a_{n-2}(16)^{n-2} + \cdots + a_1(16)^1 + a_0(16)^0$$
$$+ a_{-1}(16)^{-1} + \cdots + a_{-m}(16)^{-m}$$
$$= \sum_{i=-m}^{n} a_i(16)^i. \tag{2.7}$$

The first letters of the alphabet have been generally accepted as the additional symbols, A representing ten, B representing eleven, ..., F representing fifteen. Selected hexadecimal numbers are shown in the fourth column of Figure 2.2.

Sixteen different combinations may be represented by exactly four bits. Thus we derive the following rule for binary to hexadecimal conversion: starting at the radix point, divide the integer part and the fractional part of the binary number into groups of four bits. Each group represents a hexadecimal digit.

EXAMPLE 2.7. Convert the binary number of Example 2.6 into a hexadecimal number.

$$\begin{array}{lccccc}
\text{Binary:} & 0001 & 1110 & 1000 & . & 0111 \\
\text{Hexadecimal:} & 1 & E & 8 & & 7
\end{array}$$

Thus, $111101000.0111_2 = 1E8.7_{16}$

Notice the drastic reduction in the number of digits required to express a given quantity as the radix is increased.

Conversion from octal or hexadecimal to a binary representation requires the converse process of that shown in Examples 2.6 and 2.7. With a little practice, conversion between the three number bases can be done by inspection.

2.4   Number Base Conversion

In this section we discuss a general method for conversion of a number given in one base (radix) to an equivalent number in any other desired base. Consider the decimal number N_{10} which has an integer part of n digits and a fractional part of m digits

$$N_{10} = d_{n-1}d_{n-2} \cdots d_1 d_0 . d_{-1}d_{-2} \cdots d_{-m}, \tag{2.8}$$

where eq. (2.8) is in positional notation, d_i representing the ith decimal digit. The numerical quantity given in eq. (2.8) may also be expressed in any other radix (number base); however, the number of digits will generally differ. For example, in base 2 the number N_2 will have s integer and t fractional digits (bits); in positional notation it is expressed

$$N_2 = b_{s-1}b_{s-2} \cdots b_1 b_0 . b_{-1}b_{-2} \cdots b_{-t}. \tag{2.9}$$

Number Base Conversion of Integers.   Conversion between bases must be done separately on the integer and the fractional parts. Thus, equating the integer parts of the decimal and binary numbers we have

$$N_{10}(\text{integer}) = b_{s-1}(2)^{s-1} + b_{s-2}(2)^{s-2} + \cdots + b_1(2)^1 + b_0. \tag{2.10a}$$

Factoring out 2 from the leading $(s-1)$ terms we obtain

$$N_{10}(\text{integer}) = 2[b_{s-1}(2)^{s-2} + b_{s-2}(2)^{s-3} + \cdots + b_1] + b_0. \tag{2.10b}$$

or

$$N_{10}(\text{integer}) = 2A_1 + b_0,$$

where A_1 is the polynomial in the square brackets of eq. (2.10b) and b_0 is the *remainder*. The next bit, b_1, is obtained by factoring 2 out of A_1

$$A_1 = 2A_2 + b_1.$$

Successive factoring of 2 yields

$$A_2 = 2A_3 + b_2$$
$$A_3 = 2A_4 + b_3$$
$$\vdots$$
$$A_i = 2A_{i+1} + b_i$$

where A_i is a polynomial one degree higher than A_{i+1} and b_i is the binary remainder. The sth step completes the process yielding the remainder b_{s-1}.

Notice that the remainders b_0 through b_{s-1} are the binary digits of the integer part of N_2 in eq. (2.9).

EXAMPLE 2.8. Convert to binary the decimal integer number 53.

$$53 \div 2 = 26, \text{ remainder } 1 = b_0 \text{ (lsb)}$$
$$26 \div 2 = 13, \text{ remainder } 0 = b_1$$
$$13 \div 2 = 6, \text{ remainder } 1 = b_2$$
$$6 \div 2 = 3, \text{ remainder } 0 = b_3$$
$$3 \div 2 = 1, \text{ remainder } 1 = b_4$$
$$1 \div 2 = 0, \text{ remainder } 1 = b_5 \text{ (msb)}$$

The binary equivalent of the decimal number is obtained by reading the remainders from *bottom to top:* $53_{10} = 110101_2$.

Number Base Conversion of Fractions. To convert the fractional part of a decimal number to a fractional binary number we have

$$N_{10}(\text{fractional}) = b_{-1}2^{-1} + b_{-2}2^{-2} + \cdots + b_{-t}2^{-t}. \qquad (2.11a)$$

Multiplication of eq. (2.11a) by 2 yields

$$2N_{10}(\text{fractional}) = b_{-1} + [b_{-2}2^{-1} + b_{-3}2^{-2} + \cdots + b_{-m}2^{-t+1}], \qquad (2.11b)$$

where b_{-1} is 0 or 1, and the expression in the square brackets is smaller than 1. t successive multiplications by 2 yield $b_{-1}, b_{-2}, \ldots, b_{-t}$, in that order. The process is completed when the fractional part is 0 after multiplication by 2, or it is terminated when the desired accuracy is attained.

EXAMPLE 2.9. Convert to binary the decimal fractional number $.39_{10}$.

$$.39 \times 2 = 0.78 \qquad b_{-1} = 0$$
$$.78 \times 2 = 1.56 \qquad b_{-2} = 1$$
$$.56 \times 2 = 1.12 \qquad b_{-3} = 1$$
$$.12 \times 2 = 0.24 \qquad b_{-4} = 0$$
$$.24 \times 2 = 0.48 \qquad b_{-5} = 0$$
$$.48 \times 2 = 0.96 \qquad b_{-6} = 0$$
$$.96 \times 2 = 1.92 \qquad b_{-7} = 1$$
$$.92 \times 2 = 1.84 \qquad b_{-8} = 1$$
$$.84 \times 2 = 1.68 \qquad b_{-9} = 1$$
$$.68 \times 2 = 1.36 \qquad b_{-10} = 1.$$

Terminating the process here, a *round-off error* is incurred having a value ϵ smaller than 2^{-10}, i.e., $\epsilon < 1/1024$. The binary equivalent of the decimal fraction is obtained by reading the b_i's from *top to bottom:* $.39_{10} = .0110001111_2 + \epsilon$.

Number Base Conversion of Mixed Numbers. The method is illustrated by a decimal to hexadecimal conversion.

EXAMPLE 2.10. Convert to hexadecimal the decimal number 53.39_{10}.

(i) Integer part
$$53 \div 16 = 3, \text{ remainder } 5 = h_0 \text{ (lsd)}$$
$$3 \div 16 = 0, \text{ remainder } 3 = h_1 \text{ (msd)}$$

where lsd and msd are the least and most significant digits, respectively, and h_0 and h_1 are the hexadecimal digits.

(ii) Fractional part

$.39 \times 16 =$	6.24	$h_{-1} = 6$
$.24 \times 16 =$	3.84	$h_{-2} = 3$
$.84 \times 16 =$	13.44	$h_{-3} = D \ (= 13_{10})$
$.44 \times 16 =$	7.04	$h_{-4} = 7.$

Thus $53.39_{10} = 35.63D7_{16}$.

Check. We may use the results of Examples 2.8 and 2.9 to check the results of the last example, since binary to hexadecimal conversion may be done by inspection. Thus

$$35.63D7_{16} = (11 \quad 0101 \quad . \quad 0110 \quad 0011 \quad 1101 \quad 0111)_2$$

The round-off error $\epsilon < 16^{-4} = 2^{-16} < 1/6.4 \times 10^4$.

Number Base Conversion between Numbers in any Bases. The procedure shown earlier may be also used for conversion between numbers of any bases. In general, *if a number in base* r_1 *is to be converted to base* r_2, *the conversion arithmetic must be carried out in base* r_1. It is convenient to establish addition and multiplication tables for base r_1 to aid in the conversion.

EXAMPLE 2.11. Convert 25223_6 to a decimal number. Base 6 arithmetic tables are shown in Figure 2.4.

Conversion of the integer number requires successive division by $10_{10} = 14_6$:

$$
\begin{array}{r}
1423_6 \\
14_6 \overline{)25223_6} \\
14 \\
\hline
112 \\
104 \\
\hline
42 \\
32 \\
\hline
103 \\
50 \\
\hline
\end{array}
$$

$$\text{remainder} = 13_6 = 9_{10} = \text{least significant digit.}$$

+	0	1	2	3	4	5
0	0	1	2	3	4	5
1	1	2	3	4	5	10
2	2	3	4	5	10	11
3	3	4	5	10	11	12
4	4	5	10	11	12	13
5	5	10	11	12	13	14

×	0	1	2	3	4	5
0	0	0	0	0	0	0
1	0	1	2	3	4	5
2	0	2	4	10	12	14
3	0	3	10	13	20	23
4	0	4	12	20	24	32
5	0	5	14	23	32	41

Base 10	Base 6
10	14
20	32
30	50
40	104
50	122
60	140
70	154
80	212
90	230

FIGURE 2.4. Base 6 arithmetic tables.

Next we divide the intermediate result 1423_6 by 14_6

$$
\begin{array}{r}
101_6 \\
14_6\overline{)1423_6} \\
14 \\
\hline
23 \\
14 \\
\hline
\end{array}
$$

$$\text{remainder} = 5_6 = 5_{10}.$$

Next step:

$$
\begin{array}{r}
3_6 \\
14_6\overline{)101} \\
50 \\
\hline
\end{array}
$$

$$\text{remainder} = 11_6 = 7_{10}.$$

Final step: $3_6 \div 14_6 = 0$, remainder $= 3_{10}$ = most significant digit.
Result: $25223_6 = 3759_{10}$.

The procedure for conversion of a fraction given in base r_1 to a fraction in base r_2 requires successive *multiplication* by the value of r_2 that must be expressed in base r_1.

2.5   Negative Numbers

Thus far we have dealt with positive integer and fractional numbers only. We have also established the constraint that all circuitry shall be implemented with 2-state devices representing 0 and 1. No circuit component is, however, available to represent the signs " + " and " − ". There are three commonly accepted ways of expressing negative binary numbers: sign-and-magnitude, 1's-complement, and 2's-complement representation.

Sign-and-Magnitude Representation.   In ordinary arithmetic a number such as -345.76 is represented by the *sign* " − " and the *magnitude* 345.76. Similarly, in binary *sign-and-magnitude* representation we express a number by a sign followed by the magnitude, or absolute value, of the number as shown in the second column of Figure 2.5. The sign that is placed to the left of the most significant bit represents a " + " when 0, and a " − " when 1. Thus a binary number N of n integer and m fractional bits is expressed in the sign-and-magnitude representation by $n+m+1$ bits

$$N \text{ (sign-and-magnitude)} = \underbrace{b_n 2^n}_{\substack{\text{sign} \\ \text{bit}}} + \underbrace{\sum_{i=0}^{n-1} b_i 2^i}_{\substack{\text{integral} \\ \text{number bits}}} + \underbrace{\sum_{i=-m}^{-1} b_i 2^i}_{\substack{\text{fractional} \\ \text{number bits}}} , \quad (2.12)$$

where the sign bit, b_n, is 0 for positive numbers and 1 for negative numbers. Collecting under one summation sign the bits representing the integral and fractional parts of the number, we obtain

$$N \text{ (sign-and-magnitude)} = b_n 2^n + \sum_{i=-m}^{n-1} b_i 2^i. \quad (2.13)$$

This method of numbers representation is easily readable since the magnitude part is identical for positive and negative numbers. However, it will be seen that implementation of digital arithmetic is not as easy as with other number representations.

Binary addition and subtraction are carried out in the same way as the

Decimal	Binary Sign-and-magnitude	Binary 1's-complement	Binary 2's-complement
− 8			1, 0 0 0
− 7	1, 1 1 1	1, 0 0 0	1, 0 0 1
− 6	1, 1 1 0	1, 0 0 1	1, 0 1 0
− 5	1, 1 0 1	1, 0 1 0	1, 0 1 1
− 4	1, 1 0 0	1, 0 1 1	1, 1 0 0
− 3	1, 0 1 1	1, 1 0 0	1, 1 0 1
− 2	1, 0 1 0	1, 1 0 1	1, 1 1 0
− 1	1, 0 0 1	1, 1 1 0	1, 1 1 1
− 0	0, 0 0 0	1, 1 1 1	0, 0 0 0
+ 0		0, 0 0 0	
1	0, 0 0 1	0, 0 0 1	0, 0 0 1
2	0, 0 1 0	0, 0 1 0	0, 0 1 0
3	0, 0 1 1	0, 0 1 1	0, 0 1 1
4	0, 1 0 0	0, 1 0 0	0, 1 0 0
5	0, 1 0 1	0, 1 0 1	0, 1 0 1
6	0, 1 1 0	0, 1 1 0	0, 1 1 0
7	0, 1 1 1	0, 1 1 1	0, 1 1 1

(a)

Decimal	Binary Sign-and-magnitude	Binary 1's-complement	Binary 2's-complement
+ 11	0, 0 1 0 1 1	0, 0 1 0 1 1	0, 0 1 0 1 1
− 11	1, 0 1 0 1 1	1, 1 0 1 0 0	1, 1 0 1 0 1
+ .3125	0, . 0 1 0 1	0, . 0 1 0 1	0, . 0 1 0 1
− .3125	1, . 0 1 0 1	1, . 1 0 1 0	1, . 1 0 1 1
+ 31	0, 1 1 1 1 1	0, 1 1 1 1 1	0, 1 1 1 1 1
− 31	1, 1 1 1 1 1	1, 0 0 0 0 0	1, 0 0 0 0 1
0	0, 0 0 0 0 0	$\begin{Bmatrix} 0, 0 0 0 0 0 \\ 1, 1 1 1 1 1 \end{Bmatrix}$	0, 0 0 0 0 0

(b)

FIGURE 2.5. (a) Positive and negative binary number representations.
(b) Selected binary numbers shown in three number representations.

corresponding operations in sign-and-magnitude representation of decimal numbers. For example, adding two positive numbers we have

$$
\begin{array}{l}
\text{Sign} \\
\downarrow \\
\quad\ 0\ 1\ 0\ 0\ 1 \quad \leftarrow \text{Carry} \\
\ 0,0\ 0\ 1\ 1\ 0\ 1 \quad \leftarrow \text{Addend} \\
+0,0\ 0\ 1\ 0\ 0\ 1 \quad \leftarrow \text{Augend} \\
\hline
\ 0,0\ 1\ 0\ 1\ 1\ 0 \quad \leftarrow \text{Sum}
\end{array}
$$

Note that a comma has been added between the sign and the number bits.

When adding a positive number to a negative number we have to check their absolute magnitudes to find the appropriate sign of the result. Also, when the two numbers involved in an addition are of the same sign, their sum may exceed the bit capacity of the adding circuit. Thus a carry out from the most significant bit would be lost and would yield an erroneous result, except for an *overflow* monitor circuit that is incorporated in the adder to detect this condition. The overflow problem is also present in other representations of binary numbers.

1's-Complement Representation.   As in sign-and-magnitude representation, a number expressed in 1's-complement notation consists of a sign bit in the leftmost position followed by number bits. For positive numbers the two representations are identical. In case of negative numbers the number bits are represented by their 1's-complements. A 1's-complement of a number is obtained by inverting each bit, including the sign bit, of the corresponding positive number. For example, given the positive number $N = 0,11001_2$ $(= 25_{10})$, the 1's-complement of N is $1,00110_2$.

The 1's-complement, $\overline{N}_1$, of a number N having n integers and m fractional bits can also be expressed as

$$
\overline{N}_1 = 2^n - \left(\sum_{i=-m}^{n-1} b_i 2^i \right) - 2^{-m}, \tag{2.14}
$$

where b_i is the binary coefficient (0 or 1) of the ith bit. The expression in the parentheses equals N, thus

$$
\overline{N}_1 = 2^n - N - 2^{-m}. \tag{2.15}
$$

In case of an integer number, eq. (2.15) reduces to

$$
\overline{N}_1 = 2^n - N - 1. \tag{2.16}
$$

The 1's-complement of $\overline{N}_1$ is N, as shown below:

$$
(\overline{N}_1) = 2^n - \overline{N}_1 - 2^{-m}; \tag{2.17}
$$

substituting for $\overline{N}_1$ from eq. (2.15) we obtain

$$(\overline{\overline{N}_1}) = 2^n - (2^n - N - 2^{-m}) - 2^{-m} = N. \qquad (2.18)$$

Selected numbers in 1's-complement representation are shown in the third column of Figure 2.5. A binary number, N, of n integer and m fractional bits is thus represented by $(n + m + 1)$ bits, where one bit is employed to indicate the sign. The range of numerical values, R, covered by a number expressed in 1's-complement representation is

$$-(2^{n+m-1} - 1) \le R \le + (2^{n+m-1} - 1). \qquad (2.19)$$

The third column of Figure 2.5a shows the range of 1's-complement numbers that can be represented by four bits. Note the anomaly of two ways of representing zero, namely 00 $\cdots$ 00 and its complement 11 $\cdots$ 11. These two 0's are sometimes referred to as the "positive" and "negative" 0, respectively, which causes some inconvenience in digital systems.

Arithmetic operations with 1's-complement notation are greatly facilitated since the sign and the number bits are treated equally. No distinction between these bits is made in addition, subtraction, multiplication, and division; no special circuitry is required for subtraction which is obtained through adding to the minuend the *complement* of the subtrahend.(These are also true of the 2's-complement representation to be discussed later.) The arithmetic circuitry, however, must detect two conditions: (a) a sign change occurring when the result of an arithmetic operation exceeds the bit capacity of the adder and (b) a "carry-out" that is generated in some but not all combinations of addition of 1's-complemented numbers. A "carry-out," if generated, is added to the lsb of the sum as an "end-around carry" to yield the correct result.

EXAMPLE 2.12.   Given a 1's-complement adder of 6-bit capacity find $25_{10} - 13_{10} = 0,11001_2 - 0,01101_2$.

$$
\begin{array}{rll}
25_{10} = & 0,11001 & \leftarrow \text{ Minuend} \\
-13_{10} = & 1,10010 & \leftarrow \text{ 1's-complement of subtrahend} \\
\hline
& (1)\,0,01011 & \leftarrow \text{ Sum} \\
& 1 & \leftarrow \text{ Add end-around carry} \\
\hline
12_{10} = & 0,01100 & \leftarrow \text{ Result}
\end{array}
$$

The overflow in Example 2.12 was added as an "end-around carry" to the intermediate result to yield the correct answer: $+12_{10}$. However, when two numbers are added that exceed the bit capacity of the adder, an arithmetic overflow is generated. This is accompanied by an unwarranted change in sign which has to be detected by special circuitry.

EXAMPLE 2.13.   Given a 1's-complement adder of 6-bit capacity find $-25_{10} - 13_{10}$.

$$
\begin{array}{rl}
-25_{10}= & 1{,}00110 \\
-13_{10}= & 1{,}10010 \\
\hline
-38_{10}\neq & (1)\,0{,}11000
\end{array}
$$

Note that the sign bit has changed, which is unwarranted in the addition of two numbers having equal signs. The change in sign bit indicates that the bit capacity of the adder has been exceeded and an incorrect result has been obtained.

2's-Complement Representation.   The 2's-complement, $\overline{N}_2$, of a number N may be obtained by subtracting its absolute value from a binary number that is one order greater than the most significant bit of N. Thus

$$
\overline{N}_2 = 2^n - \sum_{i=-m}^{n-1} b_i 2^i, \tag{2.20}
$$

where the quantity under the summation represents N. Also, by comparing the 2's-and 1's-complement representations eq. (2.20) and (2.14), respectively, we obtain

$$
\overline{N}_2 = \overline{N}_1 + 2^{-m}. \tag{2.21}
$$

From eq. (2.21) it follows that the 2's-complement of N may also be generated by first obtaining its 1's-complement (i.e., inverting every bit in N) and then adding 1 in the lsb position, which may be integer or fractional.

EXAMPLE 2.14.   Find the 2's-complement of $N = 26_{10} = 0{,}11010$ utilizing the methods derived from eqs. (2.20) and (2.21).
  (a) The highest order (weight) of N, including the sign bit, is 2^5. Thus, subtracting the absolute magnitude from 2^6 we obtain from eq. (2.20)

$$
\begin{array}{rl}
2^6 = & 10\ 00000 \\
+26_{10} = & 0{,}11010 \leftarrow \text{Subtract} \\
\hline
& 1{,}00110 \leftarrow \text{2's complement, } -26_{10}
\end{array}
$$

  (b) Starting with the 1's-complement we have from eq. (2.21)

$$
\begin{array}{rl}
-26_{10} = & 1{,}00101 \leftarrow \text{1's-complement, } -26_{10} \\
& \quad\quad\ 1 \leftarrow \text{Add} \\
\hline
& 1{,}00110 \leftarrow \text{2's-complement, } -26_{10}
\end{array}
$$

A third technique for obtaining the 2's-complement is as follows: Examine the binary number to be complemented, starting with the lsb. Start complementing the higher order bits *after* the first 1.

The 2's-complement of $\overline{N}_2$ is N, as shown below:

$$(\overline{\overline{N}_2}) = 2^n - \overline{N}_2. \tag{2.22}$$

Substituting for $\overline{N}_2$ from eq. (2.20) we obtain

$$(\overline{\overline{N^2}}) = 2^n - (2^n - N) = N. \tag{2.23}$$

Selected numbers represented in 2's-complement notation are shown in the last column of Figure 2.5. We can see that negative numbers only are expressed differently in the three representations. The range, or domain, of numerical values, R, covered by a number of n integer and m fractional bits in a 2's-complement representation is

$$-2^{n+m-1} \leq R \leq +(2^{n+m-1} - 1). \tag{2.24}$$

In 1's- and 2's-complement arithmetic, subtraction is effected through addition of the *complement* of the subtrahend.

EXAMPLE 2.15. Using 2's-complement notation find $13_{10} - 25_{10} = 0{,}01101_2 - 0{,}11001_2$.

$$
\begin{array}{ll}
0{,}11001 & \leftarrow \text{Subtrahend} = +25_{10} \\
1{,}00110 & \leftarrow \text{Invert bits} \\
\underline{\qquad 1} & \leftarrow \text{Add 1} \\
1{,}00111 & \leftarrow \text{2's-complement of subtrahend} \\
\underline{0{,}01101} & \leftarrow \text{Minuend, add } +13_{10} \\
1{,}10100 & \leftarrow \text{Result}
\end{array}
$$

The sign bit of the result is 1, indicating a negative number, as expected. To obtain the absolute value we employ 2's-complementation on the result obtained above.

$$
\begin{array}{ll}
1{,}10100 & \leftarrow \text{Result} \\
0{,}01011 & \leftarrow \text{Invert bits} \\
\underline{\qquad 1} & \leftarrow \text{Add 1} \\
0{,}01100 = 12_{10} & \leftarrow \text{Absolute value of result}
\end{array}
$$

The reader may try for practice adding various combinations of positive and negative numbers. Unlike in 1's-complement arithmetic, no "end-around carry" is added when an overflow is generated in a result that is within the range which can be represented by the given number of bits.

EXAMPLE 2.16.   Using 2's-complement representation find $25_{10} - 13_{10} =$ $0,11001_2 - 0,01101_2$.

The 2's-complement of $0,01101_2$ is $1,10011_2$

$$\begin{array}{ll} 0,11001 & \leftarrow \text{ Minuend} = 25_{10} \\ \underline{1,10011} & \leftarrow \text{ Add 2's-complement of subtrahend} \\ (1)\,0,01100 & \leftarrow \text{ Result; overflow bit in ( ) is ignored} \end{array}$$

A correct result $0,01100_2 = +12_{10}$ is obtained, except for the overflow bit that is shown in parentheses and is eliminated by the circuitry.

An adder is said to be adding *modulo-2^n* (mod-2^n) if it carries out the add operation in the usual way except that it eliminates any bit of weight 2^n. The number represented by 2^n is called the *modulus* of the arithmetic; all lower weight bits are called the *residue*. For example, in mod-2 addition $1 + 1 = 0$; the overflow bit, having a value of 2, was neglected. Other examples of use of modular representation are the clock which counts in mod-12, the angle of a sine wave which may count mod-360 degrees or mod-2π radians, or an electronic counter of N digits which counts modulo-2^N. A modulo-sum M is a sum with respect to the modulus M with the carry ignored.

Radix- and Diminished Radix-Complements for Any Number Base.   Complement representation may be generalized to apply to any number of base r. Of particular interest are the 9's- and 10's-complement representations because of the BCD code that is frequently used for execution of digital arithmetic operations.

The *radix-complement*, $\overline{N}_r$ of a number of n integer and m fractional digits is

$$\overline{N}_r = r^n - N_r, \tag{2.25}$$

where N_r is the absolute value of the number. Consider the 5-digit decimal $(r = 10)$ number 123.45; the radix- or 10's-complement of this number is

$$10^3 - 123.45_{10} = 876.55_{10}.$$

The *diminished radix-complement* $\overline{N}_{r-1}$ of an n integer and m fractional number is

$$\overline{N}_{r-1} = r^n - N_r - r^{-m} = \overline{N}_r - r^{-m}. \tag{2.26}$$

The diminished radix-complement of the decimal number (also called the 9's-complement) 123.45_{10} is 876.54_{10}.

We can also obtain the 10's-complement of a decimal number by subtracting each digit from 10. Similarly, to obtain the 9's-complement of a decimal number we subtract each digit from 9.

2.6   8-4-2-1 Binary Coded Decimal (BCD) Numbers

Most computers use base 2 for internal operations but some, notably for business applications, utilize the decimal system. Since we are limited to devices that have two stable states, some coding scheme has to be introduced whereby decimal numbers may be represented by use of binary devices. Encoding and decoding is extensively discussed in Chapter 9; however, we introduce one coding scheme early in the text to serve in several examples and illustrations that are presented before we have had the opportunity to study the subject of coding in greater detail.

Four bistable devices can assume 16 possible combinations (states): $0000, 0001, \ldots, 1111$. We may choose any 10 of these 16 states to represent the 10 decimal digits. The most natural choice is to assign to the bits binary weights 8, 4, 2, and 1, as shown in Figure 2.6. This code is known as the *naturally weighted binary coded decimal (BCD)*, or the *8-4-2-1 BCD*. Each decimal digit is represented by four bits, thus six combinations of the 4-bit code are not used causing a loss of efficiency in bit utilization.

An n-digit decimal number is expressed by n groups of four bits each.

EXAMPLE 2.17.   Express 56.78_{10} in BCD notation.

$$
\begin{array}{ccccc}
5 & 6 & . & 7 & 8 \\
0101 & 0110 & & 0111 & 1000
\end{array}
$$

Thus $56.78_{10} = 0101\ 0110.0111\ 1000$ (BCD)

Decimal	BCD 8 4 2 1 ← Weights
0	0 0 0 0
1	0 0 0 1
2	0 0 1 0
3	0 0 1 1
4	0 1 0 0
5	0 1 0 1
6	0 1 1 0
7	0 1 1 1
8	1 0 0 0
9	1 0 0 1

FIGURE 2.6.   8-4-2-1 binary coded decimal (BCD) numbers.

The number of possible 4-bit BCD codes is $\approx 3 \times 10^{10}$ and several of these are discussed in Chapter 9.

2.7   Floating Point Representation and Arithmetic

In operations involving large numbers that undergo several arithmetic steps it is important to retain the maximum number of significant digits. The floating point notation preserves these significant digits and it is quite similar in form to the scientific notation. For example, the speed of light in meters/second is expressed in scientific notation as 3×10^8, which is much shorter than 300,000,000. Leading or trailing zeros do not add to the accuracy of a calculation when a number is expressed in scientific notation; their retention may introduce substantial rounding-off errors.

Floating Point Representation. A binary number expressed in floating point notation consists of an *exponent E* (also called the *characteristic*), and a *mantissa M* (also referred to as the *fraction field*). The terms exponent and mantissa, borrowed from logarithms, should not imply that floating point numbers are logarithms. Both E and M depend on the size of the computer word and on the application: E may typically have 7 to 9 bits, while M may use 16 to 60 bits. In small computers two or even three computer words may have to be utilized to express E and M with a sufficient number of bits for the desired accuracy of the computation.

There are several methods of expressing a number in floating point notation, which differ in detail only. Common to all these is that the leading bits are reserved for E, while the remaining bits represent a fractional number M, where M must be in the range

$$1/2 \leq |M| < 1. \tag{2.27}$$

Exponent and mantissa may be expressed in 2's-complement representation, and the value of E is adjusted to ensure the relation of eq. (2.27). This adjustment is called *normalization*.

EXAMPLE 2.18.   Express in floating point notation the binary number $N = -1.10011011$.

Since N is not normalized, i.e., $|M| > 1$, we require an exponent $E = +2^1$ to yield

$$N = 2^1 \times (-.110011011).$$

Assuming that eight bits are reserved for E and that 2's-complements are used for M (and E) we have

$$N \text{ (floating point)} = \underset{E}{0,0000001,} \quad \underset{M}{1,001100101,}$$

where the leading bit in E and in M indicates the respective sign ($+$ or $-$).

The value of E may also be considered as the number of shifts that have to be carried out for normalization of M to bring it within the range of $1/2 \leq |M| < 1$. For $|M| \geq 1$, the mantissa has to be shifted right and the exponent will be positive. For example, $N = 101.101$ may be written $N = 2^3 \times (.101101)$, and E is $(0,0...11)_2$. Conversely, for $|M| < 1/2$, the number is shifted left to make the leading bit 1, and E has a negative value which is equal to the number of shifts.

EXAMPLE 2.19.   Express in floating point notation the following numbers (written for convenience in base 8): $+0.653_8$, -0.732_8, $+1215_8$, -0.0074_8 and $+261.2_8$. Use 2's-complements for negative E and M and 5 bits to express E.

The results are shown in Figure 2.7.

Octal	Binary	Floating point	
		Exponent	Mantissa
+ 0.653	+ . 1 1 0 1 0 1 0 1 1	0, 0 0 0 0	0, 1 1 0 1 0 1 0 1 1
− 0.732	− . 1 1 1 0 1 1 0 1 0	0, 0 0 0 0	1, 0 0 0 1 0 0 1 1 0
+ 1215	+ 1 0 1 0 0 0 1 1 0 1	0, 1 0 1 0	0, 1 0 1 0 0 0 1 1 0 1
− 0.0074	− . 0 0 0 0 0 0 1 1 1 1	1, 1 0 1 0	1, 0 0 0 1
+ 261.2	+ 1 0 1 1 0 0 0 1 . 0 1	0, 1 0 0 0	0, 1 0 1 1 0 0 0 1 0 1

FIGURE 2.7.   Floating point representation of numbers given in Example 2.19.

One representation of floating point numbers in a 32-bit computer word is shown in Figure 2.8a. It is assumed that (a) the mantissa is expressed in sign-and-magnitude form, (b) the leading bit (bit 0) serves as the sign bit for M, (c) the exponent utilizes seven bits and does not use a sign bit, and (d) E is represented in "excess-64" notation. In other words, in order to accommodate large and small magnitudes of numbers, 64_{10} is added to the exponent obtained in the normalization process. This technique eliminates the need for a sign bit for E.

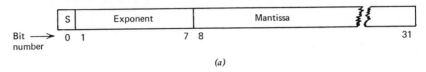

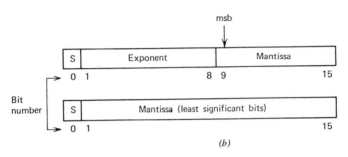

FIGURE 2.8. Floating point representation: (*a*) by a computer word of 32 bits, (*b*) by two computer words of 16 bits each.

EXAMPLE 2.20.   Determine the range of positive numbers, N, that can be expressed using the normalized floating point notation described with reference to Figure 2.8*a*.

Since exponent E has 7 bits, the total range of the exponent is 2^{127}. Since E is expressed in "excess-64" notation, the exponent of the largest number is $2^{127-64} = 2^{+63}$. The largest value of the mantissa is $(2^{24} - 1) \times 2^{-24} = 1 - 2^{-24}$. Thus $N_{max} = 2^{+63}(1 - 2^{-24})$. To find N_{min} we note that the smallest value of E in Figure 2.8*a* is 0, corresponding to 2^{-64} in "excess-64" notation. The smallest value of the *normalized* mantissa is 2^{-1}, yielding $N_{min} = 2^{-64} \cdot 2^{-1} = 2^{-65}$.

Thus
$$2^{-65} \leq N \leq 2^{+63}(1 - 2^{-24}). \tag{2.28}$$

A floating point representation using two 16-bit computer words is shown in Figure 2.8*b*. The binary value of the leftmost bit in the upper word determines the sign of the exponent, while the corresponding bit in the lower word determines the sign of the mantissa.

Floating Point Arithmetic.   Multiplication, division, addition, and subtraction of two numbers in floating point representation is presented below. Given two normalized floating point numbers, $X = 2^{E_x}(M_x)$, $Y = 2^{E_y}(M_y)$, the product is

$$X \cdot Y = (2^{E_x + E_y})(M_x \cdot M_y) = 2^{E_v}(M_v), \tag{2.29}$$

and the quotient is

$$X \div Y = (2^{E_x - E_y})(M_x \div M_y) = 2^{E_w}(M_w). \tag{2.30}$$

Normalization of the result may be required if $|M_v|$ or $|M_w|$ fall outside of the range 1/2 to 1. For example, let $X = 2^5(.101)$, $Y = 2^{12}(.1001)$, then $X \cdot Y = 2^{17}(.0101101) = 2^{16}(.101101)$.

Addition and subtraction requires alignment of the two numbers, the exponent of the larger number serving as the exponent of the sum or the difference.

EXAMPLE 2.21. Obtain the sum of $X + Y = Z$, where $X = 2^{E_x}(M_x) = 2^6(.101)$, $Y = 2^{E_y}(M_y) = 2^{11}(.101101)$.

The binary point of X has to be shifted five places to the left, since $E_y - E_x = 5$

$$X = 2^{11}(.00000101)$$
$$Y = 2^{11}(.101101)$$
$$\overline{Z = 2^{11}(.10111001)}$$

Alignment is always made with reference to the larger number. Given two numbers, one with a positive exponent and the other with a negative exponent, the binary point of the latter number has to be shifted since its value is the smaller of the two numbers.

Notice that the effect of normalization in arithmetic operation is to retain the maximum number of significant bits.

PROBLEMS

2.1 Determine the binary number corresponding to 311_{10}.

2.2 Determine the octal number corresponding to 437_{10}.

2.3 Find the decimal number corresponding to 6531_7.

2.4 Determine the base 6 number corresponding to decimal 6437.

2.5 (a) Convert $.739_{10}$ to base 6.
(b) Convert $.4233_6$ to base 10.

2.6 Convert the following decimal numbers to base 16:
(a) 27, (b) 467, (c) 511.

2.7 Given a decimal integer number of n digits, how many digits are required to represent the same number in base 3, 4, 8, and 16?

2.8 Given an integer N of n digits in base r_1, find a general expression for the number of digits, m, required to represent N in base r_2. That is, find m as a function of n, r_1, and r_2.

2.9 Express the following mathematical constants in base 8:
(a) $\pi = 3.14159_{10}$,   (b) $e = 2.718_{10}$,   (c) $c = 2.998 \times 10^8$.

2.10 Establish an addition and multiplication table for base-8 numbers and convert the following octal numbers to decimal:
(a) 1003, (b) 1602, (c) 3377, (d) 7777, (e) .077, (f) .007,
(g) .322, (h) .000777. Express the decimal fractions to six significant figures.

2.11 Find the octal equivalent of 10^n where $0 \le n \le 9$.

2.12 Determine the decimal, octal, and hexadecimal equivalents of the binary numbers 0.111101101, 11.001011, and 1101101.

2.13 Determine the binary, octal, and hexadecimal equivalents of the decimal numbers 278.5, 0.0522, 15.381, 0.1375.

2.14 (a) Establish an addition table for base-16 numbers.
(b) Make a table showing (hexadecimal) A times each of the 16 hexadecimal symbols, i.e., $A \cdot 1 = A$, $A \cdot 2 = 14$, $A \cdot 3 = 1E$, ... $A \cdot F = 96$.
(c) Find the decimal equivalent of $AE5C3_{16}$.

2.15 Show the three binary representations of the following decimal numbers: -47.3, -0.0712, and -1.023.

2.16 Find the radix-complements and diminished radix-complements of the following numbers: 1.12_{10}, 0.378_{10}, 11234_6, 102.201_3.

2.17 Convert 25.3_{10} to an octal number, allowing a conversion error $\le .001$.

2.18 Find the BCD representation of 153.75_{10}.

2.19 Express in floating point notation the following binary numbers: $+0.101001$, -0.100011, $+10111.01$, -0.00000101, $+0.0001011$. Use 2's-complements for negative exponents and mantissas.

2.20 Use the floating point notation shown in Section 2.7 and carry out the following arithmetical operations:
(a) -0.1101×11.01
(b) 1.0001×0.00001101
(c) $-0.1101 \div 11.01$
(d) $0.000001101 \div 1.0001$

(e) $0.00011011 + 0.11001$

(f) $-0.001 - 11.01$

2.21 In a particular floating point notation an exponent of zero is represented by the 7-bit pattern $1000000 = 64_{10}$, plus 1 is represented by $1000001 = 65_{10}$, and minus 1 by $0111111 = 63_{10}$. What is the largest decimal number that can be expressed with this choice of representation of the exponent? What is the smallest positive decimal number that can be expressed?

2.22 Consider the corrections on the exponents, if any, that have to be made when multiplying or dividing two numbers using the floating point convention of Problem 2.21.

CHAPTER 3

Switching Functions

In this chapter we introduce an algebra to serve as a basic tool in the design of switching functions. These functions can be manipulated to obtain simplified expressions in an analogous way to ordinary algebra. Simplified, or minimized, expressions of switching functions have an important practical implication: they enable the systems designer to realize a set of given specifications with a minimum number of components.

Emphasis will be given to map methods for simplification of algebraic expressions since these methods draw on the power of pattern recognition by the human mind and are thus easily implemented. An algorithm for minimization using a tabular method will be briefly treated.

3.1 Switching Algebra

Boole[1] developed a symbolic logic to express reasoning with help of mathematical operations. In this framework, known as *Boolean algebra*, there exist two statements: *true* and *false*, represented by numbers 1 and 0, respectively. The algebra was applied by Shannon[2] to the analysis of relay and switching circuits, and today it is used extensively by designers of logic and computer systems. These systems utilize components such as switches, transistors, or magnetic materials (core, tape) that can be characterized by two states: closed or open, conducting or nonconducting, magnetization in one direction or in the opposite direction. The states of these components can, therefore, be described by an algebra using two integers: 1 and 0. Operations in this algebra are based on a set of postulates from which theorems are developed, some of which are similar to theorems of ordinary algebra.

We shall utilize switches in various combinations as an aid in presenting postulates of the algebra. These simple electrical devices exhibit two mutually exclusive states. When the switch is closed, current is allowed to pass between the terminals. We shall refer to a closed switch as being in *state 1.* Conversely, an open switch that does not allow current to flow between the terminals will be referred to as being in *state 0.*

A switching network is made up of a number of switches connected in series, or in parallel, or in a series-parallel combination. The network will be defined to be in state 1 *iff* (if and only if) current is allowed to pass between its terminals. Otherwise the network is in state 0.

Capital letters will be used to label and to describe the states of the switches. A bar over the letter, e.g., $\overline{X}$, will indicate an open switch (state 0), while the same switch in its closed condition, i.e., in state 1, will be described by X. We shall also require descriptions of three *operators,* AND, OR, and NOT, which are given below.

The AND *operation* is performed by two or more switches connected in series as shown in Figure 3.1*a.* Current will flow between the terminals *a* and *b* of Figure 3.1*a* iff both *X and Y* are closed. The symbol "·" will be used in this text for the AND operation, sometimes also called "Boolean product," "intersection," or "conjunction." Additional symbols found in the literature for the AND operation are: " ∧ ", "∩", "&".

The OR operation is performed by two or more switches connected in parallel, as shown in Figure 3.1*b.* Current will flow between the terminals *a* and *b* of Figure 3.1*b* iff *one or more* switches are closed. No current will flow if *all* switches are open. The symbol " + " will be used in this text for the OR operation, sometimes also called "Boolean sum," "union," or "disjunction." Additional symbols found in the literature for the OR operation are: " ∨ ", "∪".

The NOT operation may be illustrated by a change of the state of a switch. If switch X was initially closed (= state 1), then the NOT operation will bring it to state 0, as shown in Figure 3.1*c.*

A bar "—" over the symbol labeling the switch will be used to indicate the operation NOT, e.g., $\overline{A}$, $\overline{X}$. The NOT operation is also referred

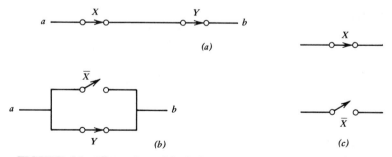

FIGURE 3.1. Illustration of logical operations with aid of switches: (a) AND, (b) OR, (c) NOT.

to in the literature as "negation," "complementation," or "inversion." Additional symbols in use are $\sim$, $-$, $'$, e.g., $(\sim A)$, $(-B)$, (C').

EXAMPLE 3.1. Referring to Figure 3.2, write down the states of the switches that will allow current to flow between terminals a and b. In other words, what are the possible combinations of the states of the switches that result in a state 1 of the network?

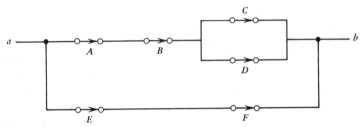

FIGURE 3.2. Switching network of Example 3.1.

Current will flow in the upper branch if A *and* B are in state 1, *and* either C *or* D are in state 1. This is represented by

$$A \cdot B \cdot (C + D).$$

The lower branch will allow a current flow between a and b iff E *and* F are in state 1, i.e., $E \cdot F$. Since current will flow between the terminals if either transmission path is completed, the network in state 1 is fully described by

$$A \cdot B \cdot (C + D) + E \cdot F.$$

Postulates. The set of postulates given below provides the basis of a switching algebra and will be used in deriving theorems that are suitable for designing switching networks with any two-valued (binary) elements. A check of the first three postulates may be obtained by reference to Figure 3.3.

> $P1a$: $0 \cdot 0 = 0$
> $P1b$: $1 + 1 = 1$.

Notice that the postulates are presented in pairs. This is due to the duality existing between each pair. For example, changing in $P1a$ each 0 to 1 and each "$\cdot$" to "$+$" we obtain $P1b$. Postulate $P1a$ may be obtained by analogous

 *(a)*

 *(b)*

 *(c)*

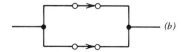 *(d)*

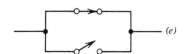 *(e)*   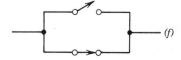 *(f)*

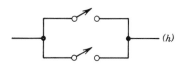 *(g)*

 (h)

FIGURE 3.3. Representation of postulates by use of switches:
(*a*) $0 \cdot 0 = 0$, (*b*) $1 + 1 = 1$, (*c*) $0 \cdot 1 = 0$, (*d*) $1 \cdot 0 = 0$, (*e*) $1 + 0 = 1$,
(*f*) $0 + 1 = 1$, (*g*) $1 \cdot 1 = 1$, (*h*) $0 + 0 = 0$.

changes of variables and operators in *P1b*. This duality also holds for the
postulates given below and for the theorems presented later.

P2a: $0 \cdot 1 = 1 \cdot 0 = 0$
P2b: $1 + 0 = 0 + 1 = 1$

P3a: $1 \cdot 1 = 1$
P3b: $0 + 0 = 0$

$P4a$: $\overline{0} = 1$: The NOT of an open circuit is a closed circuit.
$P4b$: $\overline{1} = 0$: The NOT of a closed circuit is an open circuit.

$P5a$: $X = 1$ if $X \neq 0$⎫   A circuit cannot assume
$P5b$: $X = 0$ if $X \neq 1$⎭   both states simultaneously.

Postulates $P5a$ and $P5b$ express the binary nature of the algebra and conceptually represent the statements *true* $(= 1)$ and *false* $(= 0)$.

Theorems. The theorems given below can be proved with help of the postulates developed earlier, and can easily be checked by use of switch diagrams (perfect induction method, see Problem 3.3).*

Theorem 1. Commutative law
   $Th1a$: $X \cdot Y = Y \cdot X$
   $Th1b$: $X + Y = Y + X$

Theorem 2. Associative law
   $Th2a$: $X \cdot (Y \cdot Z) = (X \cdot Y) \cdot Z$
   $Th2b$: $X + (Y + Z) = (X + Y) + Z$

Theorem 3. Distributive law
   $Th3a$: $X \cdot (Y + Z) = X \cdot Y + X \cdot Z$
   $Th3b$: $X + Y \cdot Z = (X + Y) \cdot (X + Z)$

Theorem 4. Identities
   $Th4a$: $X \cdot 1 = X$
   $Th4b$: $X + 0 = X$

Theorem 5. Null elements
   $Th5a$: $X \cdot 0 = 0$
   $Th5b$: $X + 1 = 1$

Theorem 6. Complements
   $Th6a$: $X \cdot \overline{X} = 0$
   $Th6b$: $X + \overline{X} = 1$

Theorem 7. Idempotent property
   $Th7a$: $X \cdot X = X$
   $Th7b$: $X + X = X$

Theorem 8. Absorption
   $Th8a$: $X + X \cdot Y = X$
   $Th8b$: $X \cdot (X + Y) = X$

---

* Truth tables, discussed in Section 3.2, provide an alternative method for checking the theorems by perfect induction.

Theorem 9. Involution (or double negation)

$$Th9: (\overline{\overline{X}}) = X$$

Theorem 10. DeMorgan's theorem

$$Th10a: \overline{(X \cdot Y \cdot Z \cdots)} = \overline{X} + \overline{Y} + \overline{Z} + \ldots$$
$$Th10b: \overline{(X + Y + Z + \ldots)} = \overline{X} \cdot \overline{Y} \cdot \overline{Z} \cdots$$

Simplification by Use of Theorems. The above theorems are useful in manipulation of equations in switching algebra to obtain simplified expressions. Additional theorems may also be developed and proved by use of the ten theorems.

EXAMPLE 3.2. Show that $(X + Y) \cdot (X + Z) = X + YZ$.*

From the distributive law, $Th3a$, $(X + Y) \cdot (X + Z) = X + XY + XZ + YZ$. Applying absorption, $Th8a$,
$X + XY = X$, and $X + XZ = X$; hence
$(X + Y)(X + Z) = X + X + YZ$.
From the idempotent property, $Th7b$, this reduces to $X + YZ$.

Note that we have applied Theorems 3a, 8a, and 7b, yielding Theorem 3b.

EXAMPLE 3.3. Simplify the function $f(A,B) = A + \overline{A}B$.

$A + \overline{A}B = (A + \overline{A})(A + B)$, by $Th3b$;
but $(A + \overline{A}) = 1$, by $Th6b$. Hence
$A + \overline{A}B = (A + B)1 = A + B$, by $Th4a$.

DeMorgan's theorem, $Th10$, is extensively applied in the design of switching networks and for simplification of algebraic expressions. The theorem can also be applied to obtain the complement (NOT) of any switching function. From Theorems $Th10a$ and $Th10b$ we deduce the following rule: Given a switching function $f(X,Y,Z\ldots)$; to obtain the complement of f, i.e., $\overline{f}$, change all " $+$ " operations to "$\cdot$", all "$\cdot$" operations to "$+$", and complement each variable.

EXAMPLE 3.4. Complement the switching function $f(X,Y,Z) = X \cdot [(Y\overline{Z} + W) + \overline{V}]$.

Thus, $\overline{f}(X,Y,Z) = \overline{X \cdot [(Y\overline{Z} + W) + \overline{V}]} = \overline{X} + \overline{(Y\overline{Z} + W + \overline{V})} = \overline{X} + (\overline{Y} + Z) \cdot \overline{W} \cdot V$.

---

* The symbol "$\cdot$" of the operator AND will be frequently omitted when not required to avoid ambiguity.

3.2 Venn Diagrams and Truth Tables

Venn diagrams* and truth tables provide additional methods for representing switching functions. In a Venn diagram a *class K* is defined as all possible regions within an area as, for example, the squares in Figures 3.4–3.7. An *element X* within the class is a *set* of points bounded by a closed curve (circles in the figures).

A truth table consists of two parts: a systematic listing of all the combinations of the input variables, and a column describing the result (output 0 or 1) obtained from each input combination, which depends on the logical operation performed.

These two representations, Venn diagrams and truth tables, and their correspondence will become clearer from the description below of the three basic operators: NOT, AND, and OR.

NOT. If the element X in class K is represented by all the points within the closed curve (circle), then $\overline{X}$ is represented by the area that lies within K but outside the closed curve, as shown in Figure 3.4a. The truth table

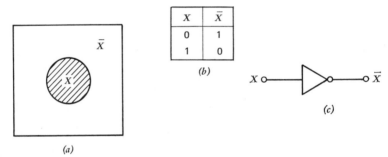

(a)

X	$\overline{X}$
0	1
1	0

(b)

(c)

FIGURE 3.4. The NOT operation: (*a*) Venn diagram, (*b*) truth table, (*c*) graphic symbol.

describing the NOT operation is shown in Figure 3.4b. The negation of X is $\overline{X}$, while the negation of $\overline{X}$ is $\overline{\overline{X}}$ (applying the involution theorem yields $\overline{\overline{X}} = X$). The circuit performing the NOT operation is called an *inverter*, its graphic symbol is shown in Figure 3.4c. We have deliberately postponed the discussion of the circuit details of Figure 3.4c since the *logical* operation performed by the circuit can be fully understood by reference to its truth table or to the respective Venn diagram. The same approach will be taken in this section when describing other operators.†

---

* Named after the 19th-century British mathematician.

† The practical implementation of operators and their circuit details are discussed in Chapter 5.

AND. The operation or function AND on elements X, Y, Z, ..., i.e., $(X \cdot Y \cdot Z \cdots)$ is represented in a Venn diagram by the largest *overlap* of the elements involved in this operation. The shaded area in Figure 3.5a represents the operation $X \cdot Y$. Note that the area outside $X \cdot Y$ is $\overline{X \cdot Y}$, which follows from the definition of the logical NOT. The truth table of $X \cdot Y$ is shown in Figure 3.5b. Two variables can be combined in four different ways, i.e., $\overline{X}\overline{Y}$, $\overline{X}Y$, $X\overline{Y}$, and XY. In general, a truth table of n variables has 2^n rows to describe all the possible combinations.

A circuit performing the AND operation is called an AND *gate*, its graphic symbol is shown in Figure 3.5c.

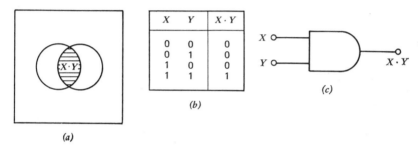

(a)   (b)   (c)

FIGURE 3.5.   The AND operation: (a) Venn diagram, (b) truth table, (c) graphic symbol.

OR. The operation or function OR on elements X, Y, Z, ..., i.e., $(X + Y + Z + ...)$, is represented in a Venn diagram by the smallest area that includes all the elements involved in this operation. The shaded area in Figure 3.6a shows the operation $(X + Y)$ and represents the statement *either X or Y or both*. For this reason, this function is also referred to as the INCLUSIVE–OR, to distinguish it from the EXCLUSIVE–OR that performs the operation *either X or Y*. The term OR will be used throughout the text to mean INCLUSIVE–OR. The truth table of $X + Y$ is shown in the second column of Figure 3.6b. A circuit performing the operation of INCLUSIVE–OR is called an OR *gate*, and its symbol is shown in Figure 3.6c.

The Venn diagram for the EXCLUSIVE–OR, $X \oplus Y$, and the graphic symbol of the gate performing the operation $X \oplus Y$ are shown in Figures 3.6d and e, respectively; its truth table is given in the last column of Figure 3.6b.

Although the logical operations of Figures 3.4 through 3.6 show two variables only, it should be understood that Venn diagrams, truth tables, and circuits (gates) can be also constructed for multivariable functions.

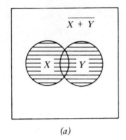

(a)

X	Y	X + Y	X ⊕ Y
0	0	0	0
0	1	1	1
1	0	1	1
1	1	1	0

(b)

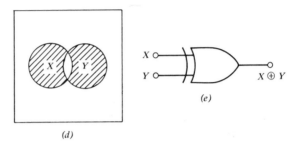

(c)

(d)

(e)

FIGURE 3.6. The OR operations: (*a*) INCLUSIVE–OR, Venn diagram; (*b*) Truth tables; (*c*) INCLUSIVE–OR, graphic symbol; (*d*) EXCLUSIVE–OR, Venn diagram; (*e*) EXCLUSIVE–OR, graphic symbol.

EXAMPLE 3.5. Draw a Venn diagram and establish a truth table for the three-variable switching function $f(A,B,C) = \bar{A}\bar{B}C + \bar{A}B\bar{C} + \bar{A}BC + A\bar{B}\bar{C} + A\bar{B}C + AB\bar{C}$.

The shaded area in the Venn diagram, Figure 3.7*a*, represents the function. The truth table of Figure 3.7*b* shows 2^3 possible combinations of

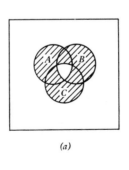

(a)

A	B	C	F
0	0	0	0
0	0	1	1
0	1	0	1
0	1	1	1
1	0	0	1
1	0	1	1
1	1	0	1
1	1	1	0

(b)

FIGURE 3.7. The 3-variable switching function of Example 3.5: (*a*) Venn diagram, (*b*) truth table.

the three-variable function. The six combinations that represent the terms of the function are shown by a "1" in the last column. In other words, since the function is an INCLUSIVE–OR of the six terms, it will be in state 1 *iff* one or more of the above terms is in state 1.

The function of Example 3.5 can be simplified by algebraic manipulation as discussed in Section 3.1, or by use of the Venn diagrams (see Problem 3.8). Neither approach is as convenient or powerful as the mapping method that will be presented in Section 3.5.

Some valid questions are appropriate at this point: (*a*) Why should we be so greatly concerned with simplification of functions? (*b*) Since the number of variables of a switching function is finite, why don't we tabulate all possible functions together with their corresponding minimized expressions for easy reference?

Switching circuits are commonly used to implement mathematical operations, communication networks, or control functions in measuring or automatic process control. These digital systems may typically utilize 10^3 to 10^6 circuits. Therefore it is important to reduce the overall number of components.

EXAMPLE 3.6. Consider the switching network of Figure 3.8*a* represented by the equation $f = AB(C + \overline{A}\overline{C})$ and requiring one OR and two AND

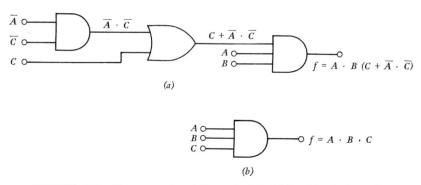

(*a*)

(*b*)

FIGURE 3.8. Two networks of Example 3.6 yielding identical results: (*a*) $f = AB(C + \overline{A}\overline{C})$, (*b*) $f = ABC$.

---

gates. Simplifying, we obtain $f = ABC + AB\overline{A}\overline{C} = ABC + (A\overline{A}) \cdot BC = ABC$, shown in Figure 3.8*b*. The networks of Figure 3.8*a* and *b* perform identical functions; Figure 3.8*b*, however, requires one AND gate only.

Tabulation of minimized expressions for all possible functions is not practical even with modern mass storage, such as a computer memory of large capacity.

EXAMPLE 3.7.   Determine the number of switching functions of n variables for $1 \leq n \leq 5$.

We have shown earlier that a function of n variables has 2^n possible combinations which can be represented in a truth table by 2^n rows. The output of each row can have a value of either 0 or 1, resulting in 2^{2^n} different output functions, as shown in the third column of Figure 3.9. This

n	2^n	2^{2^n}	N
1	2	4	3
2	4	16	6
3	8	256	22
4	16	65,536	402
5	32	$\approx 4.3 \times 10^9$	$\approx 1.23 \times 10^6$

FIGURE 3.9.   Number of output functions, 2^{2^n}, for $1 \leq n \leq 5$ variables, and the number of functions of the same type N.

number can be reduced considerably by recognizing that many functions differ only in the symbols assigned to the input variables. For example, function $f_1 = AB + C(AB + \bar{C})$ and $f_2 = CB + A(CB + \bar{A})$ are different, but one can be converted into the other through interchange of the variables A and C. Thus the simplified expressions $f_1 = AB$, or $f_2 = BC$ are of the same *type*. Classification by type reduces the number of functions, as shown in the last column of Figure 3.9. However, even tabulation by *type* becomes impractical for $n > 5$.

3.3   Analysis and Synthesis of Switching Networks

Operations NOT, AND, and OR were introduced in the previous section. It can be shown that it is possible to realize all switching functions using these three operations.* The result of an *analysis* of a switching network is an equation representing the function performed by operators such as AND, OR, NOT operating on the input variables of a network. In the example below we also show the equivalence of networks realized with (*a*) relay contacts, (*b*) gates.

* This problem will be further discussed in Chapter 4.

EXAMPLE 3.8.   Derive the equation representing the networks shown in Figure 3.10a and b.

In the relay representation of the function, Figure 3.10a, we note three networks in series: $f(A,B,C) = f_1 \cdot \bar{E} \cdot f_2$, where $f_1 = A\bar{C} + \bar{A}C$ and $f_2 = \bar{B} + \bar{C}\bar{D}$; thus $f(A,B,C) = (A\bar{C} + \bar{A}C) \cdot \bar{E} \cdot (\bar{B} + \bar{C}\bar{D})$. To analyze the gate network we start from the left and write down the result of each operation at the outputs of the respective gates, as shown in Figure 3.10b. f_1 is represented by two AND gates and one OR gate; f_2 is represented by one AND gate and one OR gate. The output is derived from a 3-input AND gate describing the function $f = f_1 \cdot \bar{E} \cdot f_2 = (A\bar{C} + \bar{A}C) \cdot \bar{E} \cdot (\bar{C}\bar{D} + \bar{B})$, as before.

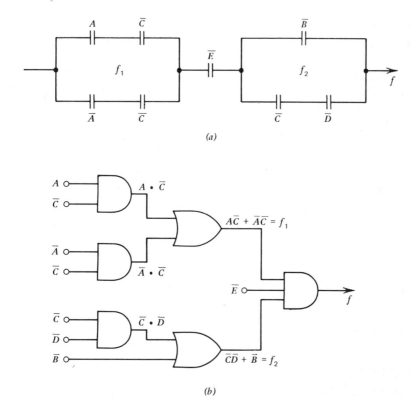

(a)

(b)

FIGURE 3.10.   Representation of the switching function $f = (A\bar{C} + \bar{A}C)$ $\cdot \bar{E} \cdot (\bar{B} + \bar{C}\bar{D})$ **(a) by relay network and (b) by AND–OR gate network.**

We have shown in Example 3.8 the analysis of a *combinational* switching network. This type of network yields at the output a logical 1 for a defined set of combinations of the input variables. Any other combinations will result in a state 0 at the output. The output of the network is also independent of its past history. That is, although the input variables may undergo an arbitrary sequence of states, the output will yield a logical 1 if one or more combinations of the input variables describing the network have been satisfied, irrespective of the sequence of their states that preceded the appropriate combinations.

In *synthesis* of a combinational circuit we start with a statement, a truth table, or an equation. We then proceed to interconnect components such as switches, relay networks, or semiconductor gates that fulfil the specifications set earlier in the statement.

EXAMPLE 3.9.   Implement the circuit that realizes the statement "Electrical engineering students shall take as an elective either switching theory or computer programming."

Assigning the variable X to "switching theory," and the variable Y to "computer programming," we obtain a truth table as in the last column of Figure 3.6b. An "either-or" statement is satisfied by the EXCLUSIVE-OR circuit, which may be synthesized from AND and OR gates as shown in Figure 3.11.

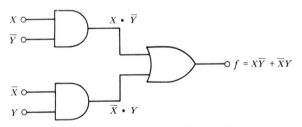

FIGURE 3.11.   Exclusive–OR circuit.

EXAMPLE 3.10.   Implement the following function, using AND and OR gates: $f(A,B,C,D) = [A + \bar{B}\bar{C}\bar{D}] \cdot [\bar{A}\bar{D} + B(\bar{C} + A)]$.

In synthesizing a network utilizing gates from a given equation, we start from the output gate. The function f can be written $f = f_1 \cdot f_2$, where $f_1 = A + \bar{B}\bar{C}\bar{D}$ and $f_2 = \bar{A}\bar{D} + B(\bar{C} + A)$. As shown in Figure 3.12, the output stage is thus implemented by an AND gate. f_1 is the OR of two terms, one

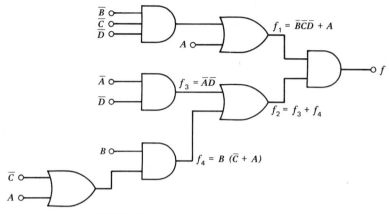

FIGURE 3.12. Synthesis of the function
$$f = [A + \overline{B}\overline{C}\overline{D}] \cdot [\overline{A}\overline{D} + B(\overline{C} + A)].$$

requiring a 3-input AND gate to realize $\overline{B}\overline{C}\overline{D}$. f_2 can be expressed as the OR of two terms: $f_2 = f_3 + f_4$, where $f_3 = \overline{A}\overline{D}$ and $f_4 = B(\overline{C} + A)$. Implementation of f_4 follows from the previous steps.

The maximum number of gates that any signal (variable) of Figure 3.12 has to traverse is four; hence this network is referred to as a 4-stage, or 4-level, network. As another example, the configuration shown in Figure 3.11 represents a 2-stage network.

3.4 Standard Forms

Any switching function can be expressed in one of two *standard*, or *canonical*, forms. These forms, which will be found useful in simplification procedures discussed in the next section, yield expressions that can be implemented with 2-stage gate networks; no general theory has been developed to date that results in a minimized expression for n-stage networks when $n > 2$. An expression written in one of the standard forms can be compared with another expression written in the same form. If both standard forms are identical, then the expressions must be equal, otherwise they must be different.

Let us first define the following terms:

A *literal* is a variable or its complement, e.g., X, $\overline{X}$, Y, $\overline{Y}$.

A *product term* is an expression consisting of literals that are connected by the function AND, e.g., $X \cdot \overline{Y} \cdot Z$, $\overline{A} \cdot \overline{B}$.

A *sum term* is an expression consisting of literals that are connected by the function OR, e.g., $(\overline{X} + Y + Z)$, $(A + \overline{B})$.

A function of n variables has 2^n *standard product terms*, also called *minterms*. Each minterm is unique, and is a product (AND) of exactly n literals. It can be shown that any switching function may be be expressed as a sum of standard product terms (also referred to as *canonical products* or *disjunctive normal form*).

EXAMPLE 3.11.   Express the following function in a standard sum-of-products form:

$$f(X_3, X_2, X_1, X_0) = (\overline{X}_1 + \overline{X}_0)(\overline{X}_3 + \overline{X}_1)(\overline{X}_3 + X_2)(X_1 + X_0).$$

Let $(\overline{X}_1 + \overline{X}_0) = f_1$, $(\overline{X}_3 + \overline{X}_1) = f_2$, $(\overline{X}_3 + X_2) = f_3$, and $(X_1 + X_0) = f_4$. From the distributive law

$$
\begin{aligned}
f &= (f_1 \cdot f_2) \cdot (f_3 \cdot f_4) \\
&= (\overline{X}_3 \overline{X}_1 + \overline{X}_3 \overline{X}_0 + \overline{X}_1 + \overline{X}_1 \overline{X}_0)(\overline{X}_3 X_1 + X_2 X_1 + \overline{X}_3 X_0 + X_2 X_0) \\
&= \overline{X}_3 X_1 \overline{X}_0 + \overline{X}_3 X_2 X_1 \overline{X}_0 + \overline{X}_3 \overline{X}_1 X_0 + \overline{X}_3 \overline{X}_1 X_0 + \overline{X}_3 X_2 \overline{X}_1 X_0 \\
&\quad + X_2 \overline{X}_1 X_0.
\end{aligned}
$$

The function has four variables: X_3, X_2, X_1, and X_0. To obtain a standard sum-of-products expression we shall apply *Th6b* $(X_i + \overline{X}_i = 1)$ to those terms of the function that have less than four literals, where X_i is the literal absent from a product term.

$$
\begin{aligned}
f(X_3, X_2, X_1, X_0) &= \overline{X}_3(X_2 + \overline{X}_2)X_1 \overline{X}_0 + \overline{X}_3 X_2 X_1 \overline{X}_0 \\
&\quad + \overline{X}_3(X_2 + \overline{X}_2)\overline{X}_1 X_0 + \overline{X}_3(X_2 + \overline{X}_2)X_1 X_0 \\
&\quad + \overline{X}_3 X_2 \overline{X}_1 X_0 + (X_3 + \overline{X}_3)X_2 \overline{X}_1 X_0.
\end{aligned}
$$

Application of *Th3a* and *Th7b* results in

$$
\begin{aligned}
f(X_3, X_2, X_1, X_0) &= \overline{X}_3 X_2 \overline{X}_1 X_0 + \overline{X}_3 \overline{X}_2 \overline{X}_1 X_0 + \overline{X}_3 X_2 X_1 \overline{X}_0 \\
&\quad + \overline{X}_3 X_2 X_1 \overline{X}_0 + \overline{X}_3 \overline{X}_2 X_1 \overline{X}_0 + X_3 X_2 \overline{X}_1 X_0.
\end{aligned}
$$

The final expression in Example 3.11 is in the form of a standard sum-of-products. It is rather cumbersome, hence we shall develop a more convenient notation as follows: each *minterm* shall be represented by m_i, where the subscript i is the decimal equivalent of the respective minterm interpreted as a binary number. Complemented variables in the minterm are read as 0's, uncomplemented variables are read as 1's.

EXAMPLE 3.12.   Express the following standard sum-of-products using the m_i notation:

$$f(W,X,Y,Z) = W\ X\ \overline{Y}\ \overline{Z} + W\ \overline{X}\ \overline{Y}\ Z + \overline{W}\ \overline{X}\ Y\ Z + \overline{W}\ \overline{X}\ \overline{Y}\ Z + \overline{W}\ X\ Y\ Z$$

	$\uparrow\uparrow\uparrow\uparrow$	$\uparrow\uparrow\uparrow\uparrow$	$\uparrow\uparrow\uparrow\uparrow$	$\uparrow\uparrow\uparrow\uparrow$	$\uparrow\uparrow\uparrow\uparrow$
Binary no.	1 1 0 0	1 0 0 1	0 0 1 1	0 0 0 1	0 1 1 1
Decimal no.	1 2	9	3	1	7

Hence $f(W,X,Y,Z) = m_{12} + m_9 + m_3 + m_1 + m_7$, which can also be written as

$$f(W,X,Y,Z) = \sum m(1,3,7,9,12).$$

A function of n variables has 2^n *standard sum terms*, also called *maxterms*. Each maxterm is unique and is a sum (OR) of exactly n literals. It can be shown that any switching function may be expressed as a product of standard sum terms (also called *canonical sums*, or *conjunctive normal form*).

EXAMPLE 3.13. Express the following function in standard product-of-sums form:

$$f(X_3,X_2,X_1,X_0) = (X_3 + \overline{X}_2 + X_0)(\overline{X}_3 + X_1 + \overline{X}_0)(\overline{X}_3 + \overline{X}_2 + X_1 + X_0).$$

We shall first add the product $X_i \cdot \overline{X}_i$ to those sum terms that have less than four literals, where X_i is the literal absent in the sum term. Since $X_i \cdot \overline{X}_i = 0$ (*Th6a*), we have not changed the function.

$$f(X_3,X_2,X_1,X_0)$$
$$= (X_3 + \overline{X}_2 + \overline{X}_1 X_1 + X_0)(\overline{X}_3 + X_2 \overline{X}_2 + X_1 + \overline{X}_0)(\overline{X}_3 + \overline{X}_2 + X_1 + X_0).$$

Applying the distributive law, *Th3b*, we have

$$(X_j + X_i \cdot \overline{X}_i) = (X_j + X_i)(X_j + \overline{X}_i),$$
$$f = (X_3 + \overline{X}_2 + X_1 + X_0)(X_3 + \overline{X}_2 + \overline{X}_1 + X_0)(\overline{X}_3 + X_2 + X_1 + \overline{X}_0)$$
$$\cdot (\overline{X}_3 + \overline{X}_2 + X_1 + \overline{X}_0)(\overline{X}_3 + \overline{X}_2 + X_1 + X_0).$$

In a shorter notation each maxterm shall be represented by M_i, where the subscript i is the decimal equivalent of a binary number obtained after each literal has been complemented, as shown below.

EXAMPLE 3.14. Express the following standard product-of-sums using the M_i notation:

$$f(W,X,Y,Z) = (\overline{W} + \overline{X} + Y + Z)(\overline{W} + X + Y + \overline{Z})(W + X + \overline{Y} + \overline{Z})(W + X + Y + \overline{Z})(W + \overline{X} + \overline{Y} + \overline{Z})$$

Complementing	$\uparrow\ \uparrow\ \uparrow\ \uparrow$	$\uparrow\ \uparrow\ \uparrow\ \uparrow$	$\uparrow\ \uparrow\ \uparrow\ \uparrow$	$\uparrow\ \uparrow\ \uparrow\ \uparrow$	$\uparrow\ \uparrow\ \uparrow\ \uparrow$
literals	(1 1 0 0)	(1 0 0 1)	(0 0 1 1)	(0 0 0 1)	(0 1 1 1)
Decimal equivalent	12	9	3	1	7

Thus $f(W,X,Y,Z) = M_1 \cdot M_3 \cdot M_7 \cdot M_9 \cdot M_{12}$, which can also be written as $f(W,X,Y,Z) = \prod M(1,3,7,9,12)$.

A listing of minterms and maxterms for a 3-variable function is shown in Figure 3.13. We have seen in Section 3.2 that a switching function can be specified by a truth table. Each row of such a table contains all the literals and hence may represent a minterm or a maxterm of a function. Note the duality in Figure 3.13 between minterms and maxterms: applying

Decimal number	X Y Z	Minterms	Maxterms
0	0 0 0	$\bar{X}\ \bar{Y}\bar{Z} = m_0$	$X + Y + Z = M_0$
1	0 0 1	$\bar{X}\ \bar{Y}Z = m_1$	$X + Y + \bar{Z} = M_1$
2	0 1 0	$\bar{X}\ Y\bar{Z} = m_2$	$X + \bar{Y} + Z = M_2$
3	0 1 1	$\bar{X}\ YZ = m_3$	$X + \bar{Y} + \bar{Z} = M_3$
4	1 0 0	$X\ \bar{Y}\bar{Z} = m_4$	$\bar{X} + Y + Z = M_4$
5	1 0 1	$X\ \bar{Y}Z = m_5$	$\bar{X} + Y + \bar{Z} = M_5$
6	1 1 0	$X\ Y\bar{Z} = m_6$	$\bar{X} + \bar{Y} + Z = M_6$
7	1 1 1	$X\ YZ = m_7$	$\bar{X} + \bar{Y} + \bar{Z} = M_7$

FIGURE 3.13. Minterms, m_i, and maxterms, M_i, of a 3-variable function.

DeMorgan's theorem to each minterm yields a maxterm with the same subscript in the same row, and vice versa.

A switching function of n variables can be represented by one and only one sum of minterms:

$$f(X_{n-1},\ldots,X_0) = \sum_{i=0}^{n-1} a_i m_i,$$ (3.1)

where a_i is 0 or 1.

EXAMPLE 3.15.   Given the truth table of Figure 3.14, express the function $f(X,Y,Z)$ in standard S-of-P (sum-of-products) form.

X	Y	Z	f
0	0	0	0
0	0	1	0
0	1	0	1
0	1	1	1
1	0	0	0
1	0	1	1
1	1	0	1
1	1	1	0

FIGURE 3.14.   Truth table of a 3-variable function.

The function has a value 1 in rows 2, 3, 5, and 6 and a value 0 in the remaining rows. Thus from eq. (3.1)

$$f(X,Y,Z)$$
$$= 0 \cdot m_0 + 0 \cdot m_1 + 1 \cdot m_2 + 1 \cdot m_3 + 0 \cdot m_4 + 1 \cdot m_5 + 1 \cdot m_6 + 0 \cdot m_7$$
$$= m_2 + m_3 + m_5 + m_6 = \sum m(2,3,5,6).$$

A switching function of n variables can be represented by one and only one product of maxterms:

$$f(X_{n-1},\ldots,X_0) = \prod_{i=0}^{n-1} (a_i + M_i), \tag{3.2}$$

where a_i is 0 or 1.

EXAMPLE 3.16.   Express the function of Figure 3.14 in standard P-of-S (product-of-sums) form.
   From eq. (3.2),

$$f(X,Y,Z)$$
$$= (0 + M_0)(0 + M_1)(1 + M_2)(1 + M_3)(0 + M_4)(1 + M_5)(1 + M_6)(0 + M_7).$$

Each sum term $(1 + M_i) = 1$ does not change the product-of-sums expression and hence can be eliminated. Thus

$$f(X,Y,Z) = M_0 \cdot M_1 \cdot M_4 \cdot M_7 = \prod M(0,1,4,7).$$

Either eq. (3.1) or (3.2) may be utilized to represent the same switching function. Generally one form will be more economical than the other, except in the case when the number of 0 entries in the truth table is equal to the number of the 1 entries. Such functions are sometimes referred to as *neutral functions*.

EXAMPLE 3.17.   Simplify the 4-variable function

$$f(W,X,Y,Z) = \sum m(0,2,3,4,5,6,7,8,9,10,11,12,13,14).$$

The standard S-of-P form has 14 terms. The same function expressed in P-of-S form has $2^4 - 14 = 2$ terms only. Thus

$$\prod M(1,15) = (W + X + Y + \overline{Z})(\overline{W} + \overline{X} + \overline{Y} + \overline{Z}).$$

A simplified expression could have been obtained from the sum-of-products form only after considerable algebraic manipulation.

3.5   Karnaugh Maps and Simplification of Functions[3]

The Karnaugh map is probably the most extensively used tool for simplification of functions with up to six variables. It is a graphic method of presenting the information contained in a truth table and is easy to apply because of the pattern recognition power of the human mind.

The map consists of 2^n squares (elements) for an n-variable function, each element representing a standard term. The assignment of the element locations is made in such a way that two adjacent terms differ in one literal only. We shall refer to such terms as *adjacencies*. Each minterm for which a function is 1 is represented by a 1 in the respective square. For example, Figure 3.15a represents XY, 3.15b represents $\overline{X}Y + XY + X\overline{Y}$, and 3.15c represents $\overline{X}Y + X\overline{Y}$.

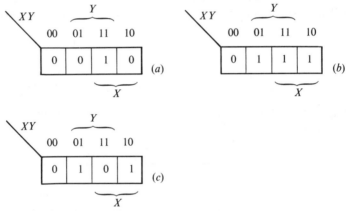

FIGURE 3.15.   Karnaugh maps for 2-variable functions: (a) XY, (b) $\overline{X}Y + XY + X\overline{Y}$, (c) $\overline{X}Y + X\overline{Y}$.

Notice that the elements at the opposite ends of the map are also adjacencies since they differ in one literal only, one representing the term $\overline{X}\overline{Y}$, the other $X\overline{Y}$.

Thus the Karnaugh map of Figure 3.15 may be viewed as a cylindrical closed surface. Each element of the surface is marked by a binary number, the most significant bit referring to the variable X, the least significant bit referring to Y. For example, an element marked on top by 01 represents the minterm $\overline{X}Y$. In addition we have marked by curly brackets the elements in which $X = 1$, and also the elements in which $Y = 1$. It follows that the variable X is in state 0 in the elements not covered by the curly bracket marked "X," and a similar statement applies to Y.

Figure 3.15 shows one Karnaugh map representation. A more common way of arranging the cells is shown in Figure 3.16a in which each cell is represented by a binary coordinate number as follows: viewing X and Y as binary digits, with X the most significant bit, we assign a decimal number to each cell that is equal to the decimal equivalent of the binary coordinates. For example, the cell marked "2" in Figure 3.16a has one coordinate $X = 1$ and the other coordinate $Y = 0$. The two coordinate numbers form a binary number 10, i.e., decimal 2. Notice that the cell numbers in a Karnaugh map are equal to the row numbers in a truth table (see for example Figure 3.13), and are also equal to the i-subscripts of the minterms.

Two 3-variable maps are shown in Figure 3.16b and c. Either cell arrangement is valid and both are in use. Again, the numbers assigned to the cells are the decimal values of the binary numbers formed from the coordinates, assuming X the most significant bit (msb), Z the least significant bit (lsb). The binary values for the coordinates, and hence the cell numbering, have been chosen to ensure that each cell has the adjacency property with respect to its neighboring cells.

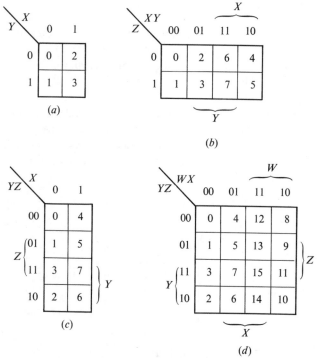

FIGURE 3.16. Karnaugh maps and cell designations: (a) 2-variable function, (b) and (c) 3-variable functions, (d) 4-variable function.

In the 4-variable map of Figure 3.16*d* the cell assignments have been made from the coordinates of the variables $WXYZ$ in which W is the msb. Each element has four adjacent elements corresponding to the four literals in which a term can differ. Also the elements in the upper row are adjacencies of the corresponding elements in the lower row. For example, cell 4 which represents the term $\overline{W}X\overline{Y}\overline{Z}$ is an adjacency to cell 6 which represents $\overline{W}XY\overline{Z}$ since the two terms differ in the literal Y only. Similarly, the four elements in the rightmost column marked by the coordinates $WX = 10$ are adjacencies of the corresponding elements in the leftmost column $(WX = 00)$.

To "map" a function given in minterms, we enter a 1 in each cell that has the same number as the subscript of the respective minterm that it represents. Similarly, to "map" a function given in maxterms, we enter a 0 in each cell that has the same number as the subscript of the respective maxterm that it represents.

EXAMPLE 3.18.   Develop the Karnaugh map for the function $f(W,X,Y,Z)$ $= \sum m(4,6,12,13,14)$.

The function is a standard sum-of-products of minterms m_4, m_6, etc. In other words, $f(X,W,Y,Z) = 1$ if any one of the five minterms exists. The subscript i of m_i has the same numerical value as the cell assignment. Therefore we enter a 1 in cell number 4 representing m_4, a 1 in cell 6 to represent m_6, etc. The five entries of the given function are shown in the Karnaugh map of Figure 3.17.

Suppose now that a function is given in a standard P-of-S form. In this case we enter a 0 in the elements corresponding to the given maxterms. Subscripts i of M_i have the same numerical values as the cell numbers in which the 0's are entered. The 0's in the map of Figure 3.17 thus represent the function

$$f(W,X,Y,Z) = \prod M(0,1,2,3,5,7,8,9,10,11,15).$$

YZ \ WX	00	01	11	10
00	0	1	1	0
01	0	0	1	0
11	0	0	0	0
10	0	1	1	0

FIGURE 3.17.   Map for the 4-variable function $f(W,X,Y,Z) =$ $\sum m(4,6,12,13,14)$.

Using DeMorgan's theorem, we can easily show that the given standard S-of-P and the P-of-S functions are equivalent.

Simplification.   Simplification of switching functions is based on recognition of adjacencies of 1's or 0's in a Karnaugh map. For example, an entry of 1 without adjacencies represents one minterm. For each two adjacent cells with "1" entries the corresponding product term will have $n - 1$ variables in an n-variable function. This follows from the application of the relation $yX + y\overline{X} = y(X + \overline{X}) = y$, where y is a product of $n - 1$ variables. For example, the terms $\overline{W}X\overline{Y}Z$ and $\overline{W}XYZ$ are adjacent since they differ in the literal Y only. Thus $\overline{W}X\overline{Y}Z + \overline{W}XYZ = \overline{W}XZ(Y + \overline{Y}) = \overline{W}XZ$.

A 4-cell adjacency represents a product term of $n - 2$ variables in an n-variable function. In general, 2^i adjacent cells represent a product term of $n - i$ variables in an n-variable function.

EXAMPLE 3.19.   Simplify the function $f_a(W,X,Y,Z) = \sum m(5,7,13,15)$.

The map for the function is shown in Figure 3.18a. The entries occupy cells for which the coordinate $X = 1$ and $Z = 1$. On the other hand, the

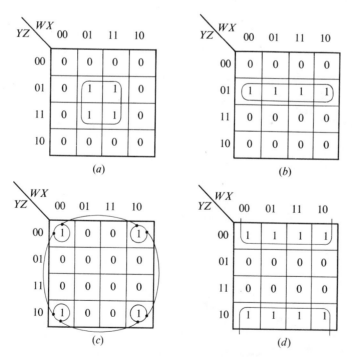

FIGURE 3.18.   Adjacencies in 4-variable functions.

coordinate W has a value 0 for m_5 and m_7, and a value 1 for m_{13} and m_{15}. The variable W is therefore eliminated from the product term. Algebraically we have

$$m_5 + m_7 + m_{13} + m_{15} = \overline{W}X\overline{Y}Z + \overline{W}XYZ + WX\overline{Y}Z + WXYZ$$
$$= X\overline{Y}Z(\overline{W} + W) + XYZ(\overline{W} + W) = X\overline{Y}Z + XYZ.$$

A similar argument shows that the coordinate Y is 0 for m_5 and m_{13}, and 1 for m_7 and m_{15}. The variable Y can therefore be eliminated from the product term, yielding

$$f_a(W,X,Y,Z) = XZ.$$

The function $f_b(W,X,Y,Z) = \sum m(1,5,9,13)$ is shown in Figure 3.18b. It can be seen that $f_b = 1$ regardless of the variables W and X. Also, the row of adjacencies has a 0 for the Y coordinate and a 1 for the Z coordinate. Thus $f_b(W,X,Y,Z) = \overline{Y}Z$.

EXAMPLE 3.20.   Simplify the function $f_c(W,X,Y,Z) = \sum m(0,2,8,10)$.

The entries of the minterms are shown in Figure 3.18c. The adjacencies here are not as obvious as in the previous examples. We have, therefore, looped the adjacent cells and marked them by connecting lines. (This technique will be most useful in dealing with 5- and 6-variable maps.) The function is independent of variables W and Y, and all the 1-entries are in cells having coordinates $\overline{X}$ and $\overline{Z}$. Thus $f_c(W,X,Y,Z) = \overline{X}\overline{Z}$.

To simplify $f_d(W,X,Y,Z) = \sum m(0,2,4,6,8,10,12,14)$ of Figure 3.18d we note that the upper and lower rows are adjacent, as indicated by the incompleted loops. In a 4-variable function, 2^3 adjacent minterms will result in a simplified expression having $4 - 3 = 1$ literal. The function is clearly independent of WX since the entries are in a row where the coordinates take on the four possible combinations of W with X. The upper row is 1 when the Y-coordinate is 0, while the lower row is 1 for a Y-coordinate = 1, i.e., the function is also not dependent on Y. Thus

$$f_d(W,X,Y,Z) = \overline{Z}.$$

The INCLUSIVE–OR of two or more functions can be obtained with aid of maps. For example, the OR of the functions of Figures 3.18a and b is the logical sum of the simplified functions $f_a + f_b = XZ + \overline{Y}Z$. When combining functions through a logical sum, care must be exercised to eliminate redundant terms. Thus the function f_c mapped in Figure 3.18c is a subset of the function f_d of Figure 3.18d, and the logical OR of $f_c + f_d = f_d = \bar{Z}$.

The problem of finding a minimum solution by map method has not been solved in a general way. The designer has to use his skill and experience to ensure a minimum, or near minimum, solution. The following rules may be used as guidelines.

1. First determine the *essential products*. These are products that contain a minterm that cannot be covered by any other set.
2. Second, combine as many minterms as possible to obtain the largest set of adjacencies.
3. Try to cover each minterm once. This may not be the best approach in every case. For example, taking the logical sum (OR) of the functions of Figures 3.18a and b, we cover minterms m_5 and m_{13} twice, obtaining $f_a + f_b = XZ + \bar{Y}Z$. Covering each minterm once only yields two alternative expressions, $XZ + \bar{X}\bar{Y}Z$ and $\bar{Y}Z + XYZ$.
4. In some cases a result with fewer terms will be obtained through selection of the 0 adjacencies of the function to obtain $\bar{f}$. Application of DeMorgan's theorem will then yield f. Alternatively, the same result may be accomplished by combining maxterms to obtain a minimized expression.

The rules were presented above for functions given in a standard S-of-P form. With minor modifications they may be applied to functions given in a standard P-of-S form.

Simplification of 5-Variable Functions.   A Karnaugh map for a 5-variable function $f(V,W,X,Y,Z)$ is shown in Figure 3.19. It is composed of two maps of 16 cells each. The left map represents all the terms for which $V = 0$, while the right map represents all the terms for which $V = 1$. This can also be seen in the assignment of the decimal numbers for the cells, since the

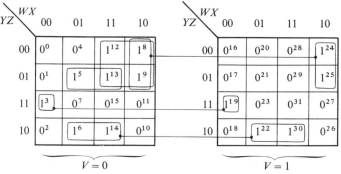

FIGURE 3.19.   Map for a 5-variable function.

variable V of the given function is assumed to carry a weight of 2^4. The cell assignments in the two maps were chosen in such a way that two cells are adjacent if they occupy the same relative positions. For example, cells 1 and 17 are adjacencies; cells 3 and 7 occupy the same relative positions as cells 19 and 23, and hence form a 4-term adjacency set: 3, 7, 19, and 23. Similarly, 0, 16, 2, 18, 8, 24, 10, and 26 are an adjacent set.

EXAMPLE 3.21.   Simplify the function

$$(V,W,X,Y,Z) = \sum m(3,5,6,8,9,12,13,14,19,22,24,25,30).$$

The minterm entries of the function are shown in Figure 3.19. One essential product is $\overline{V}X\overline{Y}Z$ in the left map. Cell 12 is covered best by including it in the 4-cell adjacency set yielding the product $\overline{V}W\overline{Y}$. Other essential products are due to adjacencies of terms in different maps, i.e., the resultant terms will be independent of V. Thus cells 3 and 19 yield the term $\overline{W}\overline{X}YZ$; cells 6, 14, 22, and 30 yield $XY\overline{Z}$; cells 8, 9, 24, and 25 yield $W\overline{X}\overline{Y}$. Note that cells 8, 9, and 13 have been covered twice. The minimized expression is thus

$$f(V,W,X,Y,Z) = \overline{V}X\overline{Y}Z + \overline{V}W\overline{Y} + \overline{W}\overline{X}YZ + XY\overline{Z} + W\overline{X}\overline{Y}.$$

Simplification of 6-Variable Functions.   The method shown earlier for construction of maps for 5-variable functions may be extended to 6-variable functions. In Figure 3.20a we have drawn four 4-variable maps. All minterms in the uppermost map contain the literals $\overline{A}\overline{B}$, in the next map the literals $\overline{A}B$, etc. An element in each map is adjacent to an element in a neighboring map if it occupies the same relative position. Notice that adjacencies may be also obtained from terms in the uppermost map $\overline{A}\overline{B}$ and the lowest map $A\overline{B}$. In Figure 3.20b we show a different cell assignment: the coordinates along the horizontal and vertical edges vary by one binary digit only. Cells that are adjacent but in different quadrants can be found from the mirror symmetry around the central two axes.

Incompletely Specified Functions.   In the previous section we have assumed that in each given problem all possible combinations have been specified. It is reasonable to assume that in some problems certain input combinations do not occur. Also, certain outputs of combinational networks may have no effect on an overall system, and hence the designer does not care about the corresponding input combinations. These *don't care conditions* can be utilized in simplifications of functions as shown below.

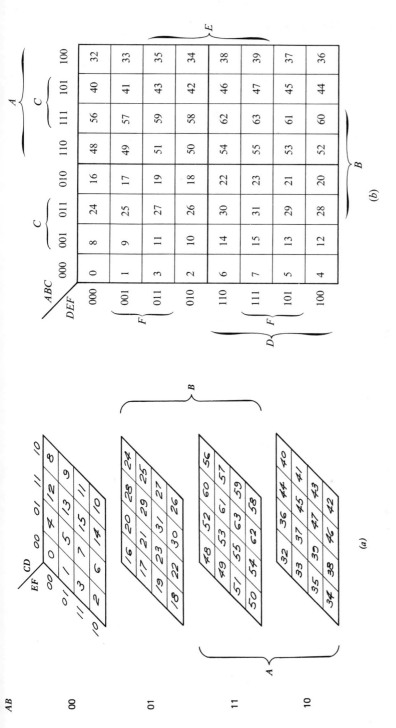

FIGURE 3.20. Maps for 6-variable functions.

EXAMPLE 3.22.   Design a minimum network that accepts a binary number and delivers ten unique outputs.

At least four variables are required to realize the network. Let us assign numbers to the outputs that are the decimal equivalents of the corresponding input combinations. The truth table of the assignment is shown in Figure 3.21a. Of the 16 available states, six are redundant. Let us assume that the combinations of these six states never appear at the

	W	X	Y	Z	Decimal no.		W	X	Y	Z
						Minimized inputs				
	0	0	0	0	0		0	0	0	0
	0	0	0	1	1		0	0	0	1
	0	0	1	0	2		—	0	1	0
	0	0	1	1	3		—	0	1	1
	0	1	0	0	4		—	1	0	0
	0	1	0	1	5		—	1	0	1
	0	1	1	0	6		—	1	1	0
	0	1	1	1	7		—	1	1	1
	1	0	0	0	8		1	—	—	0
	1	0	0	1	9		1	—	—	1
Nonexistent input combinations	1	0	1	0	10	No output				
	1	0	1	1	11					
	1	1	0	0	12					
	1	1	0	1	13					
	1	1	1	0	14					
	1	1	1	1	15					

(a)

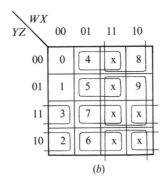

(b)

FIGURE 3.21.   Truth table and map for Example 3.22.

input of the network. We can thus utilize these states as *don't care conditions*, which are marked by x in the corresponding cells of the Karnaugh map of Figure 3.21b. (Other symbols for don't care conditions found in the literature are *d*, "$\mathbf{0}$", and "—".) Utilization of these don't care conditions in forming adjacencies is shown in Figure 3.21b. The resultant minimized input combinations are shown in the last column of the truth table. A dash "—" signifies that the particular literal has been eliminated from the product term.

The function of Figure 3.21b may be expressed in short notation as

$$f(W,X,Y,Z) = \sum m(0,1,2,3,4,5,6,7,8,9) + \sum d(10,11,12,13,14,15),$$

where $\sum d(i,j,k,...)$ represents the don't care conditions for minterms m_i, m_j, m_k, ...

When minimizing functions to obtain product-of-sums expressions, the don't care conditions may be combined with 0 entries.

EXAMPLE 3.23.   Determine the minimal P-of-S realization of the function $f(W,X,Y,Z) = \prod M(1,3,8,10) \prod d(9,11)$.

The reader can easily verify by drawing a Karnaugh map that maxterms M_1 and M_3, together with don't care conditions in cells 9 and 11, form a 4-cell adjacency yielding the sum term $X + \bar{Z}$. Combining the same don't care conditions with maxterms M_8 and M_{10} yields $\overline{W} + X$. Thus $f(W,X,Y,Z) = (X + \bar{Z})(\overline{W} + X)$.

3.6   Quine-McCluskey Minimization Algorithm

The Karnaugh map discussed in the preceding section has the advantage that it yields simplified solutions with relatively little effort. We shall apply this technique throughout the text for simplification of functions. The Karnaugh map method, however, is limited in practice to a maximum of six variables. Also, since it is an intuitive method, as distinct from an algorithmic procedure, it does not ensure minimality of the results. For the same reason it is not adaptable for use with a computer.

The Quine-McCluskey method is based on an algorithm requiring an exhaustive comparison of all the terms of the function to be minimized and is thus adaptable to be mechanized as a computer program. We shall briefly discuss the algorithm because of its general importance; a thorough treatment of the subject may be found in references (4) and (5).

The algorithm will be explained with aid of the example shown in Figure 3.22. The function to be minimized is given by the minterms

No. of 1's		Minterms			
		A	B	C	D
0	m_0	0	0	0	0 ✓
1	m_2	0	0	1	0 ✓
1	m_8	1	0	0	0 ✓
2	m_3	0	0	1	1 ✓
2	m_5	0	1	0	1 ✓
2	m_{10}	1	0	1	0 ✓
3	m_7	0	1	1	1 ✓
3	m_{11}	1	0	1	1 ✓
3	m_{13}	1	1	0	1 ✓
4	m_{15}	1	1	1	1 ✓

(a)

Index	Result of first comparison			
0, 2	0	0	x	0 ✓
0, 8	x	0	0	0 ✓
2, 3	0	0	1	x ✓
2, 10	x	0	1	0 ✓
8, 10	1	0	x	0 ✓
3, 7	0	x	1	1 ✓
3, 11	x	0	1	1 ✓
5, 7	0	1	x	1 ✓
5, 13	x	1	0	1 ✓
10, 11	1	0	1	x ✓
7, 15	x	1	1	1 ✓
11, 15	1	x	1	1 ✓
13, 15	1	1	x	1 ✓

(b)

Index	Result of second comparison			
* 0, 2, 8, 10	x	0	x	0
* 2, 3, 10, 11	x	0	1	x
* 3, 7, 11, 15	x	x	1	1
* 5, 7, 13, 15	x	1	x	1

(c)

FIGURE 3.22.   Determination of prime implicants.

$f(A,B,C,D) = \sum m(0,2,3,5,7,8,10,11,13,15)$. (In case the function is not given in standard product form it should be expanded as shown in Section 3.4.)

As a first step, all minterms are grouped according to the number of 1's they contain, as shown in Figure 3.22a. Next, every minterm of one group is compared with every minterm of its neighboring group. Minterms under comparison that differ in one literal only are combined to eliminate that literal since

$$g \cdot A + g \cdot \bar{A} = g \cdot (A + \bar{A}) = g,$$

where g is the product term containing all literals other than the literal A. The results of all possible comparisons are listed in another table, the eliminated literals being replaced by x. For example, comparison of m_0 with m_2 yields 00x0, which is shown as the first entry in Figure 3.22b. The column marked "index" shows the subscript numbers of the minterms involved in the comparison that resulted in the corresponding entry under the column "Result of first comparison." Combining m_0 and m_8 yields x000, which completes the comparison of minterms having no 1's with minterms having one 1.

Next we compare each minterm of the group having one 1 with each minterm of the group having two 1's, etc., up to the last group having four 1's. These results are also listed in Figure 3.22b.

The same procedure is now applied to the newly obtained listing and tabulated in Figure 3.22c. Only terms with x's in the same relative position need be compared. (Why? Write down the terms involved in a comparison in which the x's do not occupy the same relative positions.) The index should now show four decimal numbers corresponding to the indices of the terms compared. Note that row (0,2) combined with row (8,10) yields the same result as comparison of row (0,8) with row (2,10), and hence only one listing of the combinations is required. This exhaustive procedure ends when no comparisons of rows are possible. The resultant terms, called *prime implicants*, may cover a minterm more than once. This is analogous to an element in a Karnaugh map being included in more than one adjacency set.

In our example all prime implicants were obtained in the last comparison. This is generally not to be expected. Some terms in the first listing may not combine with any other terms, in which case they are prime implicants by definition. An asterisk is commonly placed to the left of the index numbers to distinguish prime implicants from other terms.

The final step determines the minimum number of optional prime implicants required to cover all terms of the function. The resultant terms are called *essential prime implicants*, and their Boolean sum is the desired *minimal S-of-P form* of the given function.

To find the essential prime implicants, we draw a table as shown in Figure 3.23. The numbers in the first column represent the prime implicants

	0	2	3	5	7	8	10	11	13	15
* 0, 2, 8, 10	✓	✓				✓	✓			
2, 3, 10, 11		✓	✓				✓	✓		
3, 7, 11, 15,			✓		✓			✓		✓
* 5, 7, 13, 15				✓	✓				✓	✓

FIGURE 3.23.   Determination of essential prime implicants.

obtained previously. (In our example all prime implicants happen to be in the last listing, i.e., in Figure 3.22c.) Each heading of the remaining columns represents one minterm of the original function. We now examine each row to see which minterms it covers. For example, row (0,2,8,10) covers minterms m_0, m_2, m_8, and m_{10}. This can be checked as follows. The listing in Figure 3.22c shows the expression x0x0 opposite the index (0,2,8,10). This expression represents the product $\bar{B}\bar{D}$, the other two literals having been removed in the comparison process. We now have to show that the minterms m_0, m_2, m_8, and m_{10} are indeed covered by the product $\bar{B}\bar{D}$. Expanding $\bar{B}\bar{D}$, we have

$$\bar{B}\bar{D} = (A + \bar{A})\bar{B}(C + \bar{C})\bar{D} = A\bar{B}C\bar{D} + \bar{A}\bar{B}C\bar{D} + \bar{A}\bar{B}\bar{C}\bar{D} + A\bar{B}\bar{C}\bar{D}$$

$$= m_{10} + m_2 + m_0 + m_8 .$$

Check marks are placed in row (0,2,8,10) under columns 0, 2, 8, and 10, indicating the respective minterms that have been covered by this prime implicant. The same procedure is repeated in all rows. One minterm may be covered by several rows. We first look for the minterm that is checked in *one row* only. The corresponding number (0,2,8,10) in the leftmost column represents an *essential prime implicant*, which we mark with an asterisk. Columns 5 and 13 show one check mark each; thus (5,7,13,15), i.e., *BD*, is also an essential prime implicant.

At this point columns 3 and 11 have not been covered. Either of the two remaining prime implicants may be selected as an essential prime implicant. Two equally minimal S-of-P forms of the function are thus obtained,

$$f(A,B,C,D) = \bar{B}\bar{D} + BD + \bar{B}C, \quad \text{or} \quad f(A,B,C,D) = \bar{B}\bar{D} + BD + CD.$$

The Quine-McCluskey minimization algorithm was demonstrated using an easy example. In general, the selection of essential prime implicants is not as simple. Methods for systematic selection of essential prime implicants can be found in reference (5).

PROBLEMS

3.1 In a committee consisting of three members, each member has control of one of the three switches *A*, *B*, and *C*. Voting in favor of a measure is effected by closing a switch. A "no" vote is delivered by an open switch. Design a network that will assume state 1 *iff* the measure passes by a majority vote.

3.2 Check Theorems 3a and 3b using Venn diagrams and truth tables.

3.3 Using switches, check theorems $Th1$ through $Th8$ given in Section 3.1.

3.4 Simplify the 4-variable switching function $f(W,X,Y,Z) = X + XYZ + \overline{X}YZ + \overline{X}Y + WX + \overline{W}X$ and show the theorems used in the process. Note that the steps leading to the solution are not unique. A minimum number of steps is preferred.

3.5 Using DeMorgan's theorem, complement the function

$$X \cdot (\overline{Y} + Z\overline{W} + \overline{V}S).$$

3.6 Simplify the switching function $f(X,Y) = X + \overline{X}Y$, applying complementation as the first step.

3.7 Draw a Venn diagram and show the truth table for the switching function $(\overline{X}Y + Z)$.

3.8 Simplify the switching function given in Example 3.5 using (a) algebraic manipulation, and (b) Venn diagrams. (Hint: Draw several diagrams, each representing one or preferably two terms of the function.)

3.9 Using a truth table, show that $XY + YZ + \overline{X}Z = XY + \overline{X}Z$.

3.10 Simplify the function $f = (A + B)(A + BC) + \overline{A}B + \overline{A}\overline{C} + \overline{A}BC$. Is the output dependent on the states of the input variables? Explain.

3.11 Simplify the network of Figure 3.10.

3.12 Write the complements of the functions
(a) $f = [A + \overline{B}\overline{C}\overline{D}] \cdot [\overline{A}\overline{D} + B(\overline{C} + A)]$
(b) $f = A\overline{B}C + (\overline{A} + B + D) \cdot (AB\overline{D} + \overline{B})$.

3.13 (a) Implement the function $f(A,B,C,D) = (AD + \overline{A}C) \cdot [\overline{B}(C + B\overline{D})]$.
(b) Simplify the function given in Problem 3.13(a) and draw the AND–OR network.

3.14 Prove the following two equations:

$$\overline{A} \cdot f(A,B,C,\ldots) = \overline{A} \cdot f(0,B,C,\ldots),$$
$$\overline{A} + f(A,B,C,\ldots) = \overline{A} \cdot f(1,B,C,\ldots).$$

3.15 (a) Construct a truth table and establish the switching function in sum-of-products form for a circuit which will be closed *iff* any two of five switches V, W, X, Y, and Z are closed.
(b) Repeat the above example to obtain a result in product-of-sums form.

3.16 Using a Karnaugh map, simplify the result obtained in Problem 3.15. Derive the simplified terms in S-of-P and P-of-S forms.

3.17 (a) Given a function in standard S-of-P form, derive a method to convert it to a standard P-of-S form.

(b) Derive a corresponding conversion procedure when the given function is in standard P-of-S form.

3.18 Use the Karnaugh map to simplify the functions (a), (b), and (c), below. Find the simplest S-of-P and P-of-S expressions.

(a) $f(X,Y,Z) = \prod M(0,1,6,7)$,

(b) $f(W,X,Y,Z) = \prod M(1,3,7,9,11,15)$,

(c) $f(V,W,X,Y,Z) = \prod M(0,4,18,19,22,23,25,29)$.

3.19 A manufacturing company has five plants V, W, X, Y, and Z in different locations. The company maintains a fleet of six planes A, B, C, D, E, and F to deliver supplies to the five plants. Each plane can supply a limited number of plants as shown in Figure 3.24.

Plane	$	Plants
A	12,000	V, X, Z
B	13,000	W, X, Z
C	10,000	V, W, Y
D	14,000	Y, Z
E	13,000	W, Y
F	14,000	V, X

FIGURE 3.24

The cost to send a plane also varies, as shown in the second column. Find which planes will supply the plants at least expense.

3.20 Using the Karnaugh map simplify the switching functions given below. Find the minimal S-of-P expressions and the minimal P-of-S expressions.

(a) $f(A,B,C,D) = \sum m(0,2,4,6,)$

(b) $f(A,B,C,D) = \sum m(0,1,4,5,12,13)$

(c) $f(A,B,C,D,E) = \sum m(0,4,18,19,22,23,25,29)$

(d) $f(A,B,C,D,E,F) = \sum m(3,7,12,14,15,19,23,27,28,29,31,35,39,$
$44,45,46,48,49,50,52,53,55,56,57,59)$.

3.21 Determine minimal P-of-S realizations for the functions given in Problem 3.20(a), (b), and (c) using the Quine-McCluskey method.

3.22 Determine minimal P-of-S realizations for the following functions:

(a) $f(W,X,Y,Z) = \sum m(0,2,8,9) + \sum d(1,3)$

(b) $f(W,X,Y,Z) = \sum m(1,7,11,13) + \sum d(0,5,10,15)$

(c) $f(V,W,X,Y,Z) = \sum m(2,8,9,10,13,15,16,18,19,23) + \sum d(3,11,17,22)$

(d) $f(V,W,X,Y,Z) = \sum m(0,1,2,9,13,16,18,24,25) + \sum d(8,10,17,19)$.

3.23 A switching function of nine variables has many unspecified (don't care) conditions. The function is to be in state 1 for standard products m_{187} and m_{205}. The function is to be in state 0 for standard products m_{17}, m_{89}, m_{251}. Determine the minimal P-of-S realization of the above specification.

CHAPTER 4

Combinational Logic Networks

In the previous chapter we have discussed the analysis and synthesis of logic networks using operations AND, OR, and NOT. Several methods for simplification of logic expressions were demonstrated. In this chapter we introduce additional logic operations, concentrating on those that have been implemented in integrated circuit form.

A physical device that realizes a logic operation is called a "gate." The availability of a large variety of gates is of importance to the logic designer who has to select a particular set of building blocks for an efficient solution of a given problem.

Techniques for combinational network design with NAND and NOR gates are shown to be simple extensions of the design procedures developed in the preceding chapter. Three-stage combinational design with NAND and NOR gates is demonstrated to have advantages where complements of the variables are not available. Multiple output switching functions are briefly treated and logic design with AND–OR–INVERT gates, EXCLUSIVE–OR gates, and "emitter-coupled" logic is discussed.

We also show how a digital multiplexer, which typically is a medium-scale integrated (MSI) circuit, can be utilized in the synthesis of combinational networks. Such application of an MSI circuit does not generally result in a high efficiency of gate utilization as compared with alternative solutions using small-scale integration (SSI). However, the economics of integrated circuit usually favors the MSI solution. Furthermore, LSI circuits may be especially designed and optimized to perform a given function. This aspect is further discussed in Chapter 11 in connection with "programmable logic arrays."

4.1 Logic Functions of Two Variables

Two binary variables have 2^2 possible combinations $\overline{X}\overline{Y}$, $\overline{X}Y$, $X\overline{Y}$, and XY, which can be represented in a truth table by four columns; the 4 combinations of 2 variables generate $2^4 = 16$ unique functions, as shown in Figure 4.1. Several functions, also called *operations*, are of special interest because they have been implemented in integrated circuit form. These are: AND (f_1), OR (f_7), NAND (f_{14}), NOR (f_8), EXCLUSIVE–OR (f_6), and EQUALITY (sometimes also called COINCIDENCE or EXCLUSIVE–NOR), f_9.

The algebra developed in the previous chapter used a set of three operations NOT, AND, and OR. It can be shown that all possible combinations of an n-variable function can be synthesized using the two operations NOT and AND; another possibility is to use NOT and OR. A set of operations is considered functionally complete *iff* (if and only if) all possible functions can be synthesized using the set. Operation NAND is complete by itself, and so are operations NOR, IMPLICATION (f_{11} or f_{13}), and INHIBIT (f_2 or f_4). Other complete sets are EXCLUSIVE–OR with AND, and EQUALITY with AND. If a set is functionally complete, then it should be possible to synthesize other operations using the given

f	X 0 0 1 1 Y 0 1 0 1	Name	Expression	Other names	Other symbols
f_0	0 0 0 0	ZERO	0	NULL FUNCTION	
f_1	0 0 0 1	AND	$X \cdot Y$	CONJUNCTION	$\wedge$
f_2	0 0 1 0	INHIBIT	$X \cdot \overline{Y}$		$\not\subset$
f_3	0 0 1 1	IDENTITY X	X		
f_4	0 1 0 0	INHIBIT	$\overline{X} \cdot Y$		$\not\subset$
f_5	0 1 0 1	IDENTITY Y	Y		
f_6	0 1 1 0	EXCLUSIVE–OR	$X \oplus Y$		$\neq$
f_7	0 1 1 1	OR	$X + Y$	DISJUNCTION	$\vee$
f_8	1 0 0 0	NOR	$\overline{X + Y}$	PIERCE FUNCTION	$\vee$,↓
f_9	1 0 0 1	EQUALITY	$X \odot Y$	(COINCIDENCE (EXCLUSIVE–NOR	$\equiv$
f_{10}	1 0 1 0	NOT Y	$\overline{Y}$		Y', $\sim Y$
f_{11}	1 0 1 1	IMPLICATION	$X + \overline{Y}$		$\subset$
f_{12}	1 1 0 0	NOT X	$\overline{X}$		X', $\sim X$
f_{13}	1 1 0 1	IMPLICATION	$\overline{X} + Y$		$\supset$
f_{14}	1 1 1 0	NAND	$\overline{X \cdot Y}$	SHEFFER STROKE	⋏,↑
f_{15}	1 1 1 1	ONE	1	IDENTITY FUNCTION	

FIGURE 4.1. The sixteen functions of two variables.

set (see Problem 4.13). Conversely, given a set of operations we can check for the completeness of the set by synthesizing other operations, such as NOT and AND, which we know to be a complete set: this is illustrated in the example that follows.

EXAMPLE 4.1. Show that the NAND operation $f = \overline{X \cdot Y}$ is functionally complete. Assume that logic levels corresponding to 0 and 1 are given.

The NAND operation is shown in Figure 4.1 as $f_{14} = \overline{X \cdot Y}$. To obtain the NOT operation substitute 1 for Y, thus $\overline{X \cdot 1} = \overline{X} + \overline{1} = \overline{X} + 0 = \overline{X}$. The AND operation is obtained from a NOT operating on a NAND; thus $(\overline{\overline{X \cdot Y}}) = X \cdot Y$.

We could also check each operator for its algebraic properties such as associativity, distributivity, etc. These properties would enable us to establish a set of theorems that could be used in synthesizing and minimizing switching functions. Fortunately this rather complicated approach is not required in practice. For example, in digital network synthesis with NAND and NOR gates we can utilize, with minor modifications, all the tools for synthesis and minimization that were developed in Chapter 3 using AND, OR, and NOT operations.

4.2   Positive and Negative Logic

A truth table can be used to describe logic operations without reference to actual voltage levels with respect to ground potential. A variety of logic gates will be described in Chapter 5, some requiring positive, other negative voltages.

Assume that the higher potential, H, (with respect to ground) of a logic gate represents the logic "true," while the lower potential, L, represents the logic "false." With this convention, which is called *positive logic*, we obtain a truth table as shown in Figure 4.2.

Assume next that our convention is reversed, and L represents the logic "true" while H represents the logic "false." The table for the resultant *negative logic* is shown in Figure 4.3. Comparing the two tables we find the following:

1. Positive AND is identical to negative OR.
2. Positive OR is identical to negative AND.
3. Positive NAND is identical to negative NOR.
4. Positive NOR is identical to negative NAND.
5. EXCLUSIVE–OR and EQUALITY operations are not affected by the convention representing the logic "true" or "false." This follows from the definition of these operations.

Inputs X Y	Outputs					
	AND	OR	NAND	NOR	EXCL. OR	EQUA-LITY
L L	L	L	H	H	L	H
L H	L	H	H	L	H	L
H L	L	H	H	L	H	L
H H	H	H	L	L	L	H

FIGURE 4.2. Truth table, positive logic.

Inputs X Y	Outputs					
	AND	OR	NAND	NOR	EXCL. OR	EQUA-LITY
L L	L	L	H	H	H	L
L H	H	L	L	H	L	H
H L	H	L	L	H	L	H
H H	H	H	L	L	H	L

FIGURE 4.3. Truth table, negative logic.

EXAMPLE 4.2. In a logic circuit (gate) having two inputs and one output, the voltage levels at the terminals are 0 or $+3$ V. The voltage level on the output terminal is $+3$ V *iff* both inputs are $+3$ V. If the convention of Figure 4.2 is used, that is, $+3$ V represents "true" and 0 V represents "false," the output is "true" $(+3$ V) *iff* both inputs are "true" $(+3$ V); hence, the circuit acts as an AND circuit. If, however, the convention of Figure 4.3 is used, that is, 0 V represents "true" and $+3$ V represents "false," the output is "true" (0 V) if either one or both inputs are "true" (0 V); hence the circuit performs the OR function.

The terms *positive* and *negative* logic should not be taken to imply positive or negative potentials with respect to ground. For example, truth values in some logic gates (RTL, DTL, or TTL*) are represented by 0 V and $\approx +3$ V, while for emitter-coupled logic* the respective values are -0.75 V and -1.5 V. We shall use the term *positive logic* when the more positive of the two potentials represents the logic 1. Conversely, in *negative logic* the more negative (or the lower) of the two potentials represents the

---

* For a detailed description of various logic gates see Chapter 5.

logical 1. As we have seen in Example 4.2, a given circuit can be used in either logic convention; however, its logic operation has to be redefined.

These results have practical implications as they add substantially to design flexibility. Since the same physical gate may be used to perform two logic functions, in many cases it would be convenient to represent this in a graphic symbol. A circle at the input(s) or output(s) of a gate shall indicate that the lower of the two potentials is present at the respective point(s). With this convention and from the duality of gate functions shown in Figures 4.2 and 4.3 we can draw the graphic symbols for positive and negative logic as shown in Figure 4.4. Notice that *each row* shows the *identical physical device*, e.g., an integrated circuit (or part of it), and the

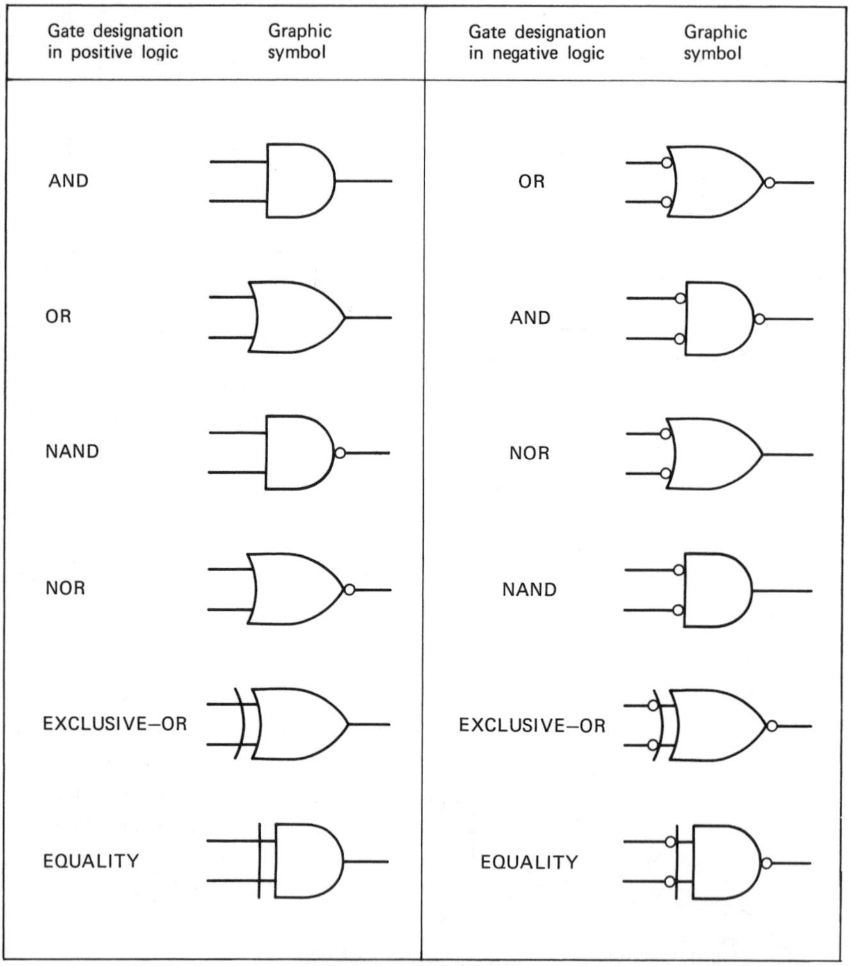

Gate designation in positive logic	Graphic symbol	Gate designation in negative logic	Graphic symbol
AND		OR	
OR		AND	
NAND		NOR	
NOR		NAND	
EXCLUSIVE–OR		EXCLUSIVE–OR	
EQUALITY		EQUALITY	

FIGURE 4.4. Graphic symbols for positive and negative logic.

logic functions that it performs. The duality of logic functions in the upper four rows can be easily proved by DeMorgan's theorem.

Several manufacturers of integrated circuits specify the duality of the operations by designating them as AND/OR or NAND/NOR, etc., or by using the term "positive logic." The majority of the specifications and some of the discussions later in this text ignore the complementary aspect and present the gates as AND, OR, NAND, NOR, etc.

4.3   Combinational Networks Using AND–OR Gates

2-Stage AND–OR Networks.   Realization of combinational networks using AND and OR gates follows directly from the expressions derived by the minimization techniques of Sections 3.5 and 3.6. The procedure using Karnaugh maps can be summarized as follows:

1.   Establish a truth table describing the function to be realized.
2.   Draw a Karnaugh map representing the function.
3.   Simplify the function to obtain a 2-stage expression in either sum-of-products (S-of-P) form, or product-of-sums (P-of-S) form. Implementation of the function with gates follows directly from step 3.

EXAMPLE 4.3.   Realize a circuit of three variables in which the output always agrees with the majority of the inputs. This circuit is also referred to as a "majority voter" circuit. It is used in applications of high reliability triple modular redundancy (TMR) circuits, in which all functions are tripled and the "majority voter" circuit is incorporated to resolve any discrepancies in the outputs in case of a failure of one channel of the TMR circuit.

The three steps leading to the implementation of the network are shown in Figure 4.5. The S-of-P realization yields $f(X,Y,Z) = XY + XZ + YZ$.

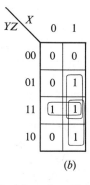

X	Y	Z	f
0	0	0	0
0	0	1	0
0	1	0	0
0	1	1	1
1	0	0	0
1	0	1	1
1	1	0	1
1	1	1	1

(a)

(b)

FIGURE 4.5.   "Majority voter" circuit: (a) truth table, (b) Karnaugh map. (*See next page for* (c) *and* (d).)

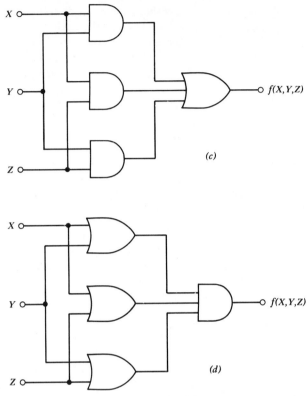

FIGURE 4.5. (c) S-of-P realization with AND–OR gates, (d) P-of-S relization with OR–AND gates.

The P-of-S expression derived from the 0 entries of the Karnaugh map yields $\bar{f}(X,Y,Z) = \overline{X}\overline{Y} + \overline{Y}\overline{Z} + \overline{X}\overline{Z}$, thus, $f(X,Y,Z) = \overline{\bar{f}}(X,Y,Z) = (X + Y)(Y + Z)(X + Z)$.

3-Stage AND–OR Networks.   Thus far we have discussed the realization of logic functions with 2-stage gate networks. In many cases, however, realization of networks consisting of three or more stages may reduce the total number of logic gates. The higher the number of stages, however, the longer is the resulting propagation time. This disadvantage can be usually tolerated if speed requirements are not stringent. The design procedure also becomes more time consuming for networks of three or more stages, which may be justified if the resulting savings are significant, e.g., if the network is manufactured in large quantities or if space or weight limitations are severe. The problem is quite complex, and we present in the following a simple outline only.

In many design problems, 3-stage combinational networks can be easily obtained by algebraic manipulation of solutions for 2-stage networks, as illustrated in the example that follows.

EXAMPLE 4.4.   Consider the function $f(W,X,Y,Z) = \sum m(1,5,9,10,11,13,14)$. It can be shown that one minimal S-of-P solution of the function is $f(W,X,Y,Z) = \overline{Y}Z + W\overline{X}Y + WY\overline{Z}$. Since the Boolean product WY is

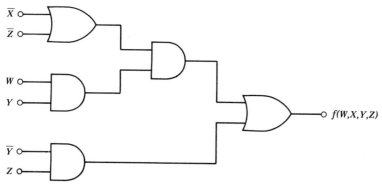

FIGURE 4.6. 3-stage AND–OR realization of the function of Example 4.4.

common to two terms it can be factored out and the function may be written as $f(W,X,Y,Z) = \overline{Y}Z + WY(\overline{X} + \overline{Z})$. Implementation of the function in this form requires a 3-stage network as shown in Figure 4.6 where the assumption was made that variables and their complements are available.

Also, 3-stage and multistage expressions can be derived from Karnaugh maps by using the following procedure:

1. Map the largest possible adjacency sets of 1-terms allowing elements with 0 entries in the adjacency loops.
2. Treat the 0-terms initially as though they were 1-terms and derive the appropriate implicants. The expression derived in this step has more minterms than it should because elements with 0 entries were included in the adjacency sets.
3. To remove the superfluous minterms draw Karnaugh submaps, each submap being formed from a loop that contained one or more 0's.
4. Loop the adjacencies (1's) in each submap to obtain the additional minterms. Since the submap was defined by the variables of the loop in step 1, these variables do not appear in the minterms of step 4.

5. Form a logical AND of the terms obtained in step 4, with the minterms defining the respective submaps. This step concludes the 3-stage minimization procedure using map methods.

Further simplifications are possible when after step 1 there are terms that are common to two or more adjacency sets.

6. A 1 that is present in more than one adjacency set can be treated as a "don't care" condition in a submap provided it has been covered at least once in an adjacency set of another submap.
7. A 0 that is present in more than one adjacency set must be treated as a 0 in every set.

EXAMPLE 4.5. Realize a 3-stage AND–OR network of the function $f(W,X,Y,Z) = \sum m(2,3,7,8,10,11,12,13,14,15)$.

The first two steps of the design procedure outlined earlier are shown in Figure 4.7a and yield $f_1 = W + Y$. Next (step 3), we draw the submap defined by literal $W\,(=1)$ as shown in Figure 4.7b involving literals X, Y, and Z only. Similarly the submap defined by literal $Y\,(=1)$, as shown in Figure 4.7c, involves literals W, X, and Z only. Note that four terms in Figure 4.7a have been covered twice by adjacency sets. Thus from step 6

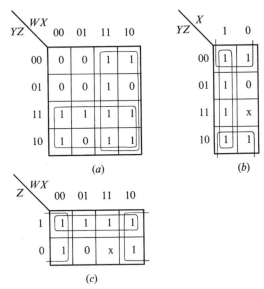

(a)   (b)

(c)

FIGURE 4.7. 3-stage AND–OR realization of the function of Example 4.5: (a) Karnaugh map for steps 1 and 2, (b) submap defined by $W = 1$, (c) submap defined by $Y = 1$.

of the procedure we may treat each such 1 entry as a "don't care condition" in one submap provided that it has been covered by an adjacency set of the other submap. This is shown in Figures 4.7*b* and 4.7*c* yielding, respectively, $f_2 = X + \bar{Z}$, and $f_3 = \bar{X} + Z$. Finally (step 5), we form the logic AND from each literal defining a submap, with the minimized expression derived from the respective submap

$$f(W,X,Y,Z) = W \cdot f_2 + Y \cdot f_3 = W(X + \bar{Z}) + Y(\bar{X} + Z).$$

The procedure does not guarantee a minimized expression since in many cases there is no clear-cut choice of implicants.

4.4 Multiple Output Combinational Networks

Thus far we have synthesized a class of networks in which a *single* output is determined by the states of the input variables. There are digital applications in which several outputs are simultaneously dependent on the states of several input variables. Such networks are referred to as *multiple output combinational networks* and their general block diagram is shown in Figure 4.8. The full adder, to be described in Section 8.2, is one simple

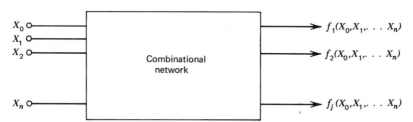

FIGURE 4.8.   Multiple output combinational network.

example of such a network, since three input variables determine the states of two outputs: "sum" and "carry-out."

A solution for multiple output combinational networks can be easily obtained treating each function independently of the others. Intuitively we know, however, that this may not be a minimal solution in the general case, since such an approach does not take into consideration the existence of terms that are common to the solution of more than one function.

EXAMPLE   4.6.   Implement   the   3-output   combinational   network $f_1(W,X,Y,Z) = W\bar{Y} + \bar{W}X$, $f_2(W,X,Y,Z) = W\bar{Y} + Y\bar{Z}$, $f_3(W,X,Y,Z) = \bar{W}X + Y\bar{Z}$.

Implementation of the three functions independent of each other would require six AND gates and three OR gates. Recognizing that each term in the above equations is common to two functions we implement the network using three AND gates and three OR gates.

The solution of Example 4.6 was arrived at by inspection. This may not be possible in the general case since the use of prime implicants common to two or more functions does not necessarily result in a minimized expression. In the following example we shall use a map method to obtain a near optimum solution with relatively little effort.

EXAMPLE 4.7. Using Karnaugh maps derive an economical solution in S-of-P form for the multiple output functions $f_1(W,X,Y,Z) = \sum m(0,1,2,3,6, 13,15)$ and $f_2(W,X,Y,Z) = \sum m(3,6,7,8,11,12,13,15)$.

First we map functions f_1 and f_2 in Figures 4.9a and 4.9b, respectively. Next we obtain $f_1 \cdot f_2 = \sum m(3,6,13,15)$, (see Figure 4.9c). From these three

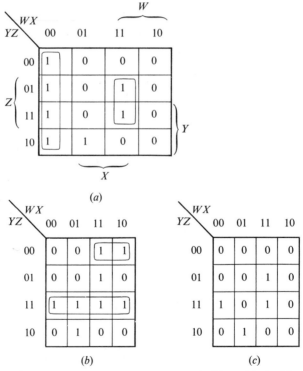

FIGURE 4.9.   Karnaugh maps for Example 4.7: (a) f_1, (b) f_2, (c) $f_1 \cdot f_2$.

maps we select prime implicants for the multiple output function that are *essential prime implicants* of one of the *three* functions. In addition the selected prime implicants in f_1 or f_2 should include at least one minterm which is not covered by any prime implicant of $f_1 \cdot f_2$ that is smaller (covering a smaller number of terms) than the selected prime implicant.

Considering f_1 we select the product term $\overline{W}\overline{X}$ and WXZ as essential prime implicants: $\overline{W}\overline{X}$ contains minterms m_0 and m_1 which are not covered by other prime implicans of f_1 or $f_1 \cdot f_2$, while WXZ contains m_{13} and m_{15} which are not covered by another prime implicant of $f_1 \cdot f_2$ that is *smaller* than WXZ. In the product term $\overline{W}Y\overline{Z}$, however, m_2 already has been covered by $\overline{W}\overline{X}$, while m_6 is a prime implicant of $f_1 \cdot f_2$ that is smaller than $\overline{W}Y\overline{Z}$.

Similar arguments applied to f_2 in Figure 4.9b yield the essential prime implicants $W\overline{Y}\overline{Z}$ and YZ. m_6 is not selected since it is covered by a smaller prime implicant in $f_1 \cdot f_2$, while m_{13} which appears in all three maps is not essential.

Terms in f_1 and f_2 that have not been covered will now be selected with help of auxiliary maps in which all the previously covered terms can be considered as don't care conditions as shown in Figure 4.10. All possible solutions must be checked since no algorithm exists for an optimal selection. Considering the two maps we notice that m_6 has not been covered in either map. For f_1 we may select $\overline{W}Y\overline{Z}$, and for f_2 $\overline{W}XY$, requiring two 3-input AND gates. A better choice is the term $\overline{W}XY\overline{Z}$ which covers m_6 in both maps and requires one 4-input gate only.

The last term to be covered is m_{13} in f_2. Two possible solutions exist: $WX\overline{Y}$ or WXZ. Since the latter term has been previously selected as an essential prime implicant in f_1 we use it for implementation of f_2. Thus

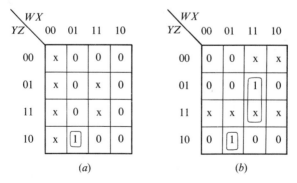

(a)   *(b)*

FIGURE 4.10. Karnaugh maps for Example 4.7: *(a)* **terms not covered in** f_1, *(b)* **terms not covered in** f_2. (*See next page for (c).*)

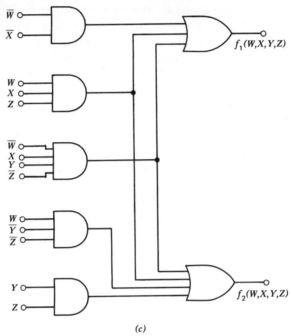

(c)

FIGURE 4.10. (c) AND–OR implementation. Terms WXZ and $\overline{W}XY\overline{Z}$ are common to both functions.

$f_1(W,X,Y,Z) = \overline{W}\,\overline{X} + WXZ + \overline{W}XY\overline{Z}$ and $f_2(W,X,Y,Z) = W\overline{Y}\overline{Z} + WXZ + \overline{W}XY\overline{Z} + YZ$. Implementation of the 2-output AND–OR network is shown in Figure 4.10c.

4.5   Combinational Networks Using NAND Gates

The NAND operation represents one of the most widely used logical functions. It has the property of functional completeness, i.e., all possible Boolean functions can be realized using the NAND gate only. It is also easily adapted to simplification procedures with Karnaugh maps which were developed for AND–OR gates in Section 3.5.

2-Stage NAND Networks.   The design procedure with NAND gates starts with a Karnaugh map from which the simplified AND–OR realization is obtained. As a next step the AND–OR network is changed to a NAND–NAND network. Any variables that are applied directly to the output NAND gate must be complemented.

EXAMPLE 4.8.   Using NAND gates, obtain a minimal realization of the function $f(W,X,Y,Z) = \sum m(1,2,6,8,9,10,11,12,13,14,15)$.

The Karnaugh map for the function is shown in Figure 4.11a, yielding $f(W,X,Y,Z) = W + Y\bar{Z} + \bar{X}\bar{Y}Z$, as shown implemented with AND–OR gates in Figure 4.11b. The AND–OR network is transformed to an equivalent NAND–NAND network, as shown in Figure 4.11c. Variable W, passing one stage only, is applied in *complemented* form to gate 1.

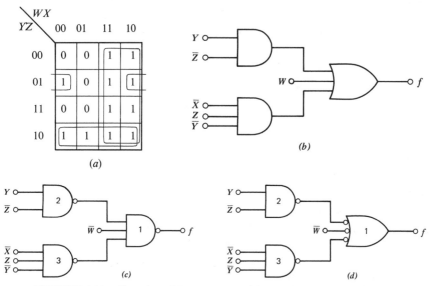

FIGURE 4.11.   Function of Example 4.8: (*a*) Karnaugh map, (*b*) implementation with AND–OR gates, (*c*) NAND–NAND implementation, (*d*) representation of the solution of Example 4.8 using graphic symbols for positive NAND and negative NOR.

We shall now apply DeMorgan's theorem to prove the validity of our procedure: The output at gate 2 is $\overline{Y\bar{Z}}$. The output at gate 3 is $\overline{\bar{X}\,\bar{Y}Z}$. Thus the output at gate 1 is $f(W,X,Y,Z) = \overline{\bar{W}(\overline{Y\bar{Z}})(\overline{\bar{X}\,\bar{Y}Z})} = W + Y\bar{Z} + \bar{X}\bar{Y}Z$.

Notice that we have used the graphic symbol of a positive NAND gate for the output stage (gate 1) of Figure 4.11c. This gate actually performs the function of a negative NOR gate, since the output will be "high" if either or both of its inputs are "low." This duality of functions performed by the *same* physical device was discussed in Section 4.2 and illustrated in Figure 4.4. It follows, therefore, that the solution of Example 4.8 shown in Figure 4.11c may also be represented by the graphic symbols as shown in Figure 4.11d. Both graphic representations are in use. In more complicated networks involving many stages, the use of graphic symbols that distinguish between positive and negative logic greatly enhances readability of the diagram, since the logic level at any given point can easily be determined by inspection: a circle indicates the lower potential while absence of a circle indicates the higher potential.

3-Stage NAND Networks.   In Example 4.8 we have presented a NAND–NAND solution for a function of four variables. An assumption was made that complements of all the four variables are available resulting in the 2-stage network of Figure 4.11. Such an assumption is appropriate in limited cases only. Modern integrated circuits contain a great number of functions within a physical package that has a limited number of output connections. These outputs most often deliver the uncomplemented variable only. Thus for realization of Figure 4.11 we would have to add four inverters.

In this section we demonstrate design techniques for realization of combinational networks with NAND gates when the complements of input variables are not given.[1] The result is a 3-stage NAND network. Note that propagation delay has not been increased in comparison with the NAND-plus-inverter realization, since delay times through NAND gates and inverters are, as a rule, comparable for a given type of logic family.* The design procedure will be illustrated by an example.

EXAMPLE 4.9.   Design a 3-variable network that delivers a logical 1 if the three inputs do not agree. This network is also known as "dissent circuit." Assume that complements of the input variables are not available.

In a function of three variables X, Y, and Z there are eight combinations. In two of these combinations all variables agree: XYZ and $\overline{X}\overline{Y}\overline{Z}$. The remaining six combinations represent the function to be implemented: $f(X,Y,Z) = \overline{X}\overline{Y}Z + \overline{X}Y\overline{Z} + \overline{X}YZ + X\overline{Y}\overline{Z} + X\overline{Y}Z + XY\overline{Z}$. A step-by-step procedure is presented next.

1. Draw a Karnaugh map of the function and mark the *minterm* that represents the three given variables, i.e., m_7 as shown in Figure 4.12a.
2. Draw loops around the adjacency sets. Each loop *must* include the marked minterm (m_7), even if such loop must enclose one or more 0's.
3. Derive implicants from the map ignoring temporarily the presence of 0's in some loops. This results in $f(X,Y,Z) = X + Y + Z$, which is

---

* For further discussion of propagation delay times see Section 5.1.

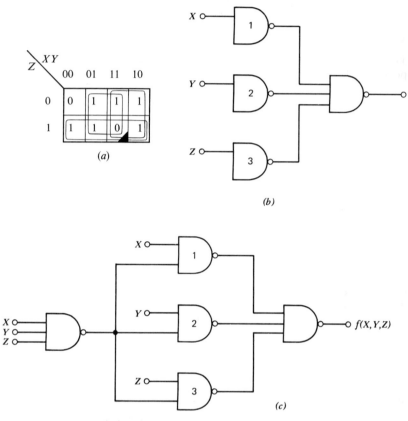

FIGURE 4.12.   Dissent circuit: (a) Karnaugh map, (b) intermediate step, $f(X,Y,Z) = X + Y + Z$, (c) 3-stage implementation with NAND gates.

implemented with four NAND gates as shown in Figure 4.12b. Note that one input only in each of the 2-input gates has been specified thus far. The other input will be used to inhibit the 0-terms that were included in the adjacency sets.

4. Derive implicants from the looped 0's only of the map. Each term in this step represents a 0 entry which we shall utilize to remove the superfluous minterms that were covered by the original adjacency loops. The product term(s) thus obtained have to be inverted to inhibit the unwanted terms. A NAND gate is ideally suited since it represents an AND–NOT function. In our example there is one looped 0 only, representing the term XYZ. Using a NAND gate we obtain at its output $\overline{XYZ}$ which we connect to the uncommitted inputs of gates 1, 2, and 3 yielding the result as shown in Figure 4.12c.

In a straightforward 2-stage NAND–NAND realization we would implement the simplified function $f(X,Y,Z) = X\overline{Y} + Y\overline{Z} + \overline{X}Z$, which requires three 2-input NAND gates, one 3-input NAND gate, and three inverters. The solution in Figure 4.12c is more economical and also more convenient for circuit layout and interconnection because it requires only two different types of gate circuits.

4.6   Combinational Networks Using NOR Gates

The NOR operation is functionally complete; thus we can realize all possible Boolean functions using the NOR gate only.

2-Stage NOR Networks.   The design procedure with NOR gates starts with a Karnaugh map from which the minimal P-of-S realization is obtained. The resultant OR–AND network is then changed to a NOR–NOR network by application of the following rules.

1. Change all OR and AND gates to NOR gates.
2. Complement all literals that are applied directly to the output stage.

EXAMPLE 4.10.   Using NOR gates obtain a simplified realization of a "majority voter" circuit that produces a logic 1 if at least two out of its three inputs are 1.

The Karnaugh map of the function is shown in Figure 4.13a from which we obtain the P-of-S expression $f(X,Y,Z) = (X + Z)(X + Y)(Y + Z)$ which is realized with an OR–AND network as shown in Figure 4.13b. The equivalent network implemented with NOR–NOR gates is shown in Figure 4.13c.

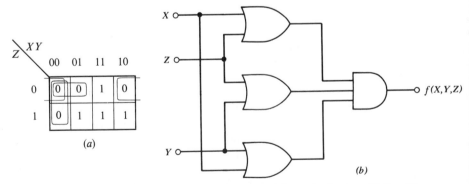

FIGURE 4.13.   "Majority voter" circuit: (a) Karnaugh map, (b) P-of-S implementation with OR–AND gates.

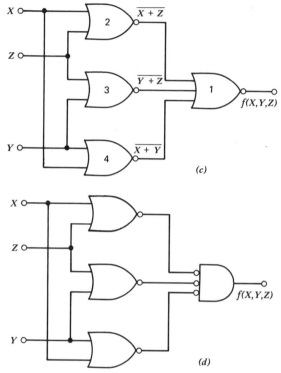

FIGURE 4.13. (*c*) NOR–NOR implementation, (*d*) representation of the "majority voter" circuit using graphic symbols for positive NOR and negative NAND.

To check the validity of the procedure note that the output at gate 2 is $\overline{X + Z}$, at gate 3 it is $\overline{Y + Z}$, and at gate 4 it is $\overline{X + Y}$. Thus the output at gate 1 is $f(X,Y,Z) = \overline{(\overline{X + Z}) + (\overline{Y + Z}) + (\overline{X + Y})} = (X + Z)(Y + Z)(X + Y)$. The circuit may also be represented by "positive" NOR gates followed by a "negative" NAND, see Figure 4.13*d*.

3-Stage NOR Networks.   The discussion preceding the design example of 3-stage NAND networks is relevant to 3-state NOR networks. When complements of input variables are not given we may add inverters where necessary. This produces a 3-stage network that may not be as economical as a 3-stage realization using NOR gates only.

EXAMPLE 4.11.   Utilizing NOR gates only, implement the 3-variable function $f(X,Y,Z) = \sum m(0,3,4,5,6,7)$. Complements of the input variables are not given.

The procedure is similar to that used in the design of 3-stage NAND networks.

1. Draw a Karnaugh map of the function and mark the *maxterm* that represents the given variables. In our example the maxterm is M_7 as shown in Figure 4.14*a*.

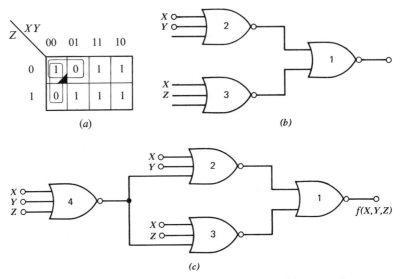

(a)   (b)

(c)

FIGURE 4.14.   3-Stage NOR network: (*a*) Karnaugh map, (*b*) intermediate step, $f(X,Y,Z) = (X + Y)(X + Z)$, (*c*) 3-stage NOR implementation.

2. Draw loops around adjacency sets of the 0-terms. Each loop *must* include the marked maxterm, even if such loop must enclose one or more 1's. (Why adjacencies of 0-terms?).
3. Derive implicants from the map, ignoring temporarily the presence of 1's in some loops. This gives an intermediate solution $f(X,Y,Z) = (X + Y)(X + Z)$ which is implemented with three NOR gates as shown in Figure 4.14*b*. Note that 2 inputs only in each of the 3-input gates have been specified thus far. The third inputs will be used to inhibit the 1's that were included in the adjacency sets.
4. Derive implicants from the looped 1's only of the map. Each term in this step represents a 1 entry which we shall utilize to remove the superfluous maxterms that were covered by the adjacency loops in step 2. The sum terms thus obtained have to be inverted to inhibit the undesired terms. In our example one maxterm only has been included in both adjacency sets. A NOR, gate 4, provides the requisite logic operation since it produces at its output the comple-

ment of that maxterm; the output at this NOR gate is $\overline{(X + Y + Z)}$ which we connect to the uncommitted inputs of gates 2 and 3, resulting in the network shown in Figure 4.14c.

To check the results we have $f_4 = \overline{(X + Y + Z)} =$ output at gate 4. Similarly, $f_2 = \overline{X + Y + \overline{(X + Y + Z)}}$ and $f_3 = \overline{(X + Z) + \overline{(X + Y + Z)}}$. Thus $f_1 = f(X,Y,Z) = \overline{f_2 + f_3}$. Substituting for f_2 and f_3 from above we obtain

$$f(X,Y,Z) = \overline{\overline{[X + Y + \overline{(X + Y + Z)}]} + \overline{[(X + Z) + \overline{(X + Y + Z)}]}}$$
$$= [(X + Y) + \overline{(X + Y + Z)}] \cdot [(X + Z) + \overline{(X + Y + Z)}]$$
$$= (X + Y + \overline{X}\,\overline{Y}\,\overline{Z})(X + Z + \overline{X}\,\overline{Y}\,\overline{Z}).$$

It can be easily shown that $X + Y + \overline{X}\,\overline{Y}\,\overline{Z} = X + Y + \overline{Z}$, and similarly $X + Z + \overline{X}\,\overline{Y}\,\overline{Z} = X + \overline{Y} + Z$. Thus $f(X,Y,Z) = (X + Y + \overline{Z})(X + \overline{Y} + Z)$.

The above expression, which is a product of maxterms $M_1 \cdot M_2$, can also be obtained directly from the Karnaugh map of Figure 4.14a and would require two 3-input NOR gates, one 2-input NOR gate, and two inverters.

3-Stage networks have the advantage that one type of circuit only is required to implement a given function. This results in simpler gate interconnections as compared with NOR-plus-inverter realizations. Stray capacitances and inductances due to interconnections are thus reduced in many cases, resulting in better high-frequency performance.

4.7 AND–OR–INVERT (AOI) Gates

The AOI gate may be considered, for the purpose of logic network analysis, an AND gate followed by a NOR gate. In practice, however, there is no series connection of an AND and NOR circuit; the AOI function is achieved through suitable interconnection of active components (transistors) keeping the propagation delay time at a low value comparable to propagation delays of other gate functions of the same logic family, while maintaining high fan-out and noise margin. These parameters are further discussed in Chapter 5.

AOI gates are specified as m-wide, n-input, where m represents the number of AND gates and n the number of inputs to each AND gate. Commonly the range of m is between 2 and 4 while n is 2 or 3. Some AOI gates are "expandable," i.e., nodes are provided at the NOR section of the AOI, at which point additional AND gates may be connected. *Expander* elements are available in form of dual 4-input AND gates, triple 3-input, 3-2-2-3 inputs, etc. The logic diagrams of an AOI gate and a gate expander are shown in Figure 4.15*b*, the internal structure of a gate in Figure 5.15. Note that the AOI gate has two *expansion nodes e* and $\bar{e}$ which are connected to the corresponding points of the expander elements. In some logic families one node only is provided for each expansion point, and the expander elements also have one node. The manufacturers' data sheets should be consulted for suitable combinations of AOI gates and expander elements.

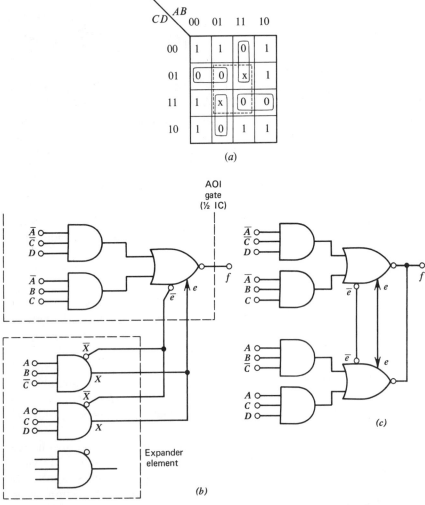

FIGURE 4.15. Use of AOI gates: (*a*) Karnaugh map, (*b*) realization with an AOI gate and an expander, (*c*) realization with a dual 2-wide 3-input AOI gate.

A 2-wide, 2-input AOI gate realizes the logic operation $f(A,B,C,D) = \overline{AB + CD} = (\overline{A} + \overline{B})(\overline{C} + \overline{D})$, which is equivalent to the OR–AND operation on complemented variables. Complementing the output of an AOI gate yields $\overline{f}(A,B,C,D) = \overline{\overline{AB + CD}} = AB + CD$, i.e., the AND–OR operation.

A 4-wide, 3-2-2-2 AOI gate with an expander performs the function

$$f = \overline{ABC + DE + FG + HJ + X},$$

where X represents the AND'ed input at the expander node.

Procedures for synthesis of combinational networks with AOI gates follow the techniques used for AND–OR implementation with the following modification: since the output function is the *complement* of an S-of-P expression we select adjacencies, from the *0 entries* in the Karnaugh map. Don't care conditions may be combined with 0 entries if they produce larger adjacency sets. Alternatively they may be combined with 1 entries if the simplified expression is desired in P-of-S form.

EXAMPLE 4.12.   Using AOI gates, realize the 4-variable function $f(A,B,C,D) = \sum m(0,2,3,4,8,9,10,14) + \sum d(7,13)$. Variables and their complements are given.

The Karnaugh map of the function is shown in Figure 4.15a. At first glance we might be tempted to select an implicant as shown by the dashed loop. However, each term in that loop is covered by an essential prime implicant eliminating the implicant shown in the dashed loop.

The resulting function

$$f(W,X,Y,Z) = \overline{\overline{A}\overline{C}D + \overline{A}BC + AB\overline{C} + ACD}$$

is shown in Figure 4.15b using a 2-wide, 3-input AOI gate with a 3-3 expander element. AOI gates are available in which the outputs may be paralleled. The above function can be implemented with such gates using a dual 2-wide, 3-input AOI gate in one integrated circuit package as shown in Figure 4.15c.

4.8   Combinational Logic with Multiplexers

Thus far we have developed design procedures for implementation of simplified combinational networks with a variety of operators. The following criteria were used leading to the desired solutions: (1) the number of gates shall be minimal, and (2) the number of gate inputs shall be minimal. Propagation delay time entered as an optional constraint and had to be considered for each alternative solution. The above criteria are valid for design with discrete components, e.g., transistor or transistor-diode circuits, or with integrated circuits (IC's) which contain a limited number of gates in one package. Such *SSI* (small scale integrated) circuits contain less than 12 gates in one package, and they commonly have 14 or 16 connections for input and output signals; their logic capability is thus limited.

The economics of integrated circuits does not favor SSI's since the manufacturing cost of a pin connection is considerably higher than that of a solid-state gate. This economic fact, partly due to an ever increasing yield in manufacture of complex solid-state circuitry, gave impetus to the development of more sophisticated logic blocks in which the ratio of gates to pin connections is substantially increased. The term *MSI* (medium scale integration) is applied in the IC industry to a physical device that contains more than 12 and less than 100 gates. When the number of gates in one package exceeds 100, then the circuit is referred to as *LSI* (large scale integration). The latter is widely used in recurrent structures, as for example, in solid-state memories. They can also be applied to combinational network implementation as discussed in Chapter 11.

Design criteria of minimality used previously for SSI's do not strictly apply to MSI's nor to LSI's, since a non-utilized gate most often does not increase the total "IC package count." It would be advantageous for the logic designer to have at his disposal a universal logic unit, *ULU*, with which he could synthesize a variety of logic functions depending on the states of some control input signals. A general discussion of combinational logic design with ULU's can be found in reference (2); the *digital multiplexer* is one form of its implementation.[3] (For combinational network synthesis using *decoders* see Problem 4.19.)

A digital multiplexer is conceptually the equivalent of a single pole multiposition switch. It is a combinational network with 4, 8, or 16 input or "data" lines, one output line (sometimes also provided with its complement), and i control, or "data select" lines, where 2^i is the number of input lines. The state of the "data select" determines which input line is connected to the output. Combinational logic is realized through judicious application of the input variables or binary constants to the "data" and "data select" inputs.

EXAMPLE 4.13. Realize the function $f(A,B,C) = \sum m(2,3,5,6)$ using a 4-input multiplexer.

The truth table of the function is shown in Figure 4.16a. Assume that the variables A and B are applied to the "data select" lines S_0 and S_1. Variable C, its complement, or binary constants (logic 1 and 0) are next selectively applied to the "data" inputs I_0 through I_3, as shown in Figure 4.16b. Variables A and B provide a transmission path for the signals applied to I_0 through I_3 depending on their combinations: $\bar{B}\bar{A}$

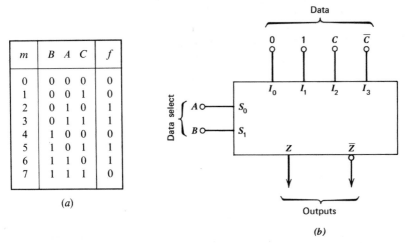

m	B	A	C	f
0	0	0	0	0
1	0	0	1	0
2	0	1	0	1
3	0	1	1	1
4	1	0	0	0
5	1	0	1	1
6	1	1	0	1
7	1	1	1	0

(a)

(b)

FIGURE 4.16. Realization of a 3-variable function with a 4-input multiplexer: (a) truth table, (b) circuit.

provides a transmission path for I_0, $\bar{B}A$ for I_1, etc. Since the given function is 0 for m_0 and m_1, independent of C, we connect I_0 to logical 0. For m_2 and m_3 the function is 1, I_1 is thus connected to logical 1. Both m_4 and m_5 require a transmission path from I_2 to the output the path being controlled by $B\bar{A}$. From the truth table, $f = 0$ (m_4) when $C = 0$ while $f = 1$ (m_5) when $C = 1$. We thus connect C to I_2. Using a similar argument for the last two minterms of the function we connect $\bar{C}$ to I_3.

The function of Example 4.13 was realized with one 4-input multiplexer. Since two such circuits are usually provided in one IC package, we have utilized one half of a package for implementation of the function. In comparison we would require two 3-input and two 2-input NAND's to implement the same function with SSI gates.

Any 4-variable function can be realized with an 8-input multiplexer in which three variables are applied to the "data select" lines while the fourth variable or the binary constants 0 and 1 are selectively applied to the input lines I_0 through I_7 as shown in Figure 4.17a. Functions of five or more variables are implemented with 2-stage multiplexer networks. A function of six variables, for example, may be realized with eight 4-input multiplexers the outputs of which feed a single 8-input multiplexer, as outlined in Figure 4.17b.

Karnaugh maps can be used to aid in the synthesis of switching functions with multiplexers. Thus using an 8-input multiplexer for realization of a 4-variable function we would require eight 4-variable maps. Each such map

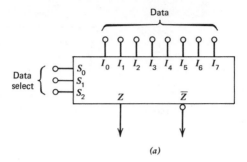

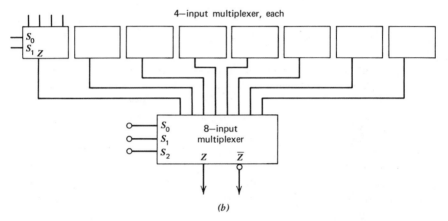

FIGURE 4.17. (*a*) Realization of a 4-variable function with an 8-input multiplexer. (*b*) Realization of a 6-variable function using digital multiplexers.

is defined by the three "data select" lines while the fourth variable is represented by one of the "data" lines.

A step-by-step design procedure for a 2-stage solution of a 6-variable function $f(A,B,C,D,E,F)$ is given below:

1. Divide the function into two parts, $g_1 = g_1(A,B,C)$ and $g_2 = g_2(D,E,F)$, reducing the 6-variable problem to two 3-variable problems.
2. In each minterm for which the function is 1 factor out the variables D, E, and F to obtain the function $g_1(A,B,C)$ of three variables.
3. Collect all terms of $g_1(A,B,C)$ corresponding to each unique input combination of D, E, and F.

4. Implement each term of $g_1(A,B,C)$ as described in Example 4.13. This step realizes the input stage (upper part) of Figure 4.17b.
5. Feed the eight outputs of the input stage to the eight inputs of the output stage that is controlled on the "data select" lines by variables D, E, and F.

In step 1 we have arbitrarily assigned variables A, B, C to one function, D, E, F to the other function. Other assignments are possible but no algorithm has been devised for selecting the best permutation for any function. General guidelines for selecting "good" permutations are:

1. Maximize the number of inputs to the output multiplexer stage which require either 0 or 1.
2. Utilize multiplexed signals that are used in more than one input to the output stage.
3. Select a permutation that maximizes pairs of multiplexed functions that are complements of each other, e.g., $g_1(A,B,C)$ and $\bar{g}_1(A,B,C)$. There are multiplexers available that provide outputs and their complements. Thus one multiplexer may be eliminated for each complementary pair of functions. If $\bar{g}_1$ is not available at the output then a simple inverter may be used to replace a multiplexer.

Each of the conditions above eliminates one multiplexer in the input stage.

EXAMPLE 4.14.[3] Implement the 6-variable function of Figure 4.18a using a 2-stage multiplexer network. Minterms in the truth table are shown only for $f(A,B,C,D,E,F) = 1$.

The function of six variables is divided into two functions of three variables each. The minterms in the truth table are sorted in groups of identical $g_2(D,E,F)$ terms. Variables A and B are applied to the data select lines of the input stage multiplexers in the upper row of Figure 4.18b, while the variable C and Boolean constants 0 and 1 are applied to the data input (I) terminals. Eight 4-input multiplexers (shown in Figure 4.18b as dual units in four IC packages) and one 8-input multiplexer are required to implement the function.

A more judicious selection of the permutations may be obtained through algebraic manipulation. Refer to Figure 4.19a: with g_1 being a function of C, D, B and g_2 a function of E, F, A (in that order) we note that the inputs to I_0 and I_1 of the output stage are identical, eliminating one

| 4-input multiplexers | 8-input multiplexer | |
A B C	I_0 I_1 I_2 I_3 I_4 I_5 I_6 I_7	D E F
0 0 0	1 0 0 0 0 0 0 0	0 0 0
0 0 1	1 0 0 0 0 0 0 0	0 0 0
0 1 1	1 0 0 0 0 0 0 0	0 0 0
1 0 0	1 0 0 0 0 0 0 0	0 0 0
1 1 0	1 0 0 0 0 0 0 0	0 0 0
0 0 0	0 1 0 0 0 0 0 0	1 0 0
0 0 1	0 1 0 0 0 0 0 0	1 0 0
0 1 1	0 1 0 0 0 0 0 0	1 0 0
1 0 1	0 1 0 0 0 0 0 0	1 0 0
1 1 1	0 1 0 0 0 0 0 0	1 0 0
0 0 0	0 0 1 0 0 0 0 0	0 1 0
0 0 1	0 0 1 0 0 0 0 0	0 1 0
0 1 1	0 0 1 0 0 0 0 0	0 1 0
1 1 1	0 0 1 0 0 0 0 0	0 1 0
0 0 0	0 0 0 1 0 0 0 0	1 1 0
0 0 1	0 0 0 1 0 0 0 0	1 1 0
0 1 1	0 0 0 1 0 0 0 0	1 1 0
1 0 0	0 0 0 1 0 0 0 0	1 1 0
0 0 1	0 0 0 0 1 0 0 0	0 0 1
0 1 0	0 0 0 0 1 0 0 0	0 0 1
0 1 1	0 0 0 0 1 0 0 0	0 0 1
1 0 0	0 0 0 0 1 0 0 0	0 0 1
0 0 0	0 0 0 0 0 1 0 0	1 0 1
1 0 0	0 0 0 0 0 1 0 0	1 0 1
0 0 1	0 0 0 0 0 1 0 0	1 0 1
1 0 1	0 0 0 0 0 1 0 0	1 0 1
0 1 1	0 0 0 0 0 1 0 0	1 0 1
0 0 1	0 0 0 0 0 0 1 0	0 1 1
1 0 1	0 0 0 0 0 0 1 0	0 1 1
1 1 0	0 0 0 0 0 0 1 0	0 1 1
1 1 1	0 0 0 0 0 0 1 0	0 1 1
1 1 0	0 0 0 0 0 0 0 1	1 1 1
0 0 1	0 0 0 0 0 0 0 1	1 1 1
0 1 1	0 0 0 0 0 0 0 1	1 1 1
1 1 1	0 0 0 0 0 0 0 1	1 1 1

(a)

FIGURE 4.18. Realization of the 6-variable function of Example 4.14 with a 2-stage multiplexer network: (a) truth table (minterms shown are only for "true" combinations of the function).

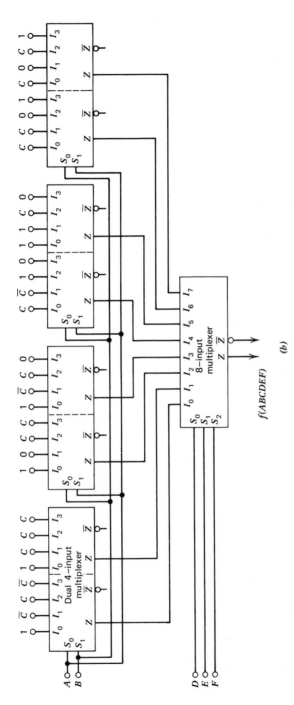

FIGURE 4.18. (*b*) 2-stage multiplexer network.

4-input multiplexers			8-input multiplexer										
C	D	B	I_0	I_1	I_2	I_3	I_4	I_5	I_6	I_7	E	F	A
0	0	0	1	0	0	0	0	0	0	0	0	0	0
1	0	0	1	0	0	0	0	0	0	0	0	0	0
0	1	0	1	0	0	0	0	0	0	0	0	0	0
1	0	1	1	0	0	0	0	0	0	0	0	0	0
1	1	0	1	0	0	0	0	0	0	0	0	0	0
1	1	1	1	0	0	0	0	0	0	0	0	0	0
0	0	0	0	1	0	0	0	0	0	0	1	0	0
1	0	0	0	1	0	0	0	0	0	0	1	0	0
0	1	0	0	1	0	0	0	0	0	0	1	0	0
1	0	1	0	1	0	0	0	0	0	0	1	0	0
1	1	0	0	1	0	0	0	0	0	0	1	0	0
1	1	1	0	1	0	0	0	0	0	0	1	0	0
1	0	0	0	0	1	0	0	0	0	0	0	1	0
0	1	0	0	0	1	0	0	0	0	0	0	1	0
1	1	0	0	0	1	0	0	0	0	0	0	1	0
0	0	1	0	0	1	0	0	0	0	0	0	1	0
1	0	1	0	0	1	0	0	0	0	0	0	1	0
1	1	1	0	0	1	0	0	0	0	0	0	1	0
1	0	0	0	0	0	1	0	0	0	0	1	1	0
1	1	0	0	0	0	1	0	0	0	0	1	1	0
1	1	1	0	0	0	1	0	0	0	0	1	1	0
0	0	0	0	0	0	0	1	0	0	0	0	0	1
1	1	0	0	0	0	0	1	0	0	0	0	0	1
0	0	1	0	0	0	0	1	0	0	0	0	0	1
1	1	1	0	0	0	0	1	0	0	0	0	0	1
0	1	0	0	0	0	0	0	1	0	0	1	0	1
1	0	1	0	0	0	0	0	1	0	0	1	0	1
0	0	0	0	0	0	0	0	0	1	0	0	1	1
0	1	0	0	0	0	0	0	0	1	0	0	1	1
1	1	0	0	0	0	0	0	0	1	0	0	1	1
1	0	0	0	0	0	0	0	0	0	1	1	1	1
0	0	1	0	0	0	0	0	0	0	1	1	1	1
1	0	1	0	0	0	0	0	0	0	1	1	1	1
0	1	1	0	0	0	0	0	0	0	1	1	1	1
1	1	1	0	0	0	0	0	0	0	1	1	1	1

(a)

FIGURE 4.19. Preferred solution to the function of Example 4.14: (a) truth table.

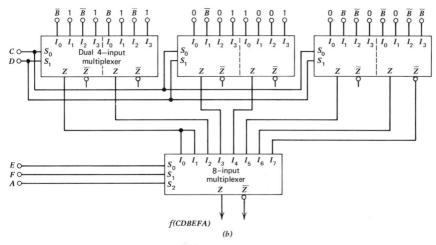

FIGURE 4.19. (*b*) 2-stage multiplexer network.

4-input multiplexer. Also, the input I_6 is a complement of I_7. Thus, utilizing the outputs Z and $\bar{Z}$ from the input stage eliminates another multiplexer. The implementation is shown in Figure 4.19*b*.

A 2^n-input multiplexer has n "data select" lines, which together with one input line can generate all possible functions of $(n + 1)$ variables. The number of combinations for $(n + 1)$ variables is $(2)^{2^{n+1}}$. Thus a 4-input multiplexer $(n = 2)$ can generate $(2)^{2^3} = 256$ different combinations, an 8-input multiplexer $(n = 3)$ can generate 65,536 combinations, etc. The potential power of multiplexers for logical design follows from the exponential increase in the possible functions that can be realized.

4.9 Miscellaneous Logic Functions and Their Realizations

EXCLUSIVE–OR Function. The EXCLUSIVE–OR (XOR) realizes the function $f(X,Y) = X\bar{Y} + \bar{X}Y = X \oplus Y$ and its symbol is shown in Figure 4.20*a*. It can be implemented with three NAND gates if complements of input variables are available; otherwise four NAND gates are required, as shown in Figure 4.20*b*. One MSI package usually contains four EX-CLUSIVE–OR gates that do not require complemented inputs.

XOR gates can be cascaded to achieve a multivariable XOR function $f(K,L,M...Z) = K \oplus L \oplus M \oplus \cdots \oplus Z$. Such cascading may be done in series as in Figure 4.21*a*, or as a symmetrical fan-in as in Figure 4.21*b*. Both configurations require an equal number of gates; Figure 4.21*b*, however, has shorter propagation delay times.

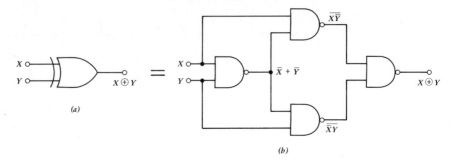

FIGURE 4.20. EXCLUSIVE–OR (XOR) gate: (a) graphic symbol, (b) realization with NAND gates.

The XOR gate performs the function of modulo-2 addition and is extensively used in arithmetical operations, code conversion, and in error detection and correction circuits, as will be discussed in Chapters 8 and 9.

EQUALITY Function. The EQUALITY function, also called COMPARATOR, COINCIDENCE, or EXCLUSIVE–NOR is a function of two variables

$$f(X,Y) = XY + \bar{X}\bar{Y} = X \odot Y,$$

i.e., $f(X,Y)$ is 1 *iff* X and Y have equal binary values. Its symbol is shown in Figure 4.22a. The complement of the EQUALITY function is the XOR function $\bar{f}(X,Y) = \overline{X \odot Y} = \overline{\bar{X}\bar{Y} + XY} = (X + Y)(\bar{X} + \bar{Y}) = X \oplus Y$.

The EQUALITY function, or EQ for short, can be realized with a XOR gate by applying the complement of either one of the two variables to an input, or through complementation of the XOR output, since $\bar{X} \oplus Y = \overline{\bar{X}}Y + \bar{X}\bar{Y} = XY + \bar{X}\bar{Y} = X \odot Y = \overline{X \oplus Y}$. Realization of the EQ function using NOR gates, when complements of variables are not given, is shown in Figure 4.22b.

The XOR and EQ operations are not "functionally complete" and each requires NAND operations for realization of all possible Boolean functions. Some useful identities for XOR and EQ gates are listed in Figure 4.23. Note the duality between some identity pairs and also between the two operations.

Simplification with aid of Karnaugh maps using the XOR or EQ functions is obtained, as with other functions, through recognition of adjacencies. The patterns are, however, not as obvious as for AND–OR networks. It is more convenient, therefore, to obtain an S-of-P expression which may then be examined for factoring possibilities to achieve XOR or EQ expressions.

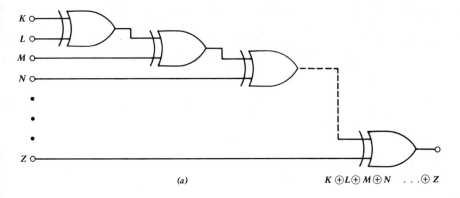

(a)

$K \oplus L \oplus M \oplus N \quad \ldots \oplus Z$

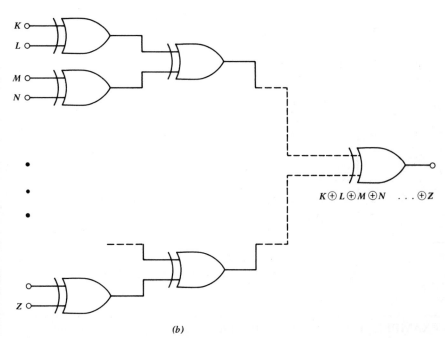

(b)

FIGURE 4.21. Multivariable XOR function: (*a*) series connection, (*b*) symmetical fan-in.

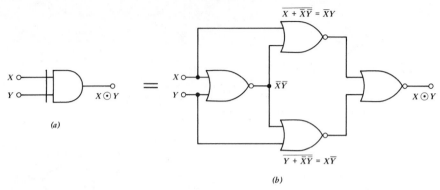

FIGURE 4.22. EQUALITY gate: (*a*) graphic symbol, (*b*) realization with NOR gates.

$X \oplus Y$	$X \odot Y$
$0 \oplus 0 = 0$	$0 \odot 0 = 1$
$0 \oplus 1 = 1$	$0 \odot 1 = 0$
$1 \oplus 0 = 1$	$1 \odot 0 = 0$
$1 \oplus 1 = 0$	$1 \odot 1 = 1$
$0 \oplus X = X$	$0 \odot X = \overline{X}$
$1 \oplus X = \overline{X}$	$1 \odot X = X$
$\overline{X} \oplus X = 1$	$\overline{X} \odot X = 0$
$X \oplus X = 0$	$X \odot X = 1$
$\overline{X} \oplus Y = X \odot Y$	$\overline{X} \odot Y = X \oplus Y$
$X \oplus \overline{Y} = X \odot Y$	$X \odot \overline{Y} = X \oplus Y$
$\overline{X \oplus Y} = X \odot Y$	$\overline{X \odot Y} = X \oplus Y$
$X \oplus Y \oplus XY = X + Y$	$X \odot Y \odot XY = X + Y$
$X \oplus XY = X \overline{Y}$	$X \odot XY = X \overline{Y}$
$X \oplus (X + Y) = \overline{X} Y$	$X \odot (X + Y) = \overline{X} Y$
$X \oplus \overline{X} Y = \overline{X + Y}$	$X \odot \overline{X} Y = \overline{X + Y}$
$X \oplus (\overline{X} + Y) = XY$	$X \odot (\overline{X} + Y) = XY$

FIGURE 4.23. Identities for XOR and EQ operations.

EXAMPLE 4.15. Implement the functions $f_1(W,X,Y,Z) = \sum m(4,5,8,9)$ and $f_2(W,X,Y,Z) = \sum m(8,11,12,15)$.

Answer:

$f_1(W,X,Y,Z) = \overline{W} X \overline{Y} + W \overline{X} \overline{Y} = \overline{Y}(W \oplus X)$ and $f_2(W,X,Y,Z) = W \overline{Y} \overline{Z} + WYZ = W(Y \odot Z)$.

The reader may enter the minterms of the two examples in Karnaugh maps to familiarize himself with some adjacency patterns. The most easily recognizable patterns are adjacency sets that are located diagonally on a map.

EXAMPLE 4.16. Consider the function $f(A,B,C,D) = \sum m(1,2,4,7,8,11, 13,14)$ as shown in Figure 4.24a. No simplification of the function is possible when AND–OR circuits are used. The diagonal adjacency sets of two minterms in a 4-variable function will each produce a term that is a Boolean product of an XOR or an EQ with the remaining two variables of the function. Terms representing the respective adjacency sets are indicated in the Karnaugh map. Thus $f(A,B,C,D) = \bar{A}\bar{C}(B \oplus D)$ $+ AC(B \oplus D) + \bar{A}C(B \odot D) + A\bar{C}(B \odot D) = (\bar{A}\bar{C} + AC)(B \oplus D)$ $+ (\bar{A}C + A\bar{C})(\overline{B \oplus D}) = (\overline{A \oplus C})(B \oplus D) + (A \oplus C)(\overline{B \oplus D})$.

The expression, which is of the form $\bar{g}h + g\bar{h} = g \oplus h$, where $g = (A \oplus C)$ and $h = B \oplus D$, reduces to $f(A,B,C,D) = A \oplus B \oplus C \oplus D$ as shown

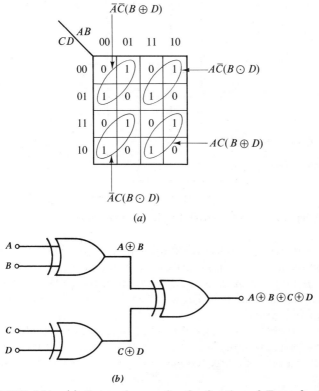

(a)

(b)

FIGURE 4.24. (a) Karnaugh map for the function of Example 4.16. (b) Realization with XOR gates.

in Figure 4.24*b*. The function is implemented with one IC package utilizing three out of four available XOR circuits. Implementation with NAND gates would require eight 4-input and one 8-input NAND, i.e., five IC packages.

OR–NOR and AND–NAND Gates.   The emitter-coupled logic family, ECL,* also referred to as current-mode logic, delivers at its output the positive OR and NOR functions. The multiple outputs that are available in some ECL gates may be interconnected as illustrated in Figure 4.25

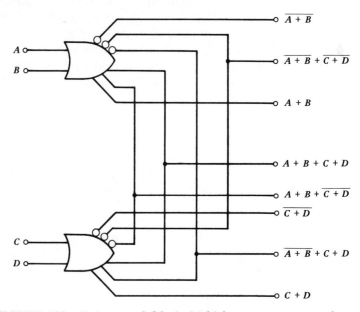

FIGURE 4.25.   Emitter-coupled logic. Multiple output gate connections.

resulting in circuit economies. Realization of the EXCLUSIVE–OR function, when complements of input variables are not given, is shown in Figure 4.26*a*. Note that the direct connection between the NOR outputs performs the OR function. This may be indicated by a small graphic OR symbol at the junction point.

Some multifunction ECL gates provide the positive AND functions when the non-inverting outputs are interconnected. The logic function (AND or OR) obtained through direct connection depends on the circuit details of a

* For a circuit description of ECL gates see Section 5.4.

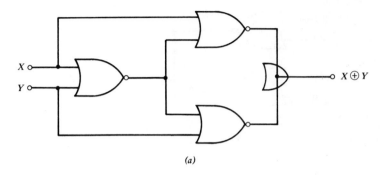

(a)

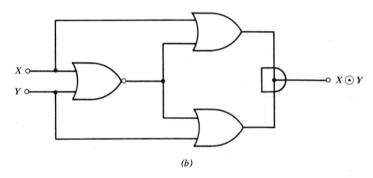

(b)

FIGURE 4.26. (*a*) ECL realization of the EXCLUSIVE–OR function.
(*b*) ECL realization of the EQUALITY function.

given ECL family; manufacturers' data should be consulted to make the correct selection.

The EQUALITY gate of Figure 4.26*b* is implemented with an ECL gate that provides the AND function when its OR outputs are interconnected. A small AND symbol may be added at the tie point, as shown.

PROBLEMS

4.1   Derive the Boolean expressions represented by Figures 4.27*a* and *b*. Comment on the results you obtained.

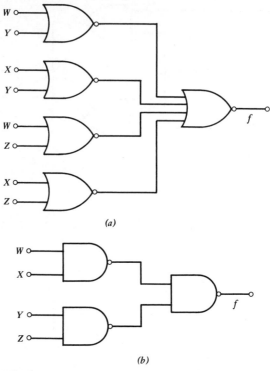

(a)

(b)

FIGURE 4.27.   Logic networks for Problems 4.1 and 4.2.

4.2 (a) Establish a truth table of the function performed by the gate network of Figure 4.27a. (b) Implement the function using AND–OR–INVERT gates only. (c) Implement the function using a multiplexer circuit.

4.3 Derive the Boolean expression in minterms and maxterms for the network of Figure 4.28.

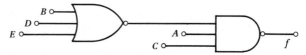

FIGURE 4.28.   Logic networks for Problem 4.3.

4.4 Design a logic network that has two "data" inputs A and B, and three "control" inputs C_0, C_1, and C_2. The network shall produce the following Boolean functions depending on the eight possible logic states of the "control" inputs: 1, $(A + B)$, $\overline{AB}$, $(A \oplus B)$, $(A \odot B)$, AB, $(\overline{A + B})$, 0.

4.5 (a) Analyze the circuit shown in Figure 4.29 and obtain the logic expressions for F_1 through F_4. (b) Connect F_1 to B_1, F_2 to C_1, and F_3 to D. What are the logic expressions at the outputs F_2, F_3 and F_4?

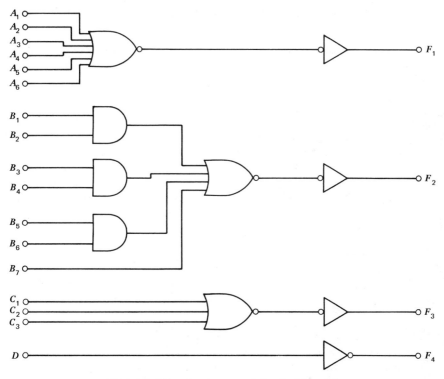

FIGURE 4.29.   Logic network for Problem 4.5.

4.6 Implement the function $f(W,X,Y,Z) = \sum m(0,1,6,8,9,14,15)$ using a minimum number of (a) 2-input AND and OR gates, (b) 2-input NAND gates and (c) 2-input NOR gates.

4.7. Obtain a minimal realization of the functions below using (i) NAND gates (ii) NOR gates. Assume that variables and their complements are available.

(a) $f_1(W,X,Y,Z) = \sum m(0,1,4,5,12,13)$
(b) $f_2(V,W,X,Y,Z) = \sum m(0,4,18,19,22,23,25,29)$
(c) $f_3(W,X,Y,Z) = \prod M(1,3,7,9,11,15)$
(d) $f_4(W,X,Y,Z) = \sum m(1,2,6,8,9,10,11,12,13,14,15)$

Compare the results of NAND and NOR implementations of each function with respect to their complexity.

4.8 Implement the multiple output network described by the functions

$$f_1(W,X,Y,Z) = W\bar{X}Y + W\bar{Z} + \bar{W}\bar{X}\,\bar{Y}Z + Y\bar{Z}$$
$$f_2(W,X,Y,Z) = \bar{W}X + \bar{W}\bar{Y}Z + W\bar{X}YZ$$
$$f_3(W,X,Y,Z) = \bar{W}\bar{X}\bar{Y} + \bar{W}XYZ + \bar{W}\bar{Z} + \bar{Y}\bar{Z}.$$

4.9 Using NAND gates only, obtain the minimal realization of a 4-variable function that delivers a logic 1 *iff* the number of "1" inputs is (a) even, and (b) odd. Repeat the problem using NOR gates only.

4.10 A logic network has four inputs I_3, I_2, I_1, and I_0 and two outputs O_1 and O_0. The state of the output lines shows which input line has a logic 1 applied to it, $O_1O_0 = 11$ corresponding to I_3, etc. If two or more input lines have a logic 1, the state of the output lines shall correspond to the input of highest priority, where I_3 has the highest priority, I_0 the lowest. Show the minimal NAND gate realization of the network that performs the function described above.

4.11 The function described by $f(X,Y,Z) = \sum m(1,2,4,7)$ is called the "odd circuit" because it produces a logical "true" when an odd number of input variables is "true." Obtain a minimal 3-stage realization of the function, assuming no complements of variables are available.

4.12 A "*half-adder*" circuit has two inputs, X and Y, and two outputs (i) $(X + Y)$ mod-2 (read "X plus Y modulo-2") and (ii) a "carry" that is 1 *iff* $X = Y = 1$. Assume that no complements of input variables are available and design a network that implements the function using a minimum number of NAND gates.
(a) How many NAND gates are required? (b) After how many gate delays is the sum output obtained? The carry output?
(c) What are the consequences of the sum and carry delay times?

4.13 Synthesize operations NOT, AND, and OR using the sets of operators given below, and draw the corresponding logic circuits.
(a) NAND, (b) NOR, (c) EXCLUSIVE–OR with AND,
(d) EQUALITY with AND.

4.14 Refer to the AND operations on 0 and 1, etc., of Figure 4.30, and establish truth tables for corresponding logic operations for the following: (a) NAND, (b) NOR, (c) IMPLICATION $(X + \bar{Y})$, (d) INHIBIT $(X\bar{Y})$, (e) EXCLUSIVE–OR, (f) EQUALITY.

X Y	$X \cdot Y$
0 0	0
0 1	0
1 0	0
1 1	1
0 X	0
1 X	X
$\overline{X}$ X	0
X X	X
X Y	$X \cdot Y$

FIGURE 4.30.   Truth table for Problem 4.14.

4.15 Given the operations f_1, f_4, f_6, f_7, f_8, f_9, and f_{14} of Figure 4.1, check which of the above operations have the algebraic (Boolean) properties of (a) distributivity, (b) associativity, (c) commutativity.

4.16 Derive a simplified solution for the multiple output network

$$f_1(A,B,C,D) = \sum m(2,3,10,13,15)$$
$$f_2(A,B,C,D) = \sum m(4,5,10,13,15)$$
$$f_3(A,B,C,D) = \sum m(8,9,10).$$

Compare the results with a realization obtained when the three functions are simplified independently of each other.

4.17 Implement the function

$$f(W,X,Y,Z) = \sum m(1,2,4,5,6,9,10,11,13)$$

using one 8-input multiplexer. How many IC packages would be required to implement the function with NAND gates?

4.18 Design a 2-stage multiplexer network that realizes the function $f(A,B,C,D,E,F) = \sum m(3,7,12,14,15,19,23,27,28,29,31,35,39,44,45,46,48,$ $49,50,52,53,55,56,57,59)$.

4.19 A *decoder* is a combinational network having n inputs and 2^n outputs where each decoded output is unique. (a) Show that such a device together with an OR gate connected to its output terminals may be utilized in the synthesis of combinational networks. (b) Implement the function $f(ABCD) = \overline{A}\overline{B}\overline{C}D + \overline{A}\overline{B}CD + \overline{A}B\overline{C}D + \overline{A}BCD + A\overline{B}\overline{C}D + A\overline{B}CD + AB\overline{C}D + ABCD$, using one 4-input decoder and one 8-input OR gate. (c) Compare the resulting number of IC's with a solution utilizing SSI gates only.

CHAPTER 5

Logic Gates

Logic gates are the most basic components in the design of modern digital systems. They realize logic functions and they also provide the basis for more complex digital circuits. The realization of logic gate circuits by semiconductor devices is the subject of this chapter. Emphasis will be placed on circuits that have attained widespread and increasing use to date; additional types are discussed in the problems at the end of the chapter.

5.1 Diode Gates

A simple realization of a 3-input AND function by a diode *AND gate* is shown in Figure 5.1*a*; a symbol for an AND gate is shown in Figure 5.1*b*, and a truth table describing its operation in Figure 5.1*c*.* It can be seen that the output is "high" (+5 V) if and only if all inputs are "high." Another gate, a 3-input diode *OR gate* is shown in Figure 5.2*a*, its symbol in Figure 5.2*b*; a truth table in Figure 5.2*c*.

In both circuits, voltage values of the outputs are shown without external loads. A resistive load from the output to ground in the AND circuit of Figure 5.1*a* would deteriorate its "high" output level (lower the +5 V output in the last line of the truth table in Figure 5.1*c*); conversely, a resistive load from the output to +5 V in the OR circuit of Figure 5.2*a* would deteriorate (raise) the output voltage of the first line in the truth table of Figure 5.2*c*. Furthermore, capacitive loads on the outputs can result in transient errors.

---

* A list of symbols is given in Figure 4.4.

104

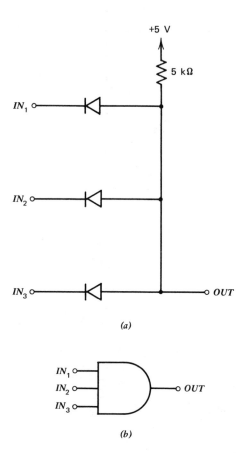

(a)

(b)

IN_1	IN_2	IN_3	OUT
0 V	0 V	0 V	+0.7 V
0 V	0 V	+5 V	+0.7 V
0	+5 V	0 V	+0.7 V
0	+5 V	+5 V	+0.7 V
+5 V	0 V	0 V	+0.7 V
+5 V	0 V	+5 V	+0.7 V
+5 V	+5 V	0 V	+0.7 V
+5 V	+5 V	+5 V	+5 V

(c)

FIGURE 5.1. A 3-input diode AND gate circuit: (*a*) circuit diagram, (*b*) AND gate symbol, (*c*) truth table showing output voltages with no external load on the output terminal.

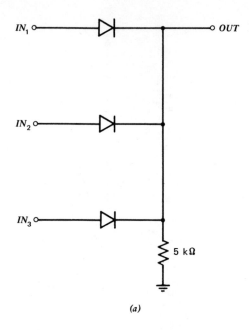

(a)

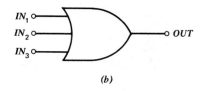

(b)

IN_1	IN_2	IN_3	OUT
0 V	0 V	0 V	0 V
0 V	0 V	+5 V	+4.3 V
0 V	+5 V	0 V	+4.3 V
0 V	+5 V	+5 V	+4.3 V
+5 V	0 V	0 V	+4.3 V
+5 V	0 V	+5 V	+4.3 V
+5 V	+5 V	0 V	+4.3 V
+5 V	+5 V	+5 V	+4.3 V

(c)

FIGURE 5.2. A 3-input diode OR gate circuit: (a) circuit diagram, (b) OR gate symbol, (c) truth table showing output voltages with no external load on the output terminal.

EXAMPLE 5.1 A 2-input AND gate circuit with a capacitive load of $C = 100$ pF is shown in Figure 5.3a. The voltages on terminals IN_1 and IN_2 are switched between 0 V and $+5$ V. It will be assumed that the voltage drops across the forward biased diodes are 0.7 V and that charges stored in the diodes are zero. When the inputs are switched from $+5$ V to

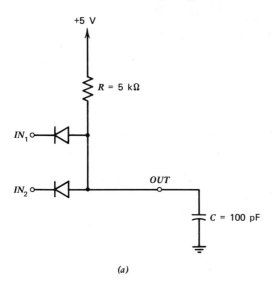

(a)

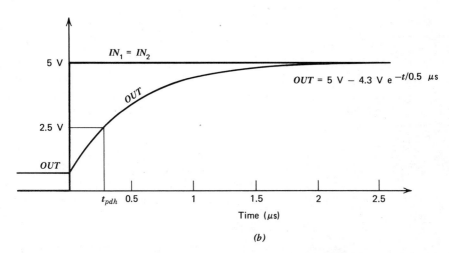

(b)

FIGURE 5.3. A 2-input diode AND gate circuit with capacitive load: (a) circuit diagram, (b) transient of the output for a positive-going input step voltage.

0 V the ouput instantaneously switches from +5 V to +0.7 V, since the impedances of the diodes and of the sources driving the inputs were neglected. When the inputs are switched from 0 V to +5 V, however, the output will rise from its initial value of +0.7 V to its final value of +5 V with a timeconstant of $RC = 5$ k$\Omega \times 100$ pF $= 0.5$ μs, passing through the "half-way" point of +2.5 V at $t_{pdh} = 0.27$ μs (see Figure 5.3b).*

In the example above, transients of a gate circuit have been described by the output waveforms for positive-going and for negative-going input steps. Frequently, however, it is sufficient to characterize the effects of these waveforms by a *propagation delay*, defined as the time required for the output signal to reach a specified level following a transition in the input signal. Propagation delays are, in general, different for positive-going and for negative-going input transitions. For this reason, both are commonly specified: one for the transition of the output to its "high" state, t_{pdh}, and one for the transition of the output to its "low" state, t_{pdl}. Since t_{pdh} and t_{pdl} depend on the external loads on the output terminal, they are usually given for specific loads.

EXAMPLE 5.2.   In the 2-input diode AND gate circuit of Figure 5.3a with an external capacitive load of $C = 100$ pF, the propagation delay to the +2.5 V level of the transient to the "high" output state, t_{pdh}, is 0.27 μs. The propagation delay to the +2.5 V level of the transient to the "low" output state, t_{pdl}, approaches zero, since the impedances of the diodes and of the sources driving the inputs have been assumed zero.

Because of their simplicity, diode gates are widely used in integrated circuits, especially in large arrays where the resulting savings in cost and space are significant. When a diode gate is followed by another diode gate, however, limitations arising from voltages across diodes and from loading have to be taken into account.

EXAMPLE 5.3.   A 2-input diode AND gate followed by a 2-input diode OR gate is shown in Figure 5.4a. When the voltage on IN_3 is zero, the output "high" voltage level is a function of resistance R_2, and is $V_h = 4.3$ V $\times R_2/(R_1 + R_2)$. With $R_1 = 5$ kΩ, this voltage is $V_h = 3.9$ V if $R_2 = 50$ kΩ; it is $V_h = 2.85$ V if $R_2 = 10$ kΩ; and it is $V_h = 2.15$ V if $R_2 = 5$ kΩ. Thus, it seems that a high value of R_2 is desirable. There are

---

* The derivation of transients is beyond the scope of this text. The subject is treated in References (1) and (2) listed at the end of the book.

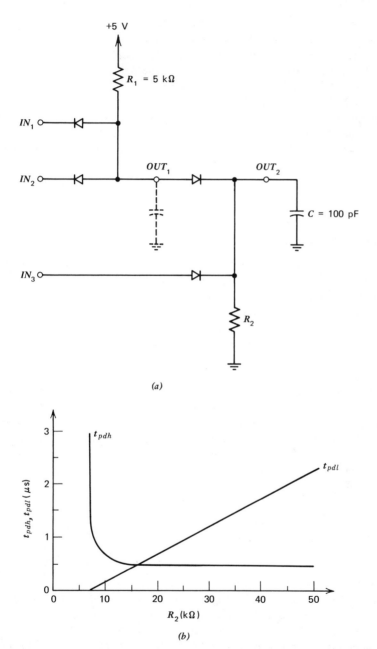

FIGURE 5.4. A 2-input diode AND gate circuit followed by a 2-input diode OR gate circuit: (a) circuit diagram, (b) propagation delays t_{pdh} and t_{pdl} as functions of the value of resistor R_2, for $C = 100$ pF.

always, however, some load capacitances present in the circuit resulting from input capacitances of subsequent circuits and from wiring capacitances. If the capacitive load on the intermediate OUT_1 (drawn by broken lines) is neglected, the propagation delays of the circuit will be governed by the values of R_1, R_2, and the value of capacitance C on terminal OUT_2. For example, if $R_2 = 5$ kΩ, t_{pdh} of the $+2.5$ V level is infinite, since $+2.5$ V is never reached at the output. If, however, R_2 is chosen large, t_{pdl} will become large. The relations for t_{pdh} and t_{pdl}, both for the level of $+2.5$ V and for $C = 100$ pF, are given without derivation in Figure 5.4b. We can see that there is no "optimum" value for R_2, but a tradeoff between t_{pdh} and t_{pdl} takes place. In practice, R_2 would be chosen in the vicinity of 17 kΩ to result in approximately equal t_{pdh} and t_{pdl}.

5.2   Diode-Transistor Logic (DTL) Inverters and Gates

In the preceding section, diode AND and OR gates were described. In order to realize the function of logic inversion, i.e., $OUT = \overline{IN}$, however, a *logic inverter* is also necessary. A simple form of such a logic inverter circuit is shown in Figure 5.5a, its symbol in Figure 5.5b. Voltages and currents in the circuit with no external load on the output are also shown in Figure 5.5a: in round parentheses for $+5$ V input voltage, and in square parentheses for 0 V input voltage.

The output voltage as function of the input voltage in the circuit of Figure 5.5a is sketched in Figure 5.5c. A significant feature is the sharp transition in the vicinity of $V_{IN} = +1.5$ V, with a result that for wide regions of input voltages (approximately between 0 V and $+1$ V, and between $+2$ V and $+5$ V) the output voltage is insensitive to changes in the input voltage. Thus, the output voltage will be a constant $+5$ V as long as the input voltage is at any value between 0 V and $+1$ V, and it will be a constant voltage of near zero when the input voltage is at any value between $+2$ V and $+5$ V. The sharp transition can be characterized by a threshold voltage, V_{TH}, which is defined as the input voltage at which the output voltage is equal to the input voltage (see Figure 5.5c). It can be seen that the threshold voltage of the circuit of Figure 5.5a is in the vicinity of $V_{TH} = +1.5$ V.

The logic inverter circuit of Figure 5.5a can be extended to a *logic gate*, as illustrated for two inputs in Figure 5.6a. It realizes the logic function $OUT = \overline{IN_1 IN_2}$, that is, the function of NOT AND, or NAND. A symbol for such a *NAND gate* is shown in Figure 5.6b, the truth table for an unloaded output in Figure 5.6c. NAND gates with up to eight inputs are available in SSI; some circuits also have provisions for a larger number of inputs.

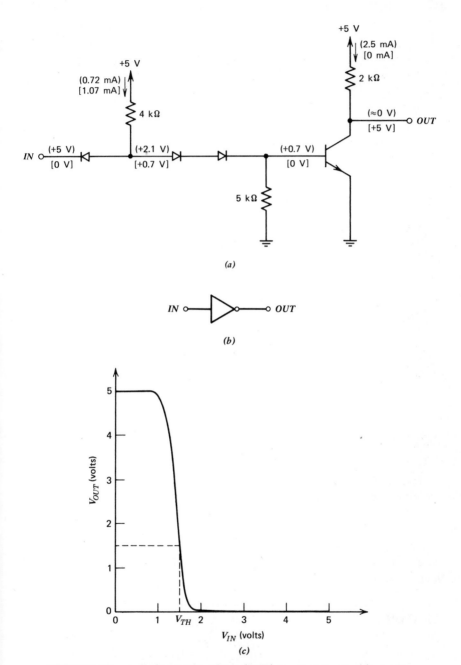

FIGURE 5.5. A diode-transistor logic (DTL) inverter circuit: (*a*) circuit diagram, (*b*) inverter symbol, (*c*) output voltage vs. input voltage.

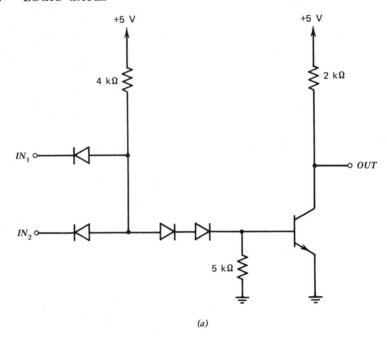

(a)

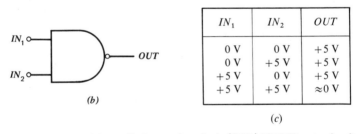

(b)

IN_1	IN_2	OUT
0 V	0 V	+5 V
0 V	+5 V	+5 V
+5 V	0 V	+5 V
+5 V	+5 V	≈0 V

(c)

FIGURE 5.6.   **A 2-input diode-transistor logic (DTL) NAND gate circuit:**
(a) circuit diagram, (b) NAND gate symbol, (c) truth table.

The DTL NAND gate circuit of Figure 5.6a, unlike diode AND and OR
gates, can be followed by identical NAND circuits without significant
loading effects.

EXAMPLE 5.4.   A logic circuit consisting of two 2-input DTL NAND
gates is shown in Figure 5.7. It can be shown that the circuit realizes
the logic function $OUT_2 = \overline{\overline{IN_1 IN_2} \cdot IN_3} = IN_1 IN_2 + \overline{IN_3}$. The voltage
levels at the output OUT_1 of the first gate circuit are still approximately
0 V or +5 V, thus there is no significant loading by the second gate circuit.

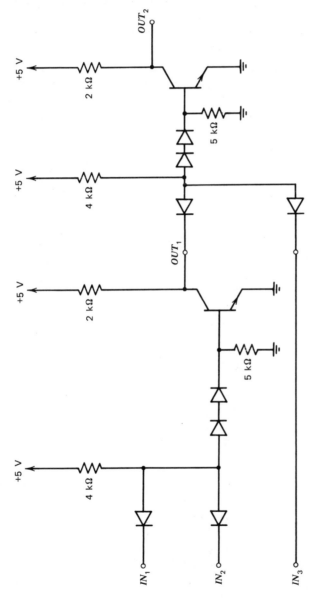

FIGURE 5.7. A logic circuit consisting of two 2-input DTL NAND gates, realizing the logic function $OUT_2 = IN_1 IN_2 + \overline{IN_3}$.

There seems to be no reason why the output of a DTL circuit could not drive several inputs simultaneously, or why a second circuit could not be followed by a third one. In fact, the output of a typical circuit can drive over half a dozen inputs simultaneously, and many circuits can be cascaded. In the latter case, however, the propagation delays of the circuits have to be added to obtain the resulting propagation delay.

EXAMPLE 5.5.   A digital circuit using four NAND gates is shown in Figure 5.8. Since some of the input signals pass through three subsequent gates, the circuit of Figure 5.8 is referred to as a 3-stage (or 3-level) circuit.

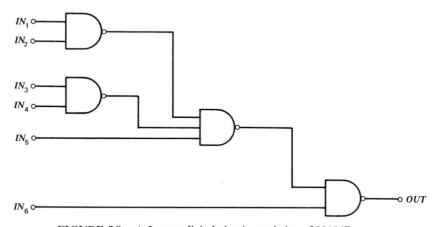

FIGURE 5.8.   A 3-stage digital circuit consisting of NAND gates.

The DTL gates used have propagation delays at the $+1.5$ V level of $t_{pdh} = t_{pdl} = 50$ ns measured with a capacitive load of $C = 50$ pF.* Thus, the propagation delay to terminal OUT of input IN_6 is 50 ns, that of IN_5 is 100 ns, and those of IN_1, IN_2, IN_3, and IN_4 are 150 ns.

When an output is used to drive several inputs simultaneously, loading limitations of the circuit have to be taken into account. These arise from the fact that the output transistor can carry only a finite current (typically 12 mA) while providing a low saturation voltage.

In general, permissible loading conditions are given for specified voltage levels. In DTL circuits it is customary to specify load conditions such that the "low" output level will always be below $+0.4$ V, and the "high" output level will always be above $+2.4$ V.

---

* 1 ns = 1 nanosecond = 10^{-9} second.

EXAMPLE 5.6.   In the DTL NAND gate circuit of Figure 5.6a, the maximum collector current in the transistor is limited to 12 mA in order to provide a saturation voltage of less than $+0.4$ V. Since at a near-zero output voltage the current through the 2 kΩ resistor is 5 V/2 kΩ = 2.5 mA, a current of 12 mA $-$ 2.5 mA = 9.5 mA flowing *into the circuit* is available for external use. The nominal input current of each circuit (see Figure 5.5a) is 1.07 mA, hence the 9.5 mA current is adequate for driving eight inputs simultaneously. Also, a "high" output level above $+2.4$ V is desired; thus it is necessary that in this state the voltage drop across the 2-kΩ resistor be less than 5 V $-$ 2.4 V = 2.6 V, i.e., the maximum current permitted to flow *out of the circuit* is 2.6 V/2 kΩ = 1.3 mA. Hence, the permitted *minimum* value of an external resistor connected between the output and ground is 2.4 V/1.3 mA = 1.85 kΩ.

The propagation delay of an inverter or gate circuit depends on the external capacitive and resistive loads on its output. For a given capacitive load, the worst case t_{pdl} occurs when the resistive current flowing *into the circuit* is maximum, that is, when the circuit is loaded by the maximum number of inputs permitted, and t_{pdl} is usually specified under such loading conditions. The worst case t_{pdh}, however, occurs when the output of the circuit is loaded by a single input only, and t_{pdh} is usually specified under such conditions.

EXAMPLE 5.7.   Propagation delays of a DTL inverter circuit are measured in the circuit of Figure 5.9a with the input waveforms of Figure 5.9b and c. The values of load resistance R and load capacitance C used in the measurements are as shown in Figure 5.9d.

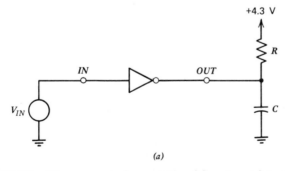

(a)

FIGURE 5.9.   **Measurement of propagation delays** t_{pdl} **and** t_{pdh}: **(a) test circuit.** (*See next page for* (**b**), (**c**), *and* (**d**).)

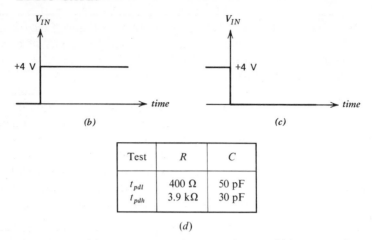

(b)   (c)

Test	R	C
t_{pdl}	400 Ω	50 pF
t_{pdh}	3.9 kΩ	30 pF

(d)

FIGURE 5.9. (*b*) input waveform for measuring t_{pdl}, (*c*) input waveform for measuring t_{pdh}, (*d*) circuit values for measuring t_{pdl} and t_{pdh}.

Wired-OR DTL Circuits. Consider the circuit of Figure 5.10 with each gate a DTL NAND gate of Figure 5.6*a*. The output will be "high" $(+5 \text{ V})$ if and only if both NAND gate outputs go "high." Hence, the circuit realizes the logic function

$$OUT = \overline{IN_1 IN_2} \cdot \overline{IN_3 IN_4} = (\overline{IN_1} + \overline{IN_2}) \cdot (\overline{IN_3} + \overline{IN_4}).$$

This circuit configuration, known as wired-OR (intrinsic-OR, wired-AND, collector dot, or collector-strapped) circuit, is especially useful when digital data are fanned in, scanned, from many parallel sources. Wired-OR

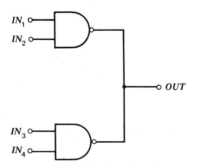

FIGURE 5.10.   A wired–OR circuit using two DTL NAND gates.

circuits are illustrated here with 2-input NAND gates; the same principle is applicable when NAND gates with more inputs are used.

EXAMPLE 5.8.   The circuit of Figure 5.11 uses four DTL NAND gate circuits in a wired-OR configuration, which are followed by an inverter.

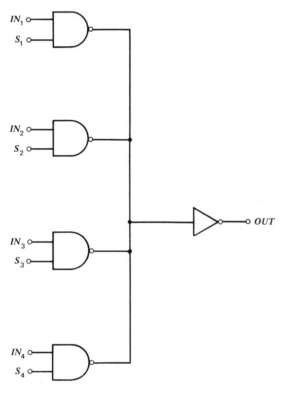

FIGURE 5.11.   A wired–OR circuit using four DTL NAND gates followed by an inverter.

Output OUT as function of the inputs can be written as $OUT = \overline{IN_1 S_1} \cdot \overline{IN_2 S_2} \cdot \overline{IN_3 S_3} \cdot \overline{IN_4 S_4} = (IN_1 S_1) + (IN_2 S_2) + (IN_3 S_3) + (IN_4 S_4)$. If inputs S_1 through S_4 are "scanned" such that only one of them is "high" at any given time, the output will equal the corresponding scanned input. This is illustrated in Figure 5.12 for the case when $IN_1 = +5$ V, $IN_2 = 0$ V, $IN_3 = +5$ V, and $IN_4 = +5$ V.

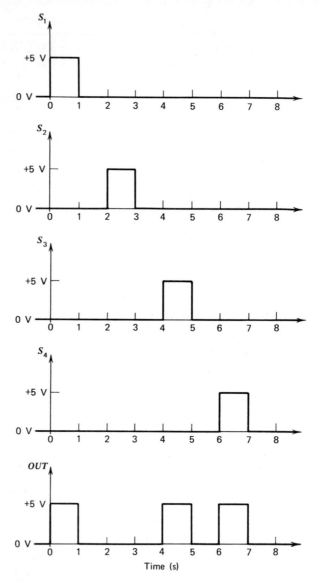

FIGURE 5.12. Operation of the wired–OR circuit of Figure 5.11 with $IN_1 = +5$ V, $IN_2 = 0$ V, $IN_3 = +5$ V, $IN_4 = +5$ V, and S_1 through S_4 scanned as shown.

Particular care must be taken in a wired-OR circuit to assure that the output loading limitations of the gates are not exceeded.

EXAMPLE 5.9.   The wired-OR circuit of Figure 5.11 utilizes four DTL NAND gates of Figure 5.6a and an inverter of Figure 5.5a. Each transistor is capable of carrying a collector current of 12 mA. Since the maximum possible collector current of any one of the four gates is the sum of the 1.07 mA load current of the inverter input and of 4×5 V/2 kΩ from the four resistors, i.e., a total of 11.07 mA, this limitation is never exceeded. It would not be possible, however, to extend the circuit to five inputs simply by a parallel connection of an additional NAND gate, since then the 12 mA load current limitation of the transistors could be exceeded.

In order to alleviate the loading limitations, DTL gate circuits are available with a choice of resistor values in the collector circuit, such as 2 kΩ as shown in Figure 5.6a, with a higher resistor value, or with no resistor at all (*"open collector"*). It is then up to the designer to select the gates such that the load limitations are not exceeded. In order to attain a short propagation delay t_{pdh}, however, it is advantageous to operate near the maximum permitted load current.

EXAMPLE 5.10.   In the circuit of Figure 5.11, the common junction of the four NAND gate outputs and the inverter input has a total capacitance, resulting from wiring and circuit capacitances, of $C = 500$ pF (rather high). Propagation time t_{pdh} of the $+1.5$ V point, neglecting the intrinsic propagation delays of the gates, can be shown to be

$$t_{pdh} = \left[500 \text{ pF} \middle/ \left(\frac{4}{2 \text{ k}\Omega} + \frac{1.07 \text{ mA}}{5 \text{ V}} \right) \right] \ln\left(\frac{5 \text{ V}}{3.5 \text{ V}} \right) = 85 \text{ ns}.$$

If, however, only one of the four NAND gates has a 2 kΩ resistor and the remaining three have the resistors omitted, then

$$t_{pdh} = \left[500 \text{ pF} \middle/ \left(\frac{1}{2 \text{ k}\Omega} + \frac{1.07 \text{ mA}}{5 \text{ V}} \right) \right] \ln\left(\frac{5 \text{ V}}{3.5 \text{ V}} \right) = 250 \text{ ns},$$

neglecting again the intrinsic propagation delays of the gates.

5.3 Transistor–Transistor Logic (TTL) Inverters and Gates

It was seen that in order to attain a short propagation delay t_{pdh} in a DTL gate, the value of the resistor at the output had to be chosen low which resulted in increased power dissipation. The continuing quest for higher speeds and reduced power dissipation in conjunction with improved manufacturing technology led to the advent of *transistor–transistor logic* (TTL) gates. There are several versions: The "standard" TTL series has typical propagation delays of 10 ns and typical dissipations of 10 mW/gate; the "low-power" TTL series has typical propagation delays of 35 ns and typical power dissipations of 1 mW/gate; the "high-speed" TTL series has typical propagation delays of 6 ns and typical power dissipations of 22 mW/gate. Still another family of circuits, the "diode-clamped" TTL, or "Schottky-clamped" TTL, series has propagation delays in the vicinity of 3 ns and typical power dissipations of 20 mW/gate, accompanied, however, by a slight increase in the "low" output level, resulting in reduced operating margins.

A schematic diagram of a "standard" TTL inverter circuit with voltage levels for $+5$ V and 0 V input voltages is shown in Figure 5.13a; the symbol for an inverter is shown in Figure 5.13b.* It can be seen that the output has a low impedance in both logic states: transistor Q_3 pulls the output up to its "high" state, transistor Q_4 pulls it down to its "low" state. This feature makes the propagation delays of TTL circuits shorter, as compared to those of DTL circuits; also the propagation delays are less sensitive to capacitive and resistive loading on the output.

Transistors Q_3 and Q_4 are driven via Q_1 and Q_2. When input *IN* is at $+5$ V, the emitter-base junction of Q_1 is cut off; Q_2 is saturated by a current of ≈ 0.7 mA flowing into its base junction through the 4 kΩ resistor and the forward biased collector-base junction of Q_1. When input *IN* is at 0 V, Q_1 is saturated and Q_2 is cut off. Diode D_1 at the input protects the circuit against spurious transients.

The output voltage as function of the input voltage is shown for various load currents in Figure 5.13c. For input voltages of $+2$ V to $+5$ V, the "low" output voltage is below $+0.4$ V for load currents up to 16 mA flowing *into the circuit*. For input voltages of 0 V to $+0.8$ V, the "high" output voltage is in the vicinity of $+2.4$ V for a load current of 2 mA flowing *out of the circuit*. In its "high" state, the output of the circuit presents a high impedance when it is driven above $+3.6$ V by an external load; driving the output above $+5$ V is not permitted.

---

* The circuit of Figure 5.13a represents one of the simplest types of TTL circuits. With improved manufacturing technology the inclusion of a larger number of transistors and improved performance are possible.

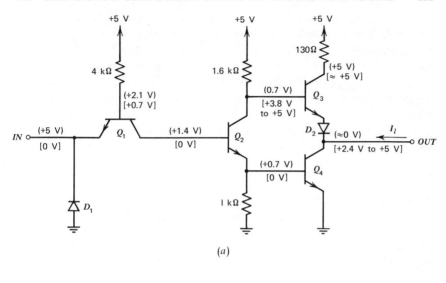

(a)

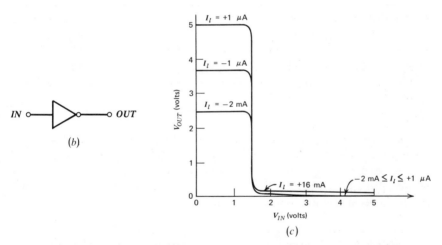

(b)

(c)

FIGURE 5.13. A standard TTL inverter circuit: (a) schematic diagram, (b) inverter symbol, (c) output voltage vs. input voltage for various load currents.

EXAMPLE 5.11. The output of the TTL circuit of Figure 5.13a is in its "low" state, i.e., at a near-zero voltage, and it is loaded by the inputs of twelve identical TTL circuits. Since at input voltages of 0 V the input current flowing out of each of the twelve inputs is about 1.1 mA, the total current flowing *into the output* of the first TTL circuit is 12×1.1 mA $=$ 13.2 mA, which is below the maximum permitted 16 mA. Because of tolerances of the circuits, however, it is usually recommended that the output of such a TTL circuit be loaded by no more than ten inputs.

When the output of the first TTL circuit is in its "high" state, it has to supply the leakage currents of the diodes at the inputs of the twelve circuits loading it. These leakage currents are in the vicinity of 50 μA each; hence the total current flowing *out of the output* of the first circuit is 12×50 μA $=$ 0.6 mA. Thus the voltage at the output of the first circuit will be between $+2.4$ V and $+3.6$ V.

The output impedance of a "standard" TTL circuit during switching is in the vicinity of 150 Ω; the intrinsic propagation delays, that is, propagation delays for zero load capacitances, are in the vicinity of 10 ns. Propagation delays are longer, however, when the output is loaded by a capacitance.

EXAMPLE 5.12.   The output of the TTL inverter circuit of Figure 5.13*a* is loaded by a capacitance of $C = 500$ pF. If the intrinsic propagation delay were neglected, t_{pdh} of the $+1.5$ V point could be written as 150 $\Omega \times 500$ pF $\ln(5 \text{ V}/3.5 \text{ V}) \approx 27$ ns. The total propagation delay can be approximated as the sum of this and the intrinsic propagation delay, i.e., $t_{pdh} \approx 27$ ns $+ 10$ ns $= 37$ ns.

Because of the availability of the large pullup current from Q_3 in the "high" output state, the circuit is capable of delivering a relatively large current into a short circuit to ground on the output; this current is limited, primarily by the 130 Ω resistor in the collector of Q_3, to a value that results in an acceptable power dissipation.*

EXAMPLE 5.13.   The output in the TTL circuit of Figure 5.13*a* is short circuited to ground while it is in its "high" state. This results in voltages of 0 V on the output, $\approx +0.7$ V on the emitter and on the collector of Q_3, and $\approx +1.4$ V on the base of Q_3. The collector current of Q_3 is $(5 \text{ V} - 0.7 \text{ V})/130 \ \Omega = 33$ mA, the base current of Q_3 is $(5 \text{ V} - 1.4 \text{ V})/$ 1.6 k$\Omega \approx 2.2$ mA. The power dissipation in the circuit, including the power dissipated in Q_1 and in the 4 kΩ resistor, is 5 V $\times$ (33 mA $+ 2.2$ mA $+ 4.3$ V/4 kΩ) ≈ 180 mW.

---

* If the output is shorted to the $+5$ V when the output is in its "low" state, it will result in a high current; the circuits are usually not protected against such short circuits.

The inverter circuit of Figure 5.13a can be extended to a NAND gate by use of a multi-emitter transistor.* A 3-input TTL NAND gate thus obtained is shown in Figure 5.14a; its symbol is in Figure 5.14b. TTL NAND gates with up to eight inputs are available; some of these can be extended to larger number of inputs by the use of additional circuitry

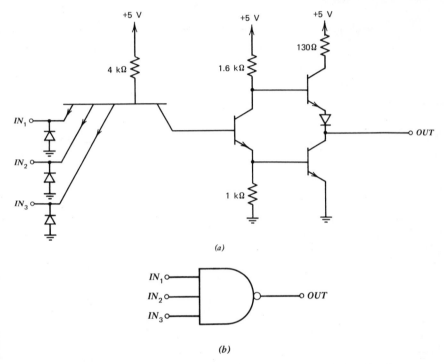

(a)

(b)

FIGURE 5.14.   A 3-input standard TTL NAND gate circuit: (a) schematic diagram, (b) 3-input NAND gate symbol.

("*expander elements*"). TTL circuits available include the AND gate (a TTL NAND gate into which an inverter stage has been incorporated), NOR and OR gates, open collector structures, and others.

AND–OR–INVERT Gates.   An important structure in TTL integrated circuits is the AND–OR–INVERT gate. One such gate realizing the logic function $OUT = \overline{IN_1 IN_2 IN_3 + IN_4 IN_5 IN_6}$ is shown in Figure 5.15. This logic function can be augmented by connecting *expander nodes* e and $\bar{e}$ to external *expander elements*. The circuit of a three-input expander element

---

* Functionally, a multi-emitter transistor may be viewed as several transistors (three in Figure 5.14a) with their bases connected together to form a single base and their collectors connected together to form a single collector.

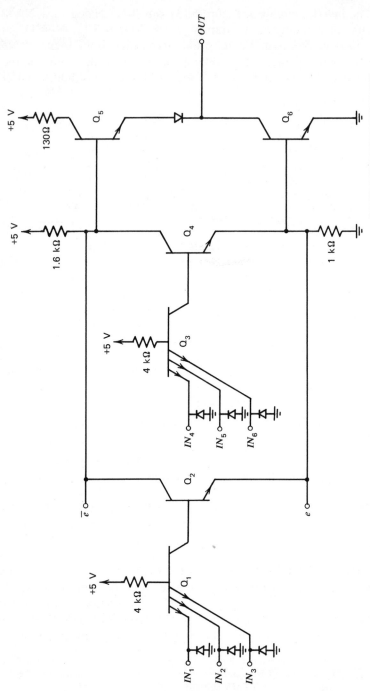

FIGURE 5.15. An AND–OR–INVERT gate circuit realizing the logic function $\overline{IN_1 IN_2 IN_3 + IN_4 IN_5 IN_6}$.

is identical to that of Q_1 and Q_2 in Figure 5.15; its use is illustrated in Figure 4.15.

Comparison of the circuit of Figure 5.15 with that of the TTL NAND gate of Figure 5.14a shows that the OR function is attained by adding transistors Q_1 and Q_2 in Figure 5.15. Thus each input signal in an AND–OR–INVERT gate passes through the same number of transistors as in a NAND gate; hence the propagation delay of the AND–OR–INVERT gate is comparable to that of the NAND gate and is superior to that of a circuit built up from individual gates. Because of this feature, and because of the potential reduction in the number of interconnections, AND–OR–INVERT gates find widespread use in digital systems.

3-State TTL Circuits.   In many cases it would be desirable to use TTL logic circuits in a wired-OR configuration similar to that of Figure 5.10 and

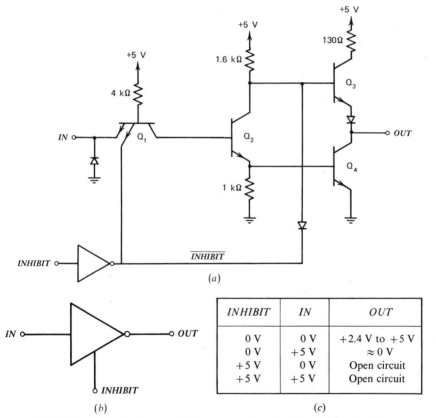

INHIBIT	IN	OUT
0 V	0 V	+2.4 V to +5 V
0 V	+5 V	≈ 0 V
+5 V	0 V	Open circuit
+5 V	+5 V	Open circuit

(b)   (c)

FIGURE 5.16.   A 3-state TTL inverter circuit: (a) simplified circuit diagram, (b) symbol, (c) truth table.

Figure 5.11. Such configurations, however, are not feasible with the TTL circuits of Figure 5.13*a*, Figure 5.14*a*, or Figure 5.15 due to their low output impedances.

In order to make fast wired-OR TTL circuits possible, special 3-state TTL (disabled TTL, Tri-state TTL) circuits have been developed. The operation of such a circuit is illustrated in Figure 5.16*a* for the simplest case of an inverter; the symbol for the circuit is shown in Figure 5.16*b*, its truth table in Figure 5.16*c*. It can be seen that as long as the *INHIBIT* input is "low," the circuit operates as an inverter and $OUT = \overline{IN}$. When the *INHIBIT* input is "high," however, Q_2, Q_3, and Q_4 are cut off and output *OUT* presents a high impedance irrespective of the state of input *IN*. The 3-state TTL inverter circuit of Figure 5.16*a* can be extended to a 3-state TTL NAND gate by use of additional emitters in transistor Q_1.

The outputs of several circuits of Figure 5.16*a* can be connected in parallel, as long as no more than one inhibit input is low.

EXAMPLE 5.14. Two 3-state TTL circuits with their outputs connected in parallel are shown in Figure 5.17*a*; the truth table is shown in Figure 5.17*b*. Note that the output is not defined in the states given in the second and third lines of the truth table; furthermore, the two circuits

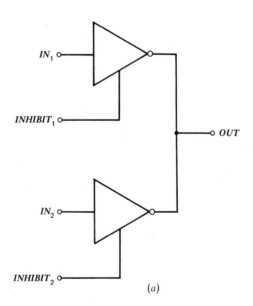

(*a*)

FIGURE 5.17. Two 3-state TTL inverter circuits in a wired-OR configuration: (*a*) circuit diagram.

$INHIBIT_1$	$INHIBIT_2$	IN_1	IN_2	OUT
0 V	0 V	0 V	0 V	+2.4 V to +5 V
0 V	0 V	0 V	+5 V	?
0 V	0 V	+5 V	0 V	?
0 V	0 V	+5 V	+5 V	$\approx$ 0 V
0 V	+5 V	0 V	0 V	+2.4 V to +5 V
0 V	+5 V	0 V	+5 V	+2.4 V to +5 V
0 V	+5 V	+5 V	0 V	$\approx$ 0 V
0 V	+5 V	+5 V	+5 V	$\approx$ 0 V
+5 V	0 V	0 V	0 V	+2.4 V to +5 V
+5 V	0 V	0 V	+5 V	$\approx$ 0 V
+5 V	0 V	+5 V	0 V	+2.4 V to +5 V
+5 V	0 V	+5 V	+5 V	$\approx$ 0 V
+5 V	+5 V	0 V	0 V	Open circuit
+5 V	+5 V	0 V	+5 V	Open circuit
+5 V	+5 V	+5 V	0 V	Open circuit
+5 V	+5 V	+5 V	+5 V	Open circuit

(b)

FIGURE 5.17.   (b) truth table.

are in conflict in trying to establish opposite logic levels at the output. These states would also result in high ($\approx$ 30 mA) currents drawn from the +5 V power supply, and for this reason they are not permitted.

5.4 Emitter-Coupled Logic (ECL) Gates

The shortest propagation delays can be attained by the use of emitter-coupled logic circuits. These circuits operate all transistors in their forward active region or in their cutoff region, hence saturation and the concomitant stored charge are avoided (see Appendix). Propagation delays in the vicinity of 1 ns can be achieved with power dissipations of 60 mW/gate, 2 ns with power dissipations of 25 mW/gate.

A circuit diagram of a 2-input emitter-coupled logic gate with propagation delays near 1 ns is shown in Figure 5.18a; levels are shown for -0.7 V and -1.6 V input voltages on IN_2 with -1.6 V on input IN_1. A symbol is shown in Figure 5.18b, a truth table in Figure 5.18c. When the -0.7 V level is assigned logical "0" and the -1.6 V level a logical "1" ("negative logic"), $OUT_1 = \overline{IN_1 IN_2}$, $OUT_2 = IN_1 IN_2$; that is, the circuit acts as an AND–NAND gate. When the -0.7 V level is assigned

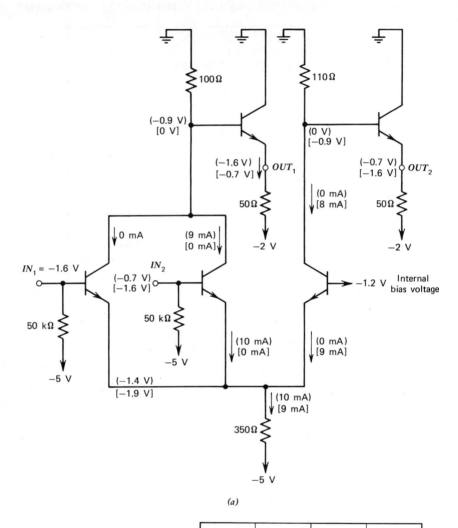

(a)

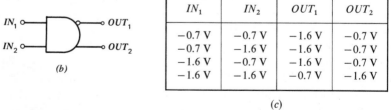

(b)

IN_1	IN_2	OUT_1	OUT_2
−0.7 V	−0.7 V	−1.6 V	−0.7 V
−0.7 V	−1.6 V	−1.6 V	−0.7 V
−1.6 V	−0.7 V	−1.6 V	−0.7 V
−1.6 V	−1.6 V	−0.7 V	−1.6 V

(c)

FIGURE 5.18. An emitter-coupled logic circuit: (a) circuit diagram, (b) symbol, (c) truth table.

logical "1" and the -1.6 V level logical "0" ("positive logic"), $OUT_1 = \overline{IN_1 + IN_2}$, $OUT_2 = IN_1 + IN_2$; and the circuit acts as an OR–NOR gate.

In order to attain short propagation delays, the outputs are loaded by 50 Ω resistors returned to -2 V, as shown. This also permits the use of terminated 50 Ω transmission lines (cables) between the circuits. Several inputs may be driven from a single output; the lengths of the interconnections may become significant, however, since their typical propagation delays can be in the vicinity of 0.1 ns/inch. Details of interconnection and load limitations are discussed in Chapter 12.

When the ultimate in operating speed is not required, outputs of several emitter-coupled circuits can be connected in parallel in a wired–OR configuration. Logic functions such as AND–OR and OR–AND can be synthesized in this manner, as illustrated in Figure 4.25.

5.5 Complementary FET Logic Inverters and Gates

We have seen in the preceding sections that there is a general tradeoff between propagation delay and power dissipation in logic circuits. Thus, propagation delays of 1 ns can be achieved by emitter-coupled logic circuits at a power dissipation of 60 mW/gate, while propagation delays of 35 ns can be achieved by low power TTL circuits at a power dissipation of 1 mW/gate. In many cases, low power dissipation is of overriding importance, and this led to the development of complementary field-effect transistor (FET) logic circuits.[3]

The gate current of an insulated-gate field-effect transistor (see Appendix) can be in the picoampere range, which may be negligible in many applications. Indeed, FET logic gates that have power dissipations of 10^{-8} W in the quiescent state are available; and they provide practical means for operation at low power consumption.

A simplified schematic diagram and the characteristics of a complementary field-effect transistor inverter are shown in Figure 5.19.* The circuit consists of a p-channel enchancement type FET Q_1 and of an n-channel enhancement type FET Q_2. With the output unloaded, when the input is at 0 V, Q_1 is turned on, Q_2 is turned off, and the output voltage is in the vicinity of $+5$ V. Conversely, when the input voltage is at $+5$ V, Q_2 is turned on, Q_1 is turned off, and the output voltage is near zero (again with unloaded output). Since only one transistor is turned on at any given time, the quiescent power dissipation is determined by leakage currents and by the load currents on the output.

* Internal protection circuits are not shown.

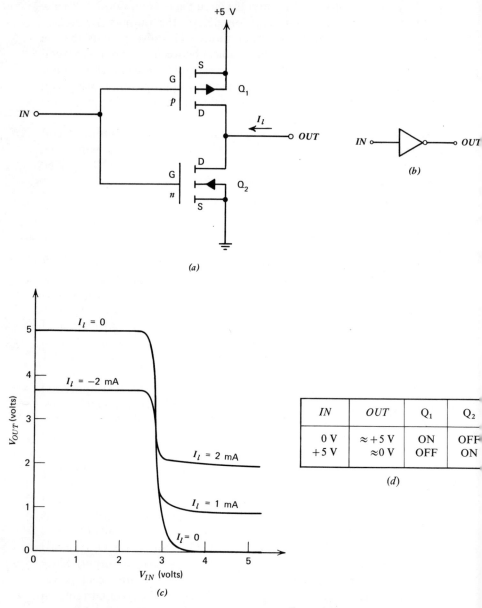

FIGURE 5.19. A complementary field-effect transistor (FET) inverter: (*a*) circuit diagram, (*b*) inverter symbol, (*c*) output voltage vs. input voltage, (*d*) truth table including states of Q_1 and Q_2.

EXAMPLE 5.15. The transistors in the inverter circuit of Figure 5.19a have leakage currents of 2 nA when they are cut off; and the quiescent power dissipation with no load on the output is 5 V × 2 nA = 10 nW.* If the output is loaded by a 100 MΩ resistor to ground, however, in the "high" output state Q_1 will conduct a current of 5 V/100 MΩ = 50 nA and dissipate a power of 5 V × 50 nA = 250 nW, resulting in a total power dissipation of 10 nW + 250 nW = 260 nW.

Because of their low input currents, the inputs of logic gates can be adequately represented by a capacitance of typically 5 pF. Assuming that there is no significant charge in the source-drain region of the field-effect transistor, propagation delays are determined primarily by capacitances and by the currents available to charge them.

EXAMPLE 5.16. In the inverter circuit of Figure 5.19a, Q_1 is capable of delivering a drain current with a magnitude of $|I_{D1}| = 1.5$ mA, Q_2 an $|I_{D2}| = 1$ mA. The threshold voltage of the circuit is $V_{TH} = +3$ V. The output of the circuit has a capacitance of 20 pF and is loaded by the input capacitances of ten circuits of 5 pF each and by wiring capacitances of 30 pF. Thus, the total capacitance loading the output is C = 20 pF + 10 × 5 pF + 30 pF = 100 pF.

In the transition to the "high" state of the output, the $V_{TH} = +3$ V level is reached in a time $t_{pdh} \cong CV_{TH}/|I_{D1}| = 100$ pF 3 V/1.5 mA = 200 ns. In the transition to the "low" state of the output, the $V_{TH} = +3$ V level is reached in a time $t_{pdl} \cong C(5\ V - V_{TH})/|I_{D2}| = 100$ pF (5 V − 3 V)/1 mA = 200 ns.

Thus, in this particular example, the propagation delays t_{pdh} and t_{pdl} are equal, $t_{pdh} = t_{pdl} = 200$ ns.

Because of the low quiescent power dissipation, ac (or transient) dissipation can become significant in field-effect transistor circuits. The energy stored in a capacitor C charged to a voltage V_c is $CV_c^2/2$. It can be shown that when the capacitor is charged from a fixed voltage via a series resistance such as the source-drain region of Q_1 in Figure 5.19a, an energy of $CV_c^2/2$ is also dissipated in Q_1. When the capacitor is discharged to 0 V via Q_2 in Figure 5.19b, the energy that was stored in the capacitor is dissipated in Q_2.

---

* 1 nA = 1 nanoampere = 10^{-9} A; 1 nW = 1 nanowatt = 10^{-9} W.

Thus, the total energy dissipated in Q_1 and Q_2 during a charge-discharge cycle is $2 \times CV_c^2/2 = CV_c^2$. If the charge-discharge cycle occurs with a frequency f, the resulting ac power dissipation is $P_{ac} = CV_c^2 f$.

EXAMPLE 5.17.   The inverter circuit of Figure 5.19a has a quiescent power dissipation of 100 nW, and the total load capacitance on its output is $C = 100$ pF. The input of the circuit is switched between 0 V and $+5$ V at a frequency of 1 MHz. Since the capacitor is switched between 0 V and $+5$ V, $V_c = 5$ V, and the ac power dissipation is $P_{ac} = CV_c^2 f = 100$ pF $\times$ $(5\text{ V})^2 \times 1$ MHz $= 2.5$ mW. Thus, the total power dissipation in the circuit is 100 nW $+$ 2.5 mW $\approx$ 2.5 mW.

The field-effect transistor (FET) inverter circuit of Figure 5.19a can be extended to a 2-input positive NAND gate as shown in Figure 5.20 (FET NAND and NOR gates can be also extended to higher number of inputs). The circuit consists of four field-effect transistors: two p-channel FET's Q_1 and Q_2 in parallel, and two n-channel FET's Q_3 and Q_4 in series. We can see that in the first three lines of the truth table at least one of Q_1 and Q_2 is turned on and at least one of Q_3 and Q_4 is turned off, hence the output is at $+5$ V. In the last line of the truth table, Q_1 and Q_2 are turned off, Q_3 and Q_4 are turned on, and the output is at 0 V.

The circuit diagram, symbol, and truth table of a 2-input field-effect transistor positive NOR gate are shown in Figure 5.21.

5.6   Complementary FET Transmission Gates

In many cases it would be desirable to connect in parallel the outputs of several FET logic gates in order to achieve a wired-OR configuration similar to the DTL wired-OR circuits of Figure 5.10 and Figure 5.11. Also, isolation between output stages is required in more complex integrated circuits. FET logic gates, however, present low (≈ 1 kΩ) output impedances both in their "low" and in their "high" states. Thus, large currents could flow if the outputs were connected in parallel.

The problem has been solved by the introduction of special circuitry that can provide a high output impedance. Such a circuit is the complementary field-effect transistor (FET) transmission gate, shown in Figure 5.22, consisting of n-channel FET Q_1 and p-channel FET Q_2. Outputs of the transmission gates can be connected in parallel as long as no more than one gate control signal G is at $+5$ V (see Figure 5.22b). This is illustrated in the example that follows.

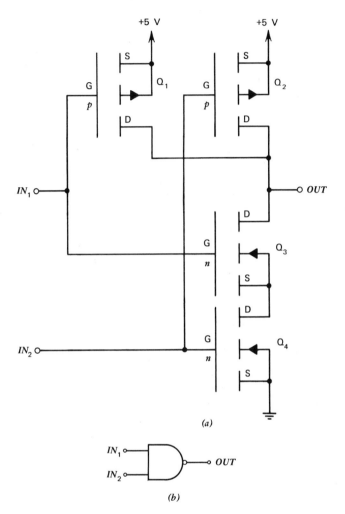

(a)

(b)

IN_1	IN_2	OUT	Q_1	Q_2	Q_3	Q_4
0 V	0 V	+5 V	ON	ON	OFF	OFF
0 V	+5 V	+5 V	ON	OFF	OFF	OFF
+5 V	0 V	+5 V	OFF	ON	OFF	OFF
+5 V	+5 V	0 V	OFF	OFF	ON	ON

(c)

FIGURE 5.20. A 2-input complementary field-effect transistor (FET) NAND gate: (a) circuit diagram, (b) NAND gate symbol, (c) truth table including the state of Q_1, Q_2, Q_3, and Q_4.

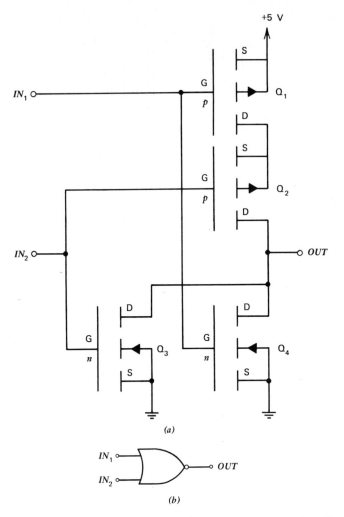

+5 V

S

G
p
Q_1

D

S

G
p
Q_2

D

IN$_1$

IN$_2$

OUT

D

D

G
n
Q_3

S

G
n
Q_4

S

(a)

IN$_1$

IN$_2$

OUT

(b)

IN_1	IN_2	OUT	Q_1	Q_2	Q_3	Q_4
0 V	0 V	+5 V	ON	ON	OFF	OFF
0 V	+5 V	0 V	OFF	OFF	ON	OFF
+5 V	0 V	0 V	OFF	OFF	OFF	ON
+5 V	+5 V	0 V	OFF	OFF	ON	ON

(c)

FIGURE 5.21. A 2-input complementary field-effect transistor (FET) NOR gate: (a) circuit diagram, (b) NOR gate symbol, (c) truth table including the states of Q_1, Q_2, Q_3, and Q_4.

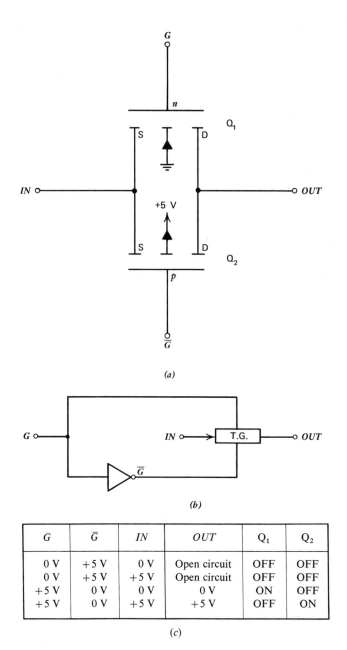

FIGURE 5.22. A field-effect transistor transmission-gate: (*a*) circuit diagram, (*b*) transmission-gate symbol, (*c*) truth table including the states of Q_1 and Q_2.

EXAMPLE 5.18.   Two field-effect transistor transmission gates with their outputs connected in parallel are shown in Figure 5.23a; the truth table is shown in Figure 5.23b. We can see that for the 14th and 15th states of the truth table the output is undetermined, and the two circuits are in conflict trying to establish the output level. These states are not allowed since they result in high (≈ 1 mA) currents drawn from the $+5$ V power supply.

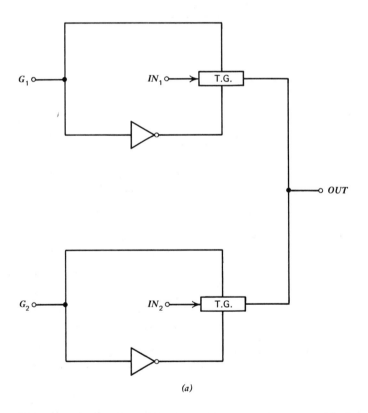

(a)

FIGURE 5.23. Two field-effect transistor transmission-gates with their outputs connected together: (a) circuit diagram.

G_1	G_2	IN_1	IN_2	OUT
0 V	0 V	0 V	0 V	Open circuit
0 V	0 V	0 V	+5 V	Open circuit
0 V	0 V	+5 V	0 V	Open circuit
0 V	0 V	+5 V	+5 V	Open circuit
0 V	+5 V	0 V	0 V	0 V
0 V	+5 V	0 V	+5 V	+5 V
0 V	+5 V	+5 V	0 V	0 V
0 V	+5 V	+5 V	+5 V	+5 V
+5 V	0 V	0 V	0 V	0 V
+5 V	0 V	0 V	+5 V	0 V
+5 V	0 V	+5 V	0 V	+5 V
+5 V	0 V	+5 V	+5 V	+5 V
+5 V	+5 V	0 V	0 V	0 V
+5 V	+5 V	0 V	+5 V	?
+5 V	+5 V	+5 V	0 V	?
+5 V	+5 V	+5 V	+5 V	+5 V

(b)

FIGURE 5.23. (b) truth table.

5.7 Tolerances, Noise Margins, Loading Rules

Transfer characteristics of the DTL logic circuit shown in Figure 5.5c indicate that for predictable operation the input voltage has to be either lower than 1 V or higher than 2 V. The situation is similar in TTL circuits and in FET circuits, as shown in Figure 5.13c and Figure 5.19c, respectively.

EXAMPLE 5.19.   The output voltage versus input voltage of a TTL inverter circuit is graphed in Figure 5.13c. We can see that the operation of the circuit is not predictable when the input voltage is in the vicinity of 1.5 V. Specifications of the circuit permit input voltages of either ≤ 0.8 V (for a high output) or ≥ 2 V (for a low output); the output is undetermined for input voltages between 0.8 V and 2 V.

In order to assure that the input voltage to a circuit is within one of the permitted regions, specifications are set on the output voltage of the circuit preceding it.

EXAMPLE 5.20.   The low output voltage of a TTL circuit is specified to be between 0 V and 0.4 V when the current into the output of the circuit is between 0 and 16 mA. The high output voltage is specified to be between 2.4 V and 5 V for a current out of the output that is between 0 and 0.6 mA. Comparing these values with the permissible inputs of ≤ 0.8 V or ≥ 2 V, we can see a margin of 0.4 V both for the low and for the high input logic levels.

The differences between the maximum permitted low input and the maximum guaranteed low output, and that between the minimum permitted high input and the minimum guaranteed high output, are designated *noise margins*. These margins (0.4 V in the example above) are available to cover undesired, but sometimes unavoidable, external interference or noise picked up along the interconnections. Specified input voltages, output voltages, and the resulting noise margins may be displayed in a *band diagram*. A band diagram for the noise margins of TTL circuits is shown in Figure 5.24.

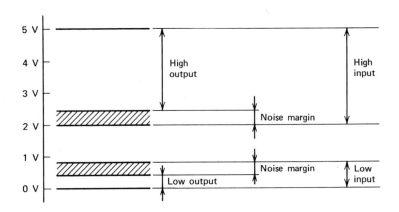

FIGURE 5.24.   A band diagram of noise margins in a TTL circuit.

Output voltages are usually specified for a given output current. Often it is desirable to express the maximum available output current in terms of the input currents of the subsequent circuits. Thus the output of a TTL circuit in Example 5.11 is capable of driving ten inputs simultaneously. Some complex circuits may have a higher input current than simpler ones. In order to simplify design, it is customary to introduce *loading rules*.

EXAMPLE 5.21. A circuit consisting of several TTL gates is shown in Figure 5.25. Inputs IN_1 and IN_4 are loaded by one gate input each, IN_2 is loaded by three gate inputs, and IN_3 by two inputs. If the currents of each input of a gate are identical, IN_1 and IN_4 may be referred to as

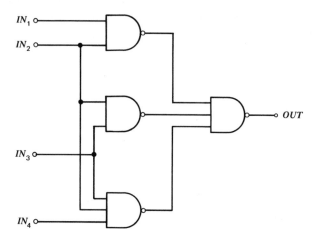

FIGURE 5.25. The circuit of Example 5.21.

representing one load each, IN_2 representing three loads, and IN_3 representing two loads. Loading rules specify that the output of the circuit may be loaded by a maximum of ten loads. Thus the output may be connected simultaneously to three, but not to four, IN_2 inputs of subsequent circuits, since three IN_2's represent nine loads, four IN_2's represent twelve loads.

5.8 Interfacing Logic Families

We have seen that there exists a general tradeoff between operating speed and power dissipation among the various logic families. Often it is necessary to use several families in order to satisfy the requirements of a particular system. In such cases, the use of *interfacing circuits* should be considered.

EXAMPLE 5.22. A digital system contains a small but fast central processing unit and an extensive but relatively slow information storage unit. In order to satisfy speed requirements, the processing unit is built using ECL circuits. Since the unit is small, the resulting power dissipation is acceptable. The large storage unit uses FET circuits that are relatively slow but draw little power, an important consideration since there are many circuits involved.

The signal levels of the ECL and FET circuits are made compatible by the use of special interfacing circuits. Outputs of the ECL circuits are connected to the inputs of the FET circuits via interfacing circuits that have a high-impedance input with a threshold of -1.2 V and an output voltage swing of 0 V to $+5$ V. Outputs of the FET circuits are connected to the inputs of the ECL circuits via interfacing circuits that have a high-impedance input with a threshold of $+3$ V and a low-impedance output with a voltage swing of -0.8 V to -1.6 V.

Under certain conditions the use of interfacing circuits is not required and two logic families can be directly connected together.

EXAMPLE 5.23.   Transfer characteristics of a TTL inverter are shown in Figure 5.13c, those of a complementary FET inverter in Figure 5.19c. When the output of a TTL inverter is connected only to the input of a FET inverter, the levels are compatible since the input current to the FET inverter is negligible. Connecting the output of a FET inverter directly to the input of a TTL inverter would be marginal, however. This is because in its low output state the output voltage of the FET inverter would be raised by the input current of the TTL inverter to about 1 V and this voltage would be too high to be used reliably as a TTL low logic level. The output of a FET inverter could be connected directly to the input of a low power TTL inverter, however, since its input draws a factor of 10 less current than the input of the standard TTL circuit of Figure 5.13a.

PROBLEMS

5.1   Sketch the output waveform of the diode AND gate of Figure 5.1a if the input waveforms are as shown in Figure 5.26 below.

5.2   Sketch the circuit diagram of 4-input diode AND gate. Prepare a truth table with input levels of 0 V and $+5$ V.

5.3   Sketch the output waveform of the diode OR gate circuit of Figure 5.2a if the input waveforms are as shown in Figure 5.26 below.

5.4   Derive the results given in Figure 5.3b.

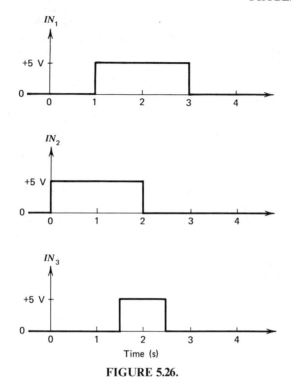

FIGURE 5.26.

5.5 Find propagation delays t_{pdh} and t_{pdl} at the $+2.5$ V level in the diode OR gate circuit of Figure 5.2a if the output of the circuit is loaded by a capacitance of $C = 100$ pF.

5.6 Derive the results given in Figure 5.4b.

5.7 Tabulate the voltages and currents in the DTL NAND gate circuit of Figure 5.6a for all four combinations of input voltages. Assume forward diode voltage drops of 0.7 V.

5.8 Sketch the output waveform of a 3-input DTL NAND gate circuit if the input waveforms are as shown in Figure 5.26 above.

5.9 Find propagation delays t_{pdh} and t_{pdl} at the $+1.5$ V level for the circuit of Figure 5.6a if the output is loaded with a capacitance of $C = 200$ pF and if the intrinsic propagation delays of the circuit are negligible.

5.10 Show that the circuit of Figure 5.7 realizes the logic function $OUT_2 = IN_1 IN_2 + \overline{IN_3}$.

5.11 Derive the logic function realized by the circuit of Figure 5.8.

5.12 Discuss the reasoning for the different values of load capacitances and load resistances in the measurements of t_{pdl} and t_{pdh} in the circuit of Figure 5.9.

5.13 Prepare a truth table for the circuit of Figure 5.10.

5.14 Draw the circuit of Figure 5.11 in detail and sketch the four collector currents as functions of time if the inputs are as given in Figure 5.12. Use component values from the circuit of Figure 5.6.

5.15 In many applications of digital integrated circuits, undesired but unavoidable "noise" of several volts is present on the input lines to the logic gates. In such cases, the $+1.5$ V threshold voltage of the DTL and TTL circuits is too low, and a higher threshold voltage is desirable. A *high-threshold NAND gate* circuit is shown in Figure 5.27. Sketch the output voltage of the circuit as a function of input voltage, assuming forward diode voltage drops of 0.7 V. What is the threshold voltage V_{TH} of the circuit?

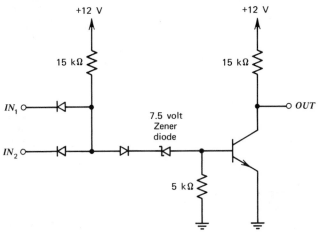

FIGURE 5.27.   High-threshold NAND gate circuit of Problem 5.15.

5.16 Tabulate the currents in the inverter circuit of Figure 5.13a for input voltages of 0 V and $+5$ V. Assume forward diode voltage drops of 0.7 V.

5.17 The TTL inverter circuit shown in Figure 5.13a has a "high" output voltage of $+3.3$ V when the output is loaded with a resistor of 8 kΩ to ground. Calculate the collector, emitter, and base voltages, and the current amplification h_{FE} (beta) of transistor Q_3.

5.18 Tabulate the voltages and currents in the 3-input TTL NAND gate circuit of Figure 5.14a for all eight combinations of the inputs, if each input can be 0 V or +5 V. Assume zero saturation voltages, diode forward voltage drops of 0.7 V, and a load resistor on the output of 8 kΩ to ground.

5.19 Draw the circuit of Figure 5.17*a* in detail and compute the currents of the circuit in the second and third states of the truth table shown in Figure 5.17*b*. Assume transistor saturation voltages of 0 V.

5.20 When the outputs of many 3-state TTL circuits are connected in parallel in a wired–OR configuration (see Figure 5.17*a*), the leakage current of each circuit in its "open" state is 40 μA. For this reason, the output current capability of each circuit is commonly made high in the high state. How many circuit outputs may be connected in parallel if the leakage current in the open state is 40 μA and the output current capability in the high state is 5.2 mA?

5.21 Sketch the voltages on OUT_1 and on OUT_2 in the ECL gate circuit of Figure 5.18*a* as functions of the voltage on IN_2 if the voltage on IN_1 is −1.6 V. What is the approximate value of the threshold voltage V_{TH}?

5.22 Extend the complementary FET NAND gate circuit of Figure 5.20 to three inputs. Draw a circuit diagram and prepare a truth table including the states of the transistors.

5.23 A 2-input *resistor-transistor logic* (RTL) circuit is shown in Figure 5.28.

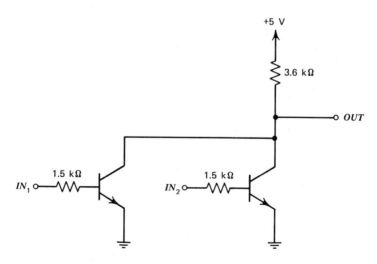

FIGURE 5.28.   Resistor-transistor logic (RTL) gate circuit of Problem 5.23.

Sketch the voltage on output OUT as a function of the voltage on input IN_1 if the voltage on input IN_2 is zero. Assume transistors with forward base-to-emitter voltage drops of 0.7 V, with $h_{FE} = \infty$ (infinite beta) and with saturation voltages of 0 V. What is the threshold voltage V_{TH}? Repeat with $h_{FE} = 10$.

5.24 An emitter-coupled logic circuit has its operating regions specified as follows. Permitted low level on the input is < -1.45 V, permitted high level on the input is > -1.15 V, guaranteed low level on the output is < -1.6 V, and guaranteed high level on the output is > -1 V. Find the noise margins.

5.25 Specify the characteristics of an interfacing circuit, including noise margins, for conversion from ECL to TTL levels, and of one for conversion from TTL to ECL levels.

CHAPTER 6

Flip-Flops

One of the most widely used storage elements in today's digital systems is the flip-flop. It is a circuit that has two stable states and transitions between these states can be effected by suitable input stimuli. For the most part flip-flops can be assembled from the logic gates described in the preceding chapter, although modern technology makes this rarely desirable. In what follows, however, flip-flops will be introduced as if they were in fact assembled from logic gates, with additional components added when required. The previous chapter on logic gates is heavily relied upon and considerations of logic levels, propagation delays, tolerances, noise margins, loading rules, and interfacing will not be discussed again in detail.

This chapter starts with the simplest flip-flop, the R-S storage flip-flop, and proceeds gradually to the master-slave flip-flop which is the most advanced type presently available. The chapter is concluded by the presentation of a summary of flip-flop characteristics suitable for later reference.

6.1   R-S Storage Flip-Flop Circuits

A circuit consisting of two cross-connected NOR gates is shown in Figure 6.1a, a state table of the circuit in Figure 6.1b. When inputs R (Reset) and S (Set) are both zero, the circuit may assume either of two stable states. In one of these, the output of the upper gate is $Q = 1$, thus the output of the lower gate is $\bar{Q} = 0$. Output $\bar{Q}$, however, is an input to the upper gate and, since $R = 0$, this results in $Q = 1$ reinforcing the original assumption. Thus, $R = 0$, $S = 0$, $Q = 1$, and $\bar{Q} = 0$ describe one stable state. Similarly, $R = 0$, $S = 0$, $Q = 0$, and $\bar{Q} = 1$ describe the other stable state.

Transition from one state to the other can be effected by inputs R or S. The $Q = 1$ state can be established by momentarily making input $S = 1$, which forces $\bar{Q}$ to 0 and hence Q to 1. Similarly, the $Q = 0$ state can be established by momentarily making input $R = 1$. When $R = 1$ and $S = 1$ (last line of Figure 6.1b), $Q = 0$ and $\bar{Q} = 0$; if R and S are both

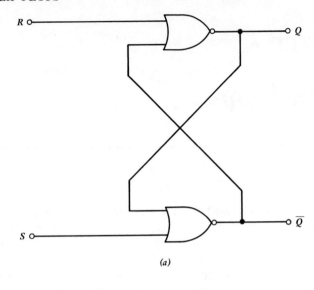

(a)

R	S	Q	$\overline{Q}$
0	0	$\begin{bmatrix} 0 \\ 1 \end{bmatrix}$	$\begin{bmatrix} 1 \\ 0 \end{bmatrix}$
0	1	1	0
1	0	0	1
1	1	0	0

(b)

FIGURE 6.1. R-S storage flip-flop using NOR gates: (*a*) circuit, (*b*) state table.

switched to 0 simultaneously, the next state of the flip-flop is undetermined. The transitions are illustrated in the timing diagram of Figure 6.2 showing signals in the circuit for a given sequence of inputs R and S, assuming gates with zero output rise and fall times and with propagation delays of t_{pd} each.

When the initial state of the flip-flop is $Q = 1$ and an input of $R = 1$ is applied to the upper NOR gate for a duration of 3 t_{pd} [see interval (1) in Figure 6.2], Q will be forced to 0 at a time delayed by t_{pd} from the rising edge of R. Output $Q = 0$ will, in turn, change the output of the lower NOR gate $\overline{Q}$ to 1 at a time delayed by an additional t_{pd} from the rising edge of R, since $S = 0$. Finally, output $\overline{Q} = 1$ will hold the output of the upper NOR gate at $Q = 0$ even after input R returns to 0.

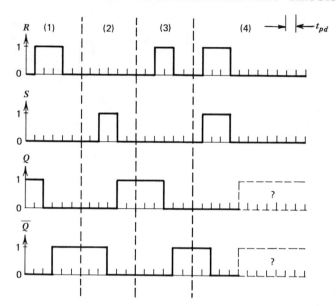

FIGURE 6.2. Timing diagram illustrating the operation of the R-S storage flip-flop circuit of Figure 6.1. For identification, four time intervals marked (1) through (4) are separated by broken lines.

When the initial state of the flip-flop is $Q = 0$ and an input of $S = 1$ is applied [interval (2) in Figure 6.2], transition to $Q = 1$ will take place in a manner similar to above. In order for the transition to take place, however, it is necessary that S remain 1 for at least 2 t_{pd} to allow time for the $Q = 1$ transition to arrive at the input of the lower NOR gate. The width of S in interval (2) is thus barely adequate to effect a transition, so is the width of R in interval (3). When both inputs $R = 1$ and $S = 1$ return to 0 simultaneously [interval (4)], the resulting state of the flip-flop is undetermined, since it can assume either of the two states equally well.

When the initial state is $Q = 1$ and an input $R = 1$ with a width of less than 2 t_{pd} is applied, the resulting state of the flip-flop is undetermined; the situation is similar when the initial state is $Q = 0$ and an input $S = 1$ with a width of less than 2 t_{pd} is applied. It can be shown that above conditions could result in a sustained oscillation in the circuit when the output rise and fall times are zero. In real circuits, however, rise and fall times are comparable to t_{pd} and such oscillations can not be sustained.

EXAMPLE 6.1. The initial state of the flip-flop of Figure 6.1 is $Q = 1$ and an input $R = 1$ is applied for a duration of t_{pd}. If the gates have zero rise and

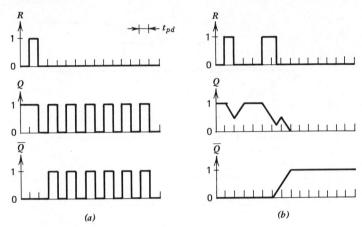

FIGURE 6.3. Signals in Example 6.1: (*a*) zero rise and fall times, (*b*) rise and fall times of $2\,t_{pd}$.

fall times, the result is a sustained oscillation (see Figure 6.3*a*). If the gates, however, have transitions that are rising and falling linearly in a time $2\ t_{pd}$, such an oscillation can not take place (see Figure 6.3*b*).

An alternate realization of the R-S storage flip-flop circuit using NAND gates and inverters is shown in Figure 6.4. Although this circuit looks more

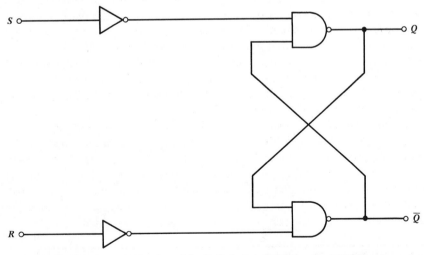

FIGURE 6.4. Realization of the R-S storage flip-flop using NAND gates and inverters.

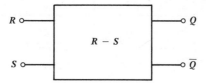

FIGURE 6.5.   A symbol of the R-S storage flip-flop.

complicated, it is used quite frequently because in many cases NAND gates are easier to build than NOR gates. Also, in many circuits the inverters at the input of the circuit of Figure 6.4 are omitted, resulting in a somewhat simpler circuit with inputs $\overline{R}$ and $\overline{S}$.

A symbol of the R-S storage flip-flop circuit is shown in Figure 6.5. A principal use of this flip-flop, as its name indicates, is as a storage element, or *latch*.

EXAMPLE 6.2.   The digital circuit of Figure 6.6. consists of four R-S storage flip-flops and four AND gates. Rise times, fall times, and propagation

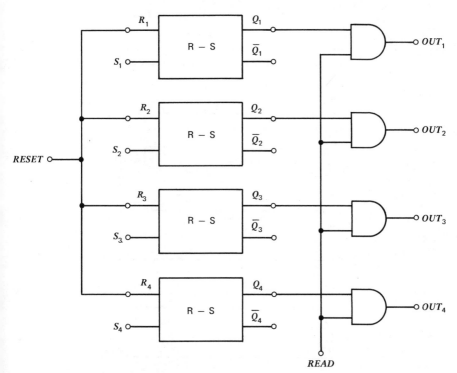

FIGURE 6.6.   Digital circuit of Example 6.2.

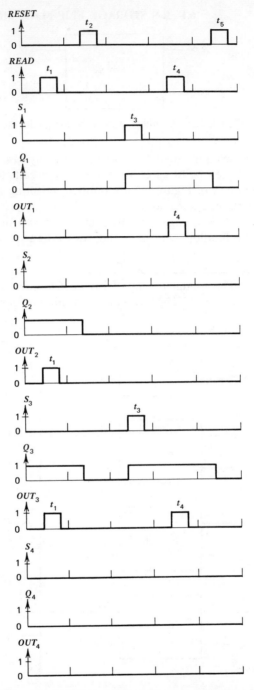

FIGURE 6.7. Timing diagram illustrating operation of the digital circuit of Figure 6.6. Rise and fall times, and propagation delays are neglected.

delays are negligible. The operation is illustrated in Figure 6.7. Initially, the states of flip-flops Q_1 through Q_4 are 0, 1, 1, 0. These states are read out at time t_1. At time t_2 the four flip-flops are reset by a common *RESET* pulse. At time t_3 the flip-flops are loaded again, this time by 1, 0, 1, 0; these data are read out at time t_4. Finally, the flip-flops are reset again at time t_5.

6.2   Clocked R-S Flip-Flop Circuits

The circuit of Figure 6.1 with two NAND gates added at its input is shown in Figure 6.8. In addition to *control inputs R* and *S* there is a *clock input C*, and the circuit is designated a clocked R-S flip-flop. The operation is illustrated in Figure 6.9 with gates that have propagation delays of t_{pd} and zero rise and fall times. The clock pulses are recurrent with widths of 2.5 t_{pd}. Inputs *R* and *S* are not allowed to change during any of the clock pulses: this restriction assures that each clock pulse leaves the circuit in a definite state; the simultaneous presence of $R = 1$ and $S = 1$, however, results in an undetermined state.

In describing transitions in clocked circuits, the state of *Q* before the clock pulse is designated as present state, Q_n, and the state of *Q* after the clock pulse as next state, Q_{n+1}. The transitions of the clocked R-S flip-flop circuit can be described by the *characteristic table* of Figure 6.10. Symbols R_n and S_n designate inputs *R* and *S* before (and during) the clock pulse; frequently these are shown simply as *R* and *S*. The first line in the table is an abbreviated notation for two transitions: a $Q_n = 0$ state to a $Q_{n+1} = 0$,

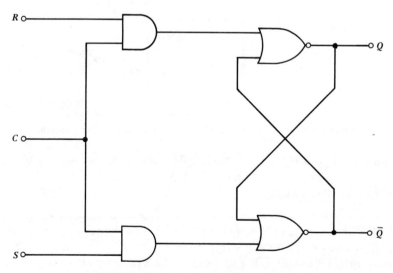

FIGURE 6.8.   Clocked R-S flip-flop circuit.

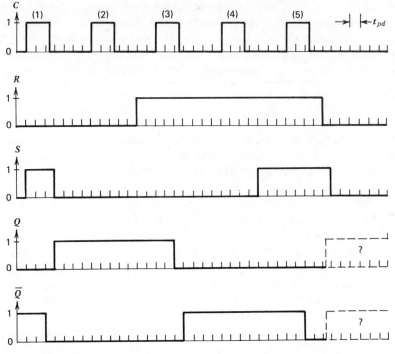

FIGURE 6.9. Timing diagram illustrating operation of the clocked R-S flip-flop circuit.

R_n	S_n	Q_{n+1}
0	0	Q_n
0	1	1
1	0	0
1	1	?

FIGURE 6.10. Characteristic table for the clocked R-S flip-flop.

and a $Q_n = 1$ to a $Q_{n+1} = 1$. The last line shows that in the clocked R-S flip-flop the simultaneous presence of $R_n = 1$ and $S_n = 1$ results in an undermined next state.

EXAMPLE 6.3. Transitions in the clocked R-S flip-flop circuit of Figure 6.8 are illustrated in the timing diagram of Figure 6.9 for five clock pulses numbered (1) through (5). For clock pulse (1), input $R_n = 0$ and input $S_n = 1$; this is a transition described by the second line in the characteristic table of Figure 6.10. Clock pulse (2) is described by the first line with $Q_n = Q_{n+1} = 1$, clock pulse (3) is described by the third line, and clock pulse (4) by the first line with $Q_n = Q_{n+1} = 0$. Finally, for clock pulse (5) $R_n = S_n = 1$, resulting in an undetermined Q_{n+1}; this case is described by the last line in the characteristic table of Figure 6.10.

A symbol of the clocked R-S flip-flop is shown in Figure 6.11. Some connections are often omitted to reduce the number of terminals required. In such circuits, clock input C is usually common to several flip-flops, and only one output, Q, is provided for each flip-flop.

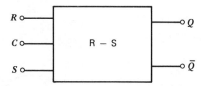

FIGURE 6.11.   A symbol of the clocked R-S flip-flop.

6.3   Extensions of the R-S Flip-Flop

Pulse-Triggered R-S Flip-Flop Circuits. The undefined state of the clocked R-S flip-flop (Figure 6.8) can be eliminated by use of additional feedback as shown in the circuit of Figure 6.12. Again, gates in the circuit are assumed to have zero rise and fall times and propagation delays of t_{pd} each. Output Q is now returned to the 3-input AND gate at input R via a non-inverting "one-input AND gate" with a delay of t_{pd}. Output $\bar{Q}$ is returned similarly to the 3-input AND gate at input S. Operation of the circuit is illustrated in Figure 6.13. When both inputs R and S are 0, signals A_R and A_S remain 0 and the state of the flip-flop is not affected by the clock, as illustrated by clock pulses C' numbered (1) and (4). When the initial state of the flip-flop is $Q = 0$, output Q via the feedback inhibits input R from passing through the input AND gate. Since an input of $R = 1$ would be of use only in establishing a $Q = 0$, this feedback does not interfere with the operation. The flip-flop can be set, however, by an input of $S = 1$ [see pulse (2)], since the feedback from $\bar{Q} = 1$ allows it to pass through the AND gate at input S. As a result, A_S is raised to 1, which sets $\bar{Q}$ to 0 and Q to 1. The feedback from $\bar{Q}$ to the input will result in reducing A_S to a width of 3 t_{pd}; this, however, is longer than 2 t_{pd}, and hence is sufficient to set the circuit. The flip-flop can be reset by R in a similar fashion when the initial state is $Q = 1$ as illustrated by pulse (3) in Figure 6.13. A significant feature of the circuit is its operation when both

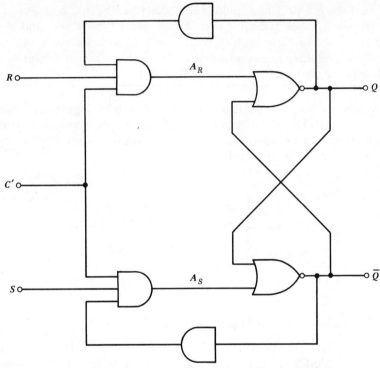

FIGURE 6.12.   Pulse-triggered R-S flip-flop circuit.

R and S are 1, as illustrated for the initial state of $Q = 0$ by clock pulse (5) in Figure 6.13. The $Q = 0$ inhibits input R via the feedback, while input S is allowed to pass, changing A_S to 1, $\overline{Q}$ to 0, and Q to 1. The feedback from Q will open the path for input R, but this will occur only after clock pulse C' is completed, thus it will have no effect. If clock pulse C' is too long, however, the circuit may end up in an undefined state: it can be shown that the width of C' has to be longer than 2 t_{pd} but shorter than 6 t_{pd} to assure correct operation.

Edge-Triggered J-K Flip-Flop Circuits.   The dependence of the operation on the width of clock pulse C' can be eliminated for long clock input pulses by incorporating a suitable pulse shaping circuit which generates a clock pulse that is always shorter than 6 t_{pd}. Such a circuit is shown in Figure 6.14; its operation is illustrated in Figure 6.15. The clock input, designated by C, is applied to a 2-input AND gate that has its other input connected to the inverted clock input signal C delayed by 3 t_{pd}. Thus, the two inputs of the AND gate are simultaneously 1 for a duration of 3 t_{pd}, and an output pulse with a width of 3 t_{pd} results at C' if the input is at least 3 t_{pd} wide.

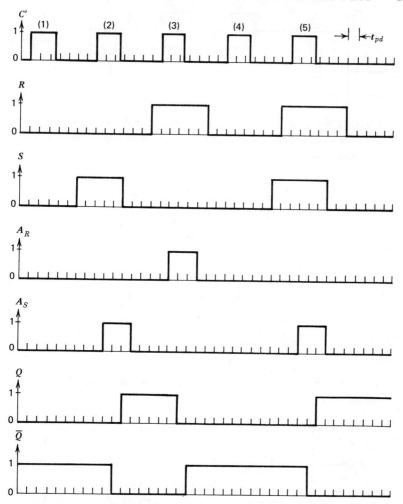

FIGURE 6.13. Timing diagram illustrating operation of the pulse-triggered R-S flip-flop circuit of Figure 6.12.

When the circuit of Figure 6.14 is included preceding clock input C' of Figure 6.12, the resulting circuit with clock input C is designated an *edge-triggered J-K flip-flop*. A symbol of this flip-flop is shown in Figure 6.16a, where input S has been relabeled J and input R has been relabeled K. The original designations S and R are not shown at all, emphasizing that the resulting flip-flop is different since it has a definite next state for $J_n = K_n = 1$; this can be also seen from the characteristic table given in Figure 6.16b.

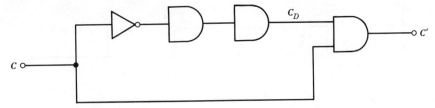

FIGURE 6.14. Pulse shaper circuit generating pulse C' of width 3 t_{pd} from the leading edge of clock input C. Propagation delay of the inverter and of each gate is t_{pd}.

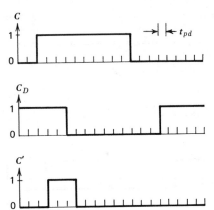

FIGURE 6.15. Timing diagram illustrating the operation of the pulse shaper circuit of Figure 6.14.

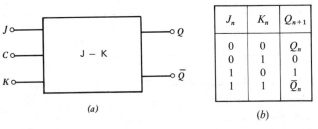

	J_n	K_n	Q_{n+1}
	0	0	Q_n
	0	1	0
	1	0	1
	1	1	$\bar{Q}_n$

(a)   (b)

FIGURE 6.16. J-K flip-flop: (a) symbol, (b) characteristic table.

A shortcoming of the edge-triggered flip-flop becomes apparent by a close examination of the circuit of Figure 6.14 for slow-rising clock pulses. If the threshold of the inverter and of the 2-input AND gate are not quite identical, for slow-rising clock pulses the output signal will depend strongly on the risetime of the clock pulse. Thus, for example, if the threshold of the 2-input AND gate is higher than that of the inverter and the difference in the thresholds is transversed by the input signal in a time longer than 3 t_{pd}, there will be no signal whatsoever on output C'. For this reason, the operation of the edge-triggered flip-flop is not reliable when risetimes of clock input C are slower than about $10\ t_{pd}$.

EXAMPLE 6.4. In the pulse shaper circuit of Figure 6.14 the inverter has a threshold voltage of $+1.25$ V and the 2-input AND gate has a threshold voltage of $+1.5$ V. Clock pulse C, shown in Figure 6.17, has rise and fall

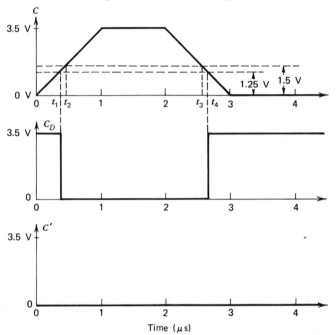

FIGURE 6.17. Signals in Example 6.4.

times of 1 μs. The propagation delays of the inverter and of the gates are $t_{pd} = 10$ ns each; hence, they are negligible. The inverter input switches at its threshold voltage of $+1.25$ V, i.e., at times t_1 and t_4, resulting in a C_D as shown in the figure. The 2-input AND gate, however, switches at its threshold voltage of $+1.5$ V, which occurs t_2 and t_3. The result is no output at all on C'.

The unpredictable operation of the edge-triggered flip-flop circuit for slow clock pulses can be overcome by inserting at clock input C a regenerative *Schmitt trigger* circuit. This circuit, described in detail in Chapter 12, generates an output signal that has fast transitions even when the transitions at its input are slow.

6.4   Master-Slave J-K Flip-Flop Circuits

The unpredictable behavior of the edge-triggered flip-flop circuit for slow-rising clock input pulses can be overcome by use of the master-slave flip-flop circuit. This circuit avoids ambiguities by incorporating two flip-flops and by setting two distinct thresholds on the clock input, resulting in four distinct timing signals as the two thresholds are crossed by the rising and falling edges of the clock input pulse.

A master-slave J-K flip-flop circuit is shown in Figure 6.18. The operation is demonstrated with gates that have threshold voltages of $+1.5$ V and nominal logic levels of 0 V and $+3.5$ V for 0 and 1, respectively. A significant feature of the circuit is the 0.7 V battery—in reality a forward-biased diode—that operates the gates connected to C_2 at levels that are different from those connected to C_1. A "one-input AND gate" is also included preceding C_1, regenerating clock input C. Figure 6.19 shows signals for the case when $J = K = 1 = +3.5$ V (clock input C is not shown, only C_1 and C_2). We can see that on the rising edge of C_1 the NOR gates connected to C_2 are activated first at time t_1, forcing B_K and B_J to zero irrespective of M and $\overline{M}$; hence, the previous state of the flip-flop

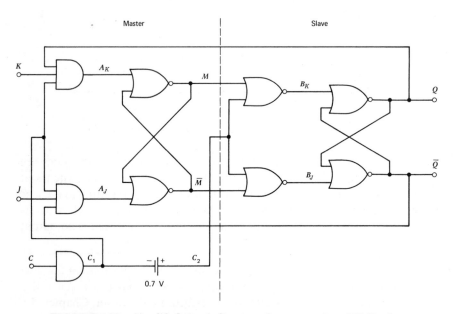

FIGURE 6.18.   Simplified circuit diagram of a master-slave J-K flip-flop. In practice the 0.7 volt battery is realized by a forward biased diode.

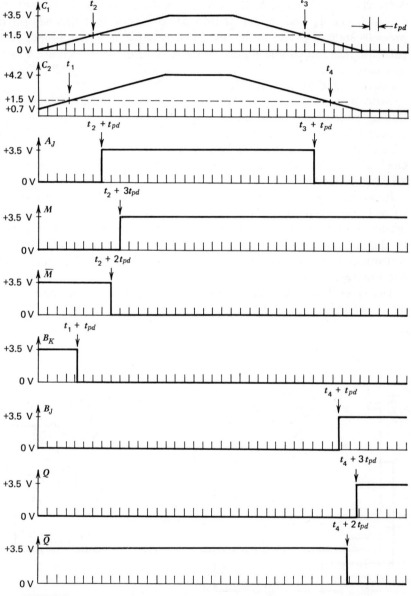

FIGURE 6.19. Timing diagram illustrating operation of the master-slave J-K flip-flop circuit of Figure 6.18 with $J = K = 1$, $Q_n = 0$, and $Q_{n+1} = 1$.

is preserved on Q and $\overline{Q}$. Next, C_1 activates the 3-input AND gates at time t_2 which via A_K and A_J set M and $\overline{M}$ to a state that depends on the previous state of Q and on inputs J and K. At time t_3 the 3-input AND gates are deactivated and inputs J and K have no further effects. Finally, commencing at time t_4, B_K and B_J, hence also Q and $\overline{Q}$, will follow the states of M and $\overline{M}$, respectively.

Note that transitions on outputs Q and $\overline{Q}$ take place following the trailing edge of the clock input pulse. This is different from the edge-triggered flip-flop, where the transition takes place following the leading edge of the clock input pulse. This, however, can be altered by replacing clock input C by its complement $\overline{C}$, with a result that the transitions will take place following the leading edge of C.

It can be shown that the operation of the master-slave J-K flip-flop is correct also when clock input C has sharp rising and falling edges. The master-slave flip-flop is a versatile circuit and finds widespread applications in digital systems. Its principal disadvantage is complexity and the concomitant increase in power consumption as compared to that of the R-S flip-flop.

The versatility of the J-K flip-flop can be enhanced by the addition of a direct set input, S_D, and a direct reset input, R_D, as shown in Figure

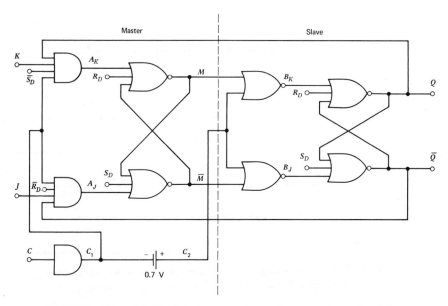

FIGURE 6.20.   J-K flip-flop incorporating a direct set input S_D and a direct reset input R_D (each one is used at three places in the circuit).

6.20. These inputs are frequently also designated as preset and clear inputs; they provide a way of setting and resetting both the master and the slave portions of the flip-flop, overriding control inputs J and K and clock C.

EXAMPLE 6.5. In the master-slave J-K flip-flop circuit of Figure 6.20, direct set input $S_D = 0$ and direct reset input $R_D = 1$; thus $\overline{S_D} = 1$ and $\overline{R_D} = 0$. The $S_D = 0$ is entered into two NOR gates where they do not influence the operation, neither does the $\overline{S_D} = 1$ entered into the upper 4-input AND gate. The $\overline{R_D} = 0$ input, however, forces A_J to 0, allowing $R_D = 1$ to reset M to 0; $R_D = 1$ also resets Q to 0.

6.5 Flip-Flop Characteristics

We have seen in the preceding sections how R-S and J-K flip-flop circuits can be built by using logic gates and how they can be described by a characteristic table. In this section other useful descriptions of flip-flops will be discussed, other types of flip-flops will be introduced, and flip-flop properties will be summarized.

A symbol of a J-K flip-flop is shown again in Figure 6.21a. It has two control inputs J and K, a clock input C, direct set and reset inputs S_D and R_D, and outputs Q and $\overline{Q}$. While S_D and R_D are important in many cases, in what follows it will be assumed that S_D and R_D are not activated, i.e., $S_D = R_D = 0$. A characteristic table is shown again in Figure 6.21b. It specifies the next state Q_{n+1} for given control inputs J_n and K_n and present state Q_n. The first line, for example, describes two transitions: $Q_n = 0$ to $Q_{n+1} = 0$ and $Q_n = 1$ to $Q_{n+1} = 1$. A complete listing of all possible transitions is given in the *state table* of Figure 6.21c, and the same information is also displayed in the Karnaugh map of Figure 6.21d; these provide a complete description of the flip-flop. Actually, the characteristic table (Figure 6.21b) provides the same information as the state table but in an abbreviated notation, which is useful when J_n, K_n, and Q_n are given.

EXAMPLE 6.6. The first line of the state table of Figure 6.21c describes the transition $Q_n = 0$ to $Q_{n+1} = 0$ taking place as a result of $J_n = K_n = 0$. The fifth line of the state table describes the transition $Q_n = 1$ to $Q_{n+1} = 1$, again taking place as a result of $J_n = K_n = 0$. The two transitions are summarized in the first line of the characteristic table of Figure 6.21b as a transition from state Q_n to state $Q_{n+1} = Q_n$.

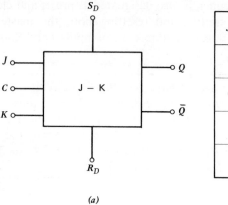

(a)

J_n	K_n	Q_{n+1}
0	0	Q_n
0	1	0
1	0	1
1	1	$\bar{Q}_n$

(b)

Q_n	J_n	K_n	Q_{n+1}
0	0	0	0
0	0	1	0
0	1	0	1
0	1	1	1
1	0	0	1
1	0	1	0
1	1	0	1
1	1	1	0

(c)

Q_n \ $J_n K_n$	00	01	11	10
0	0	0	1	1
1	1	0	0	1

Q_{n+1}

(d)

Q_n	Q_{n+1}	J_n	K_n
0	0	0	x
0	1	1	x
1	0	x	1
1	1	x	0

(e)

FIGURE 6.21. J-K flip-flop: (*a*) symbol, (*b*) characteristic table, (*c*) state table, (*d*) Karnaugh map of Q_{n+1}, (*e*) excitation table.

In the case when Q_n and Q_{n+1} are given and the possible combinations of J_n and K_n to effect the transition are sought, the use of the *excitation table* shown in Figure 6.21e may be useful.

EXAMPLE 6.7. The first line of the excitation table of Figure 6.21e describes the possible combinations of J_n and K_n that effect a transition from $Q_n = 0$ to $Q_{n+1} = 0$. These are $J_n = 0$ and $K_n = 0$, or $J_n = 0$ and $K_n = 1$. The two possibilities are summarized as $J_n = 0$ and $K_n = $ x (don't care).

A symbol of the clocked R-S flip-flop, a characteristic table, a state table, a Karnaugh map of the next state, and an excitation table are shown in Figure 6.22. As before, question mark entries denote undefined, or unknown, states and x denotes don't care condition.

The R-S and J-K flip-flops are versatile devices; in many cases, however, it is desirable to reduce the number of connections at the expense of decreased versatility. To this end, $\overline{Q}$ is often omitted; also, one or both of S_D and R_D are either omitted or made common to several flip-flops. A further possibility is to combine the two control inputs by making $K = J$ or $K = \overline{J}$. The former choice results in a *T flip-flop* described in Figure 6.23; it can be seen from the characteristic table that when $T_n = 1$, the state of the flip-flop is complemented; hence the name toggle, or T, flip-flop. The $K = \overline{J}$ choice results in the *D flip-flop* shown in Figure 6.24. Here $Q_{n+1} = D_n$; hence the name delay, or D, flip-flop.

An inspection of the excitation tables shows that for the flip-flops discussed there is always at least one combination of inputs that can effect a transition from a given state to another one. Indeed, otherwise the use of a flip-flop would be quite restricted. This is illustrated in the example that follows.

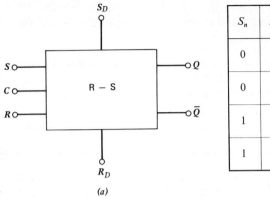

(a)

S_n	R_n	Q_{n+1}
0	0	Q_n
0	1	0
1	0	1
1	1	?

(b)

Q_n	S_n	R_n	Q_{n+1}
0	0	0	0
0	0	1	0
0	1	0	1
0	1	1	?
1	0	0	1
1	0	1	0
1	1	0	1
1	1	1	?

(c)

Q_n \ $S_n R_n$	00	01	11	10
0	0	0	?	1
1	1	0	?	1

Q_{n+1}

(d)

Q_n	Q_{n+1}	S_n	R_n
0	0	0	x
0	1	1	0
1	0	0	1
1	1	x	0

(e)

FIGURE 6.22. Clocked R-S flip-flop: (*a*) symbol, (*b*) characteristic table, (*c*) state table, (*d*) Karnaugh map of Q_{n+1}, (*e*) excitation table.

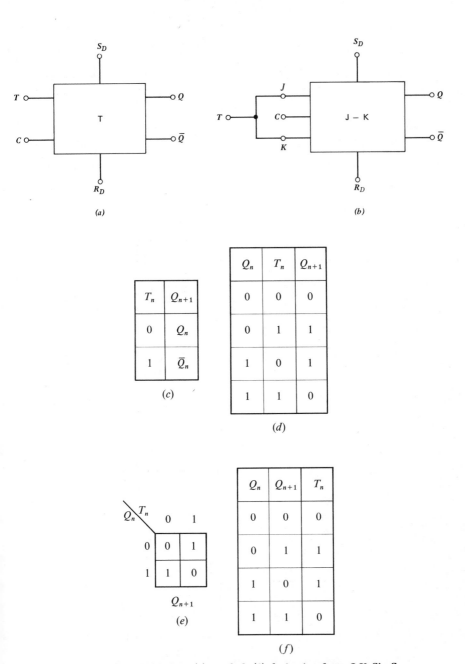

FIGURE 6.23. T flip-flop: (*a*) symbol, (*b*) derivation from J-K flip-flop, (*c*) characteristic table, (*d*) state table, (*e*) Karnaugh map of Q_{n+1}, (*f*) excitation table.

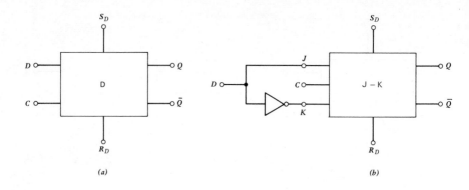

(a)

(b)

D_n	Q_{n+1}
0	0
1	1

(c)

Q_n	D_n	Q_{n+1}
0	0	0
0	1	1
1	0	0
1	1	1

(d)

Karnaugh map:

Q_n \ D_n	0	1
0	0	1
1	0	1

Q_{n+1}

(e)

Q_n	Q_{n+1}	D_n
0	0	0
0	1	1
1	0	0
1	1	1

(f)

FIGURE 6.24. D flip-flop: (a) symbol, (b) derivation from J-K flip-flop, (c) characteristic table, (d) state table, (e) Karnaugh map of Q_{n+1}, (f) excitation table.

EXAMPLE 6.8. A hypothetical "N-G" ("No Good") flip-flop is shown in Figure 6.25*a*; there is no S_D or R_D input provided. We can see in Figure 6.25*b* that there is no way to get from the $Q = 1$ state to the $Q = 0$ state.

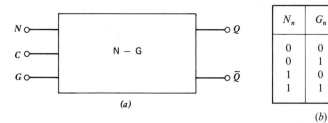

N_n	G_n	Q_{n+1}
0	0	Q_n
0	1	Q_n
1	0	Q_n
1	1	1

(*a*)

(*b*)

FIGURE 6.25. "N-G" flip-flop of Example 6.8: (*a*) symbol, (*b*) characteristic table.

Such a flip-flop would be quite useless, except if a direct reset input R_D were provided.

Any one of the J-K, R-S, T, and D flip-flops can be substituted for another one by the use of suitable logic circuitry for the control inputs.

EXAMPLE 6.9. In Figure 6.24*b*, a D flip-flop is realized by using a J-K flip-flop, an inverter for control input K, and a direct connection to control input J.

The state table and the excitation table may be utilized for systematic realization of digital circuits.

EXAMPLE 6.10. The realization of a T flip-flop using a J-K flip-flop was given in Figure 6.23*b*. Suppose that it is desired to realize a J-K flip-flop using a T flip-flop. This can be attained by programming control input T of the T flip-flop such that the transitions will follow those described in the state table of the J-K flip-flop (Figure 6.21*c*). This state table is shown again in Figure 6.26*a* with a column added for the required T_n. For each line, T_n is found from the excitation table of the T flip-flop (Figure 6.23*f*) and entered into the table. Now the table specifies T_n as a logic function of Q_n, J_n, and K_n; this is also displayed in the Karnaugh map of Figure 6.26*b* and can be written as $T_n = Q_n K_n + \bar{Q}_n J_n$. Hence a J-K flip-flop can be realized by a T flip-flop and logic gates as shown in Figure 6.26*c*.

Q_n	J_n	K_n	Q_{n+1}	T_n
0	0	0	0	0
0	0	1	0	0
0	1	0	1	1
0	1	1	1	1
1	0	0	1	0
1	0	1	0	1
1	1	0	1	0
1	1	1	0	1

(a)

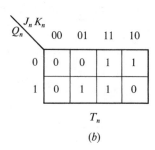

Q_n \ $J_n K_n$	00	01	11	10
0	0	0	1	1
1	0	1	1	0

T_n

(b)

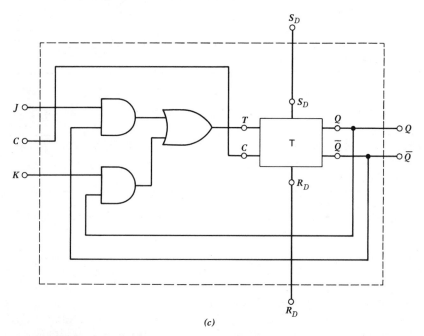

(c)

FIGURE 6.26. Realization of the J-K flip-flop using a T flip-flop and logic gates: (a) combination of the state table of the J-K flip-flop (Figure 6.21c) with the excitation table of the T flip-flop (Figure 6.23f), (b) Karnaugh map for T_n, (c) the circuit.

PROBLEMS

6.1 The timing diagram of Figure 6.2 illustrates the operation of the R-S storage flip-flop of Figure 6.1 with two NOR gates that have zero rise times and fall times, and propagation delays of t_{pd} each. Sketch a timing diagram showing Q and $\bar{Q}$ for inputs R and S as given, if the lower NOR gate has zero rise time and fall time and a propagation delay of $1.5\, t_{pd}$.

6.2 Sketch a timing diagram illustrating the operation of the circuit of Figure 6.4 if inputs R and S are as shown in the first two lines of Figure 6.2.

6.3 Realize an equivalent of the clocked R-S flip-flop of Figure 6.8 by use of NAND gates only.

6.4 Realize an equivalent of the circuit of Figure 6.12 by use of AND gates and NAND gates.

6.5 Demonstrate that in the circuit of Figure 6.12 the width of C' must be between $2\, t_{pd}$ and $6\, t_{pd}$ for correct operation. Assume that each gate has zero rise and fall time, and a propagation delay of t_{pd}.

6.6 Sketch C_D and C' in the circuit of Figure 6.14 if the inverter has a threshold voltage of $V_{TH} = +1.5$ V, the 2-input AND gate a threshold voltage of $V_{TH} = +1.25$ V, and clock C is as shown in the first line of Figure 6.17.

6.7 Sketch signals C_D and C_1 in the circuit of Figure 6.27 if input C is as shown in the first line of Figure 6.15. Assume that the inverters and the NOR gate have zero rise times and fall times, and a propagation delay of t_{pd} each.

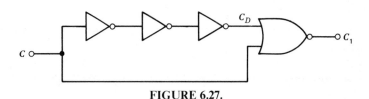

FIGURE 6.27.

6.8 Sketch the signals in the master-slave flip-flop circuit of Figure 6.18 if C_1 and C_2 are as shown in Figure 6.19. Assume $J = 1$, $K = 0$, and an initial state of $Q = 0$.

6.9 Sketch the signals in the master-slave flip-flop circuit of Figure 6.18 for a single clock input pulse with sharp rising and falling edges and with a width of 10 t_{pd}. Assume $J = K = 1$ and an initial state of $Q = 1$.

6.10 Sketch Q in the master-slave J-K flip-flop circuit of Figure 6.18 if inputs C, J, and K are as shown in Figure 6.28 and if the initial state is $Q = 0$. Assume zero rise and fall times, and zero propagation delays.

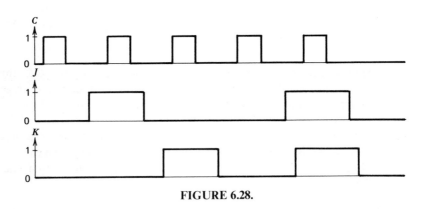

FIGURE 6.28.

6.11 Describe why inputs $\overline{S_D}$ and $\overline{R_D}$ are required in the circuit of Figure 6.20 if direct set and reset operations that are independent of J, K, and C are desired.

6.12 The master-slave J-K flip-flop with direct set and reset inputs shown in Figure 6.20 has input signals C, J, K, S_D, and R_D as shown in Figure 6.29. Sketch A_J, A_K, M, $\overline{M}$, B_J, B_K, Q, and $\overline{Q}$ if rise and fall times and propagation delays are negligible. Assume an initial state of $Q = 0$.

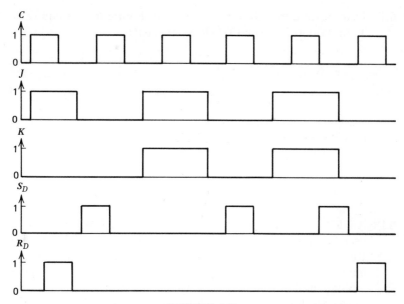

FIGURE 6.29.

6.13 Realize a D flip-flop by use of a T flip-flop and logic gates.

6.14 Prepare a state table and a transition table for the "N-G" flip-flop of Example 6.8.

6.15 The symbol and the characteristic table of a hypothetical "D-E" flip-flop is shown in Figure 6.30. Prepare a state table, a Karnaugh map for Q_{n+1}, and an excitation table. Realize the "D-E" flip-flop by use of a D flip-flop and logic gates.

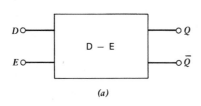

D_n	E_n	Q_{n+1}
0	0	Q_n
0	1	0
1	0	Q_n
1	1	1

(a)

(b)

FIGURE 6.30.

6.16 Establish methods for deriving the characteristic table and the excitation table from the state table, and vice versa.

6.17 Demonstrate that the circuit shown in Figure 6.31 is equivalent to a T flip-flop at its T and $CLOCK$ terminals.

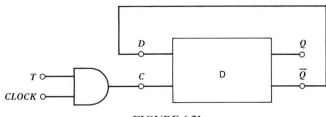

FIGURE 6.31.

6.18 An R-S flip-flop operates on logic levels of 0 V and $+5$ V. It is used as a *bounce remover*, as shown in Figure 6.32. Pushbuttons RESET and SET may bounce (disconnect) several times before settling, but the output of the flip-flop changes state only once. The operation is ambiguous, however, when the two pushbuttons are pushed simultaneously. Modify the circuit such that the RESET pushbutton overrides the SET pushbutton under such conditions.

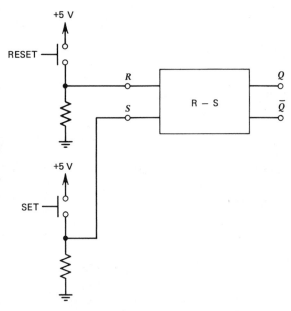

FIGURE 6.32.

CHAPTER 7

Sequential Circuits

The preceding chapter introduced flip-flops and described their operation. We have seen that these devices have two stable states and that transitions between these states can be effected by suitable combinations of the control inputs. This chapter introduces principles and examples of digital systems that, in general, consist of more than one flip-flop and have more than two states.

Most of the circuits that will be described are *synchronous*, with each transition synchronized with respect to a clock signal; an exception to this will be the *ripple counter* where transitions are initiated by the clock but are not synchronous with it.*

7.1 State Tables and Transition Diagrams

We have seen in the preceding chapter how the properties of a flip-flop could be described by a characteristic table, a state table, or an excitation table. In this section, the use of these is extended, and additional tools of analysis and design are introduced.

The state table, also known as transition table, provides a list of the states in a digital system. The maximum number of states is limited by the number of storage elements: if there are m binary storage elements (flip-flops) in a digital system, the maximum number of states is 2^m. A state table provides the following information for each state listed: (*i*) identification of the present state Q_n in binary or in decimal form, (*ii*) next states Q_{n+1} for various combinations of inputs to the system, and (*iii*) outputs resulting for each of these combinations.

---

* *Asynchronous circuits* constitute an important area of sequential circuits that is not discussed here; the reader is referred to the references listed at the end of the book.

EXAMPLE 7.1. Properties of the J-K flip-flop were described in the preceding chapter; its symbol is shown again in Figure 7.1a, and a state table in Figure 7.1b. It is assumed that no direct set input S_D and no direct reset input R_D are provided. Transitions in the state of the flip-flop are controlled by control inputs J and K, and the transitions take place at the time of the clock pulse. As a rule, it is also required that J and K do not change during the clock pulse.

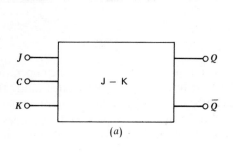

Q_n	J_n	K_n	Q_{n+1}
0	0	0	0
0	0	1	0
0	1	0	1
0	1	1	1
1	0	0	1
1	0	1	0
1	1	0	1
1	1	1	0

(b)

Present state Q_n	$J_n = 0, K_n = 0$ Next state Q_{n+1}	$J_n = 0, K_n = 1$ Next state Q_{n+1}	$J_n = 1, K_n = 0$ Next state Q_{n+1}	$J_n = 1, K_n = 1$ Next state Q_{n+1}
0	0	0	1	1
1	1	0	1	0

(c)

FIGURE 7.1. **J-K flip-flop: (a) symbol, (b) and (c): two different forms of the state table.**

The state table of Figure 7.1b lists present states Q_n, present control inputs J_n and K_n, and next states Q_{n+1} for various combinations of J_n and K_n; frequently J_n and K_n are designated simply as J and K. The state table may also be represented in the form shown in Figure 7.1c.

The number of inputs in a digital system is not related to the number of states: Thus, a J-K flip-flop has two control inputs and two states, and a T flip-flop has one control input and two states. Also, the number of outputs in a digital system is not related to the number of states. The T flip-flop of Figure 6.23 has one control input T and two outputs Q and $\overline{Q}$ (no direct set input S_D and no direct reset input R_D are provided).

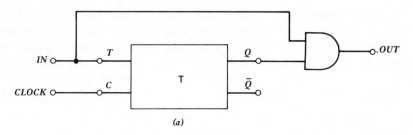

(a)

	$IN_n = 0$		$IN_n = 1$	
Present state Q_n	OUT_n	Next state Q_{n+1}	OUT_n	Next state Q_{n+1}
0	0	0	0	1
1	0	1	1	0

(b)

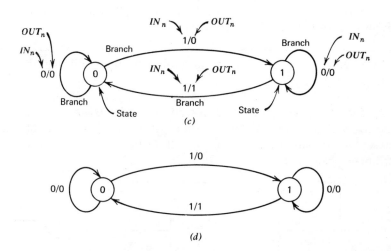

(c)

(d)

FIGURE 7.2.   Digital system using a T flip-flop: (a) circuit diagram, (b) state table, (c) transition diagram with significant features labeled, (d) transition diagram as usually represented.

Control input T, however, may be permanently connected to the "1" logic level ("$T = 1$" *flip-flop*), in which case there is no input to the system other than the clock. Figure 7.2a shows a system that uses a T flip-flop and has an output $OUT = TQ$; a state table is given in Figure 7.2b.

The information presented in a state table can also be presented in a *transition diagram*, also known as *state diagram*. In a transition diagram each state is represented by a circle, and each possible transition is represented by an oriented *branch* originating from a state and terminating at a state which may also be the original state. Inputs and outputs are identified along the branches.

EXAMPLE 7.2. Figure 7.2c shows a transition diagram of the digital circuit of Figure 7.2a. It has two states designated 0 and 1; it also has four branches with inputs and outputs identified. Thus, when the present state is 1 and the present input is $IN_n = 1$, the present output is $OUT_n = 1$ and the next state is 0, as described by the branch originating from state 1 and terminating at state 0. When the present state is 0 and the present input is $IN_n = 0$, the present output is $OUT_n = 0$ and the next state is 0, same as the present state. In practice, branches in transition diagrams are labeled as shown in Figure 7.2d.

The transition diagram can also be used to describe digital systems with several inputs and several outputs. This will be illustrated by use of the D-E flip-flop of Figure 6.30.

EXAMPLE 7.3. Figure 7.3a shows a digital circuit utilizing the D-E flip-flop of Figure 6.30. There are two inputs D and E, and two outputs OUT_1 and OUT_2. A state table is shown in Figure 7.3b, where D, E, OUT_1, and OUT_2 have been used instead of D_n, E_n, OUT_{1_n}, and OUT_{2_n}, respectively. A transition diagram is shown in Figure 7.3c. Some branches represent more than one transition; each transition is described (coded) by four binary digits as $D_n E_n / OUT_{1_n} OUT_{2_n}$.

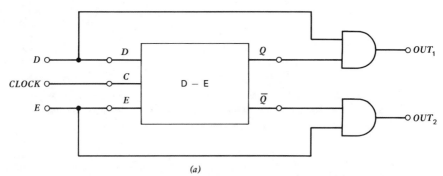

(a)

FIGURE 7.3. Digital system using the D-E flip-flop of Figure 6.30: (a) circuit diagram.

Present state Q_n	D = 0, E = 0			D = 0, E = 1			D = 1, E = 0			D = 1, E = 1		
	OUT_1	OUT_2	Next state Q_{n+1}	OUT_1	OUT_2	Next state Q_{n+1}	OUT_1	OUT_2	Next state Q_{n+1}	OUT_1	OUT_2	Next state Q_{n+1}
0	0	0	0	0	1	0	0	0	0	0	1	1
1	0	0	1	0	0	0	1	0	1	1	0	1

(b)

FIGURE 7.3. (b) state table. (See next page for (c).)

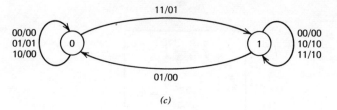

FIGURE 7.3. (c) transition diagram with each transition labeled as $D_n E_n/OUT_{1_n} OUT_{2_n}$.

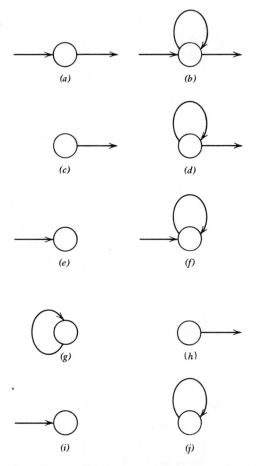

FIGURE 7.4. Various types of states and branches: (a) and (b) regular states, (c) and (d) transient states, (e) and (f) persistent states, (g) isolated state, (h) diverging branch, (i) converging branch, (j) reflecting branch.

States in a transition diagram may be classified as *regular states, transient states, persistent states,* and *isolated states.* A regular state can be entered from at least one other state and has an exit to at least one other state (Figure 7.4*a* and *b*). A transient state cannot be entered from any other state, but it has an exit to at least one other state (Figure 7.4*c* and *d*). A persistent state can be entered from at least one other state, but it has no exit to any other state (Figure 7.4*e* and *f*). An isolated state cannot be entered from any other state, and it has no exit to any other state (Figure 7.4*g*).

Branches in a transition diagram may be classified with reference to the states as *diverging branches, converging branches,* and *reflecting branches,* as shown in Figure 7.4*h, i,* and *j,* respectively. Thus, for example, in Figure 7.2*d* the branch marked by 1/0 is diverging from state 0 and is converging on state 1; also, the branches marked by 0/0 are reflecting branches.

7.2 Design of Digital Systems

The preceding section described how the operation of a given digital system can be *analyzed* by utilizing the state table and the transition diagram. In this section the opposite process will be discussed: how to *design* (synthesize) a digital system when its characteristics are specified.

The first step in the design is to translate the specified characteristics to a state table. Next, the type or types of storage elements (flip-flops) are chosen and logic circuits for the control inputs are designed using the excitation tables. Finally, the logic circuits implementing the required output or outputs are synthesized.

EXAMPLE 7.4. Design a digital system that has two states: 0 and 1. When input $IN = 0$, the system should remain in its preceding state; when $IN = 1$, it should change state. Output OUT should be 1 *iff* (if and only if) the system is in its 1 state and input $IN = 1$.

The state table of a digital system with these characteristics is given in Figure 7.2*b*; Figure 7.2*a* shows a realization using a T flip-flop. Now we shall realize the same circuit by means of a D flip-flop and also by means of an R-S flip-flop.

Realization by D flip-flop. The state table of Figure 7.2*b* is shown again in Figure 7.5*a* with two additional columns for input D_n. These columns are initially empty and will be filled in with the aid of the excitation table of the D flip-flop (Figure 7.5*b*) as follows. For $IN_n = 0$ and $^-Q_n = 0$, $Q_{n+1} = 0$; hence from the first line of the excitation table $D_n = 0$. For $IN_n = 0$ and $Q_n = 1$, $Q_{n+1} = 1$; hence from the fourth line of the excitation table $D_n = 1$. For $IN_n = 1$ and $Q_n = 0$, $Q_{n+1} = 1$; hence from the second

Present state Q_n	$IN_n = 0$: OUT_n	$IN_n = 0$: Next state Q_{n+1}	$IN_n = 0$: D_n	$IN_n = 1$: OUT_n	$IN_n = 1$: Next state Q_{n+1}	$IN_n = 1$: D_n
0	0	0	0	0	1	1
1	0	1	1	1	0	0

(*a*)

Q_n	Q_{n+1}	D_n
0	0	0
0	1	1
1	0	0
1	1	1

(*b*)

Q \ IN	0	1
0	0	1
1	1	0

$D = Q \oplus IN$

(*c*)

Q \ IN	0	1
0	0	0
1	0	1

$OUT = Q \cdot IN$

(*d*)

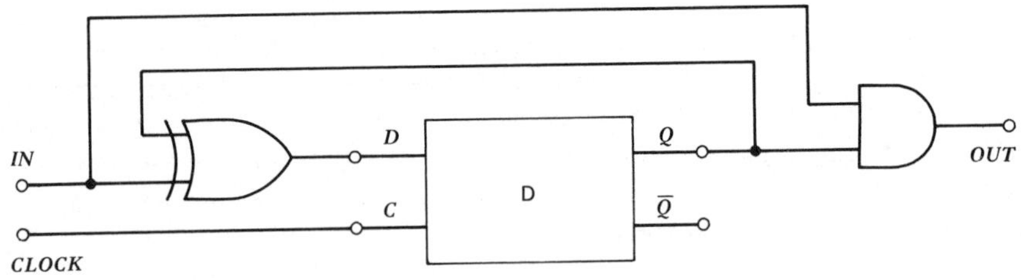

(*e*)

FIGURE 7.5. Design of a digital system specified by the state table of Figure 7.2*b* using a D flip-flop: (*a*) state table with columns added for D_n, (*b*) excitation table of the D flip-flop, (*c*) Karnaugh map for control input *D*, (*d*) Karnaugh map for output OUT, (*e*) the resulting circuit.

line of the excitation table $D_n = 1$. Finally, for $IN_n = 1$ and $Q_n = 1$, $Q_{n+1} = 0$; hence from the third line of the excitation table $D_n = 0$.

Now D_n is completely specified in Figure 7.5*a* as a function of Q_n and IN_n. This function is shown in the Karnaugh map of Figure 7.5*c*, in which the subscripts *n* have been omitted for brevity. The function given by the Karnaugh map is realized as $D = Q \oplus IN = Q \cdot \overline{IN} + \overline{Q} \cdot IN$. Output OUT is also a completely specified function of Q_n and IN_n, as shown in Figure 7.5*a* and mapped in Figure 7.5*d* with *Q*, *IN*, and OUT substituted for Q_n, IN_n, and OUT_n, respectively. The resulting digital circuit is shown in Figure 7.5*e*.

Realization by R-S flip-flop. Realization by R-S flip-flop proceeds along similar lines (see Figure 7.6). The excitation table of the R-S flip-flop, however, contains don't care conditions; hence the logic functions for the control inputs can be realized in several different ways, as can be seen from the Karnaugh maps of Figure 7.5c and d. One solution, $S = \bar{Q} \cdot IN$ and $OUT = R = Q \cdot IN$, is shown in the digital circuit of Figure 7.6d.

	$IN_n = 0$				$IN_n = 1$			
Present state Q_n	OUT_n	Next state Q_{n+1}	S_n	R_n	OUT_n	Next state Q_{n+1}	S_n	R_n
0	0	0	0	x	0	1	1	0
1	0	1	x	0	1	0	0	1

(a)

Q_n	Q_{n+1}	S_n	R_n
0	0	0	x
0	1	1	0
1	0	0	1
1	1	x	0

(b)

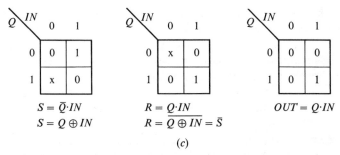

$S = \bar{Q} \cdot IN$
$S = Q \oplus IN$

$R = Q \cdot IN$
$R = \overline{Q \oplus IN} = \bar{S}$

$OUT = Q \cdot IN$

(c)

FIGURE 7.6. Design of a digital system specified by the state table of Figure 7.2b using an R-S flip-flop: (a) state table with columns added for S_n and R_n, (b) excitation table of the R-S flip-flop, (c) Karnaugh maps for control inputs S and R, and for output OUT. (*See next page for* (d).)

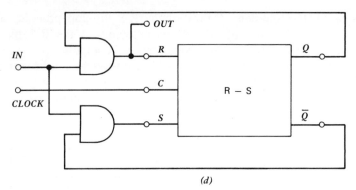

(d)

FIGURE 7.6.   (*d*) the resulting circuit.

In the preceding examples, the state tables consisted of two states 0 and 1 which were realized by the respective states of a flip-flop. It would be also possible to assign the 1 state of the flip-flop to the 0 state of the state table and the 0 state of the flip-flop to the 1 state of the state table. When the system is realized by an R-S, J-K, or T flip-flop, the changes in the assignment of the states would simply result in interchanging R for S, J for K, and Q for $\overline{Q}$ (and also S_D for R_D, if provided). In the case of a D flip-flop, however, D would have to be replaced by $\overline{D}$, which could result in a different circuit.

Thus we can see that even in the simple case of a digital system with two states there is a choice of *state assignment* that can result in different circuits. When there are more than two states, the number of available choices is further increased.

EXAMPLE 7.5.   The transition diagram of a digital system with four states is shown in Figure 7.7. The system has no input other than the clock, and the assignment of only one state is specified. There are now two independent decisions to be made: the type (or types) of flip-flops to be used has to be chosen, and also the remaining three states have to be assigned. Let us decide, for example, that we realize the system using two T flip-flops (realization by J-K flip-flops is left to the reader as an exercise). The excitation table of the T flip-flop is shown in Figure 7.8. Three possible state assignments together with the state table, Karnaugh maps, and the resulting digital circuits are shown in Figure 7.9, Figure 7.10, and Figure 7.11 on pages 183 through 188. Figure 7.9 represents a 2-bit counter counting in natural binary code, Figure 7.10 one counting in Gray code (see Chapter 9), and Figure 7.11 a down counter (see page 208).

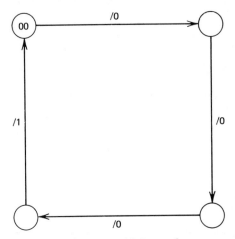

FIGURE 7.7. A digital system with incomplete state assignment.

Q_n	Q_{n+1}	T_n
0	0	0
0	1	1
1	0	1
1	1	0

FIGURE 7.8. Excitation table of the T flip-flop.

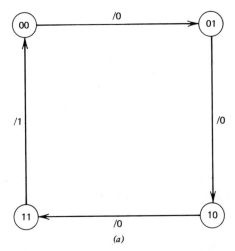

(a)

FIGURE 7.9. A possible state assignment for the digital system of Figure 7.7: (a) transition diagram. (*See next page for* (b), (c), *and* (d).)

Present states		Output	Next states		Control inputs	
Q_{1_n}	Q_{0_n}	OUT_n	$Q_{1_{n+1}}$	$Q_{0_{n+1}}$	T_{1_n}	T_{0_n}
0	0	0	0	1	0	1
0	1	0	1	0	1	1
1	0	0	1	1	0	1
1	1	1	0	0	1	1

(b)

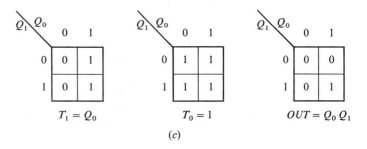

(c)

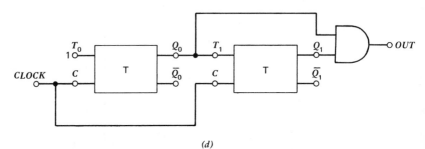

(d)

FIGURE 7.9. (b) state table, (c) Karnaugh maps of control inputs T_1 and T_0 and of output OUT, (d) the resulting circuit.

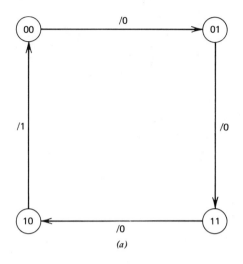

(a)

Present states		Output	Next states		Control inputs	
Q_{1_n}	Q_{0_n}	OUT_n	$Q_{1_{n+1}}$	$Q_{0_{n+1}}$	T_{1_n}	T_{0_n}
0	0	0	0	1	0	1
0	1	0	1	1	1	0
1	0	1	0	0	1	0
1	1	0	1	0	0	1

(b)

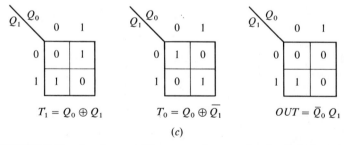

$$T_1 = Q_0 \oplus Q_1 \qquad T_0 = Q_0 \oplus \bar{Q_1} \qquad OUT = \bar{Q_0}\, Q_1$$

(c)

FIGURE 7.10. Another possible state assignment for the digital system of Figure 7.7: (*a*) transition diagram, (*b*) state table, (*c*) Karnaugh maps of control imputs T_1 and T_0 and of output OUT. (*See next page for* (*d*).)

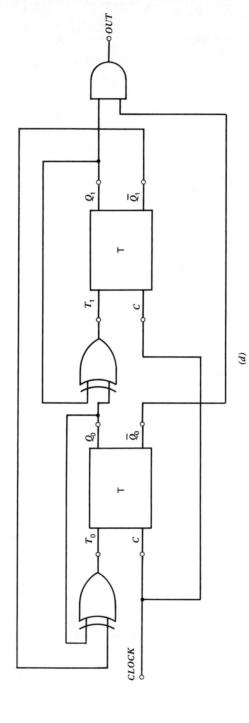

FIGURE 7.10. (*d*) the resulting circuit.

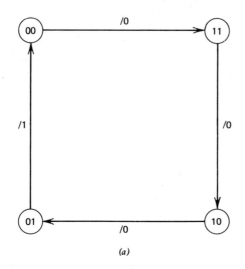

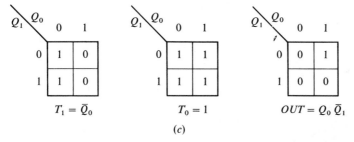

$T_1 = \bar{Q}_0$ $T_0 = 1$ $OUT = Q_0 \bar{Q}_1$

(c)

FIGURE 7.11. **A third possible state assignment for the digital system of Figure 7.7:** (a) transition diagram, (b) state table, (c) Karnaugh maps of control inputs T_1 and T_0 and of output OUT. (*See next page for* (d).)

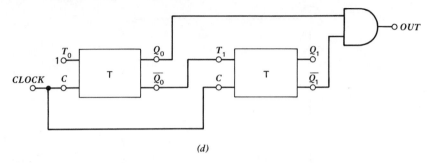

(d)

FIGURE 7.11.   (d) the resulting circuit.

Still a further degree of choice arises when the number of specified states is not an integer power of 2. In such cases the assignment of some of the states is initially unspecified, and the system is designated as *incompletely specified*.

EXAMPLE 7.6.   The transition diagram of a digital system with three specified states is shown in Figure 7.12a, the corresponding state table in

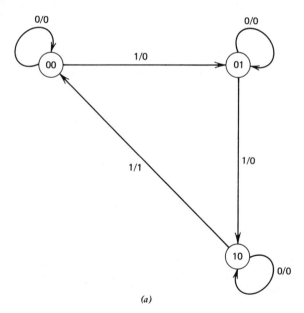

(a)

FIGURE 7.12.   An incompletely specified digital system: (a) transition diagram.

Present states		IN = 0			IN = 1		
		Output	Next states		Output	Next states	
Q_1	Q_0	OUT	Q_1	Q_0	OUT	Q_1	Q_0
0	0	0	0	0	0	0	1
0	1	0	0	1	0	1	0
1	0	0	1	0	1	0	0

(b)

Present states		IN = 0					IN = 1				
		Output	Next states		Control inputs		Output	Next states		Control inputs	
Q_1	Q_0	OUT	Q_1	Q_0	T_1	T_0	OUT	Q_1	Q_0	T_1	T_0
0	0	0	0	0	0	0	0	0	1	0	1
0	1	0	0	1	0	0	0	1	0	1	1
1	0	0	1	0	0	0	1	0	0	1	0

(c)

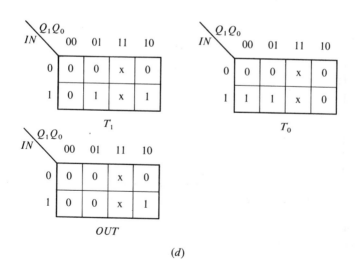

(d)

FIGURE 7.12. (*b*) **state table,** (*c*) **state table including control inputs,** (*d*) **Karnaugh maps for** T_1, T_0, **and** *OUT*. (*See next two pages for* (*e*), (*f*), (*g*), *and* (*h*).)

189

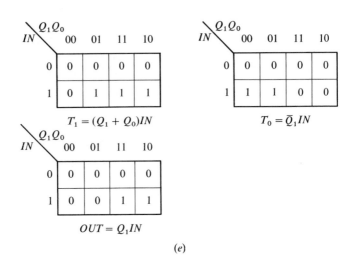

$T_1 = (Q_1 + Q_0)IN$

$T_0 = \bar{Q}_1 IN$

$OUT = Q_1 IN$

(e)

Present states		IN = 0					IN = 1				
		Control inputs		Output	Next states		Control inputs		Output	Next states	
Q_1	Q_0	T_1	T_0	OUT	Q_1	Q_0	T_1	T_0	OUT	Q_1	Q_0
0	0	0	0	0	0	0	0	1	0	0	1
0	1	0	0	0	0	1	1	1	0	1	0
1	0	0	0	0	1	0	1	0	1	0	0
1	1	0	0	0	1	1	1	0	1	0	1

(f)

FIGURE 7.12. (e) **Karnaugh maps with the don't cares resolved,** (f) **the resulting complete state table including the fourth state.**

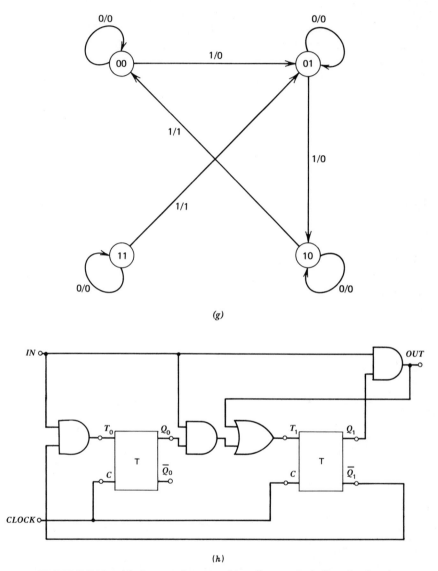

(g)

(h)

FIGURE 7.12. (*g*) the complete transition diagram including the fourth state, (*h*) the circuit.

Figure 7.12b. If the system is realized with two flip-flops, there will be an unassigned fourth state, state 11. This may be an isolated state or a transient state.

If we realize the system with two T flip-flops, the resulting state table including the control inputs is shown in Figure 7.12c, Karnaugh maps of T_1, T_0, and OUT in Figure 7.12d. We can see that because of the unspecified 11 state there are two don't care conditions on each map. A specified 11 state would remove these don't cares and replace them with 0's or 1's. We are free now, however, to choose them as we wish. One criterion in the choice can be the simplicity of the resulting logic functions: one such choice results in $T_1 = Q_1 IN + Q_0 IN$, $T_0 = \overline{Q_1} IN$, $OUT = Q_1 IN$. These choices now completely specify the system, including its 11 state, as can be seen from the Karnaugh maps of Figure 7.12e. By use of the characteristic table of the T flip-flop, we can now find the transitions of the 11 state as shown in Figure 7.12f, whence the complete transition diagram of Figure 7.12g can be derived. The resulting digital circuit is shown in Figure 7.12h.

7.3   Equivalence and Simplification

The structure of the digital system can frequently be simplified without any detriment to performance. Consider, for example, the transition diagram shown in Figure 7.13a and its realization by T flip-flops shown in Figure

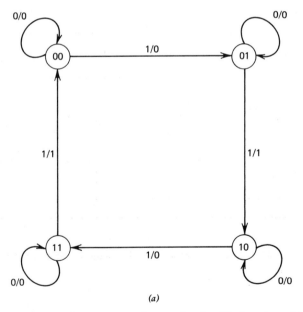

(a)

FIGURE 7.13.   A digital system that can be simplified to that of Figure 7.2: (a) transition diagram.

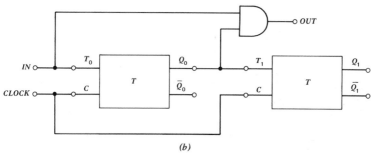

(b)

FIGURE 7.13. *(b)* **circuit.**

7.13*b*. We can see that the output of the second flip-flop in Figure 7.13*b* is not utilized, and the circuit can be reduced to the simpler circuit of Figure 7.2*a* and the transition diagram to the simpler one of Figure 7.2*d*. Such simplifications, however, are not always easily discernible.

EXAMPLE 7.7. The transition diagram of Figure 7.13*a* is shown in Figure 7.14*a*, differing in state assignments only. It can be shown that this transition diagram can be realized by the circuit shown in Figure 7.14*b*. No simple inspection would reveal, however, that the circuit, and hence the transition diagram of Figure 7.14*a*, could be simplified.

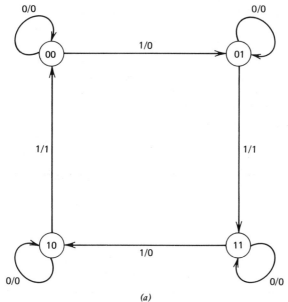

(a)

FIGURE 7.14. *(a)* **Transition diagram and digital circuit of Example 7.7.** *(See next page for (b).)*

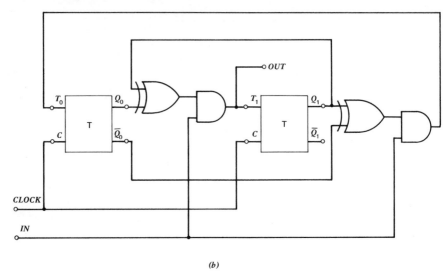

(b)

FIGURE 7.14   (b)

In many cases it is indeed difficult to determine whether a system can be simplified or not. Once there is a prospective simpler system, however, it is possible to determine whether the two systems are equivalent or distinguishable. This can be facilitated by use of two theorems which will not be proven here [see reference (3)].

The first theorem pertains to *equivalent states* and *distinguishable states*: two states are equivalent *iff* they yield identical output sequences when excited by *any* input sequence. If two states are not equivalent, then by definition they are distinguishable.

The second theorem pertains to *equivalent systems* and *distinguishable systems*: two digital systems M_1 and M_2 are equivalent *iff* each state of M_1 has at least one equivalent state in M_2 and each state of M_2 has at least one equivalent state in M_1. If M_1 and M_2 are not equivalent, then by definition they are distinguishable.

The use of these theorems will be illustrated by an example.

EXAMPLE 7.8.   Transition diagrams of two digital systems are shown in Figure 7.15*a* and Figure 7.15*b*. In the former there is no transition to state 3, also the part of the transition diagram comprising states 1 and 2 is identical to the transition diagram comprising states A and B. We can also see that states 1 and A are equivalent and so are states

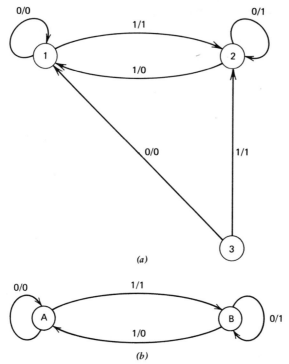

FIGURE 7.15.   Transition diagrams of two equivalent digital systems.

2 and B. Next, the first theorem will be used to show that state 3 is equivalent to state 1.

When the initial state is 3 and the input is 0, transition to state 1 with a 0 output takes place (0/0 on the transition diagram). Thus, this transition yields an output identical to that resulting from an initial state of 1. Also, the resulting next states are 1 in both cases; hence all further transitions and outputs are identical for any input sequence.

When the inital state is 3 and the input is 1, transition to state 2 with a 1 output takes place (1/1 on the transition diagram). Thus, this transition yields an output identical to that resulting from an initial state of 1. Also, the resulting next states are 1 in both cases; hence all further transitions and outputs are identical for any input sequence.

Thus, states 3 and 1 yield identical output sequences when excited by any input sequence, hence they are equivalent. Since states 1 and A are also equivalent, it follows that states 3 and A are equivalent.

The two digital systems are equivalent as a result of the second theorem, since state 1 is equivalent to state A, 2 to B, and 3 to A.

The theorems discussed above are the most basic ones on the subject. For further discussion on equivalence and simplification, the reader is referred to the references at the end of the book.

7.4   Shift Registers

Consider the digital circuit of Figure 7.16 consisting of a chain of three D flip-flops. For a D flip-flop the next state Q_{n+1} equals the present input D_n; therefore data entered at input IN are shifted one step to the right on each clock pulse. This structure is designated a *shift register.**

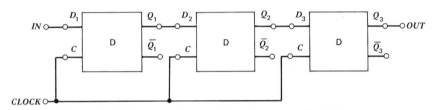

FIGURE 7.16.   A 3-stage shift-register using D flip-flops.

In a shift register circuit, data are shifted along the chain simultaneously at each flip-flop; hence the circuit must be sychronous, that is, all transitions must take place at approximately the same time. Thus all C inputs must occur simultaneously, or at least within a time that is short compared to the propagation delay of each flip-flop. This constraint may lead to some difficulties, especially in high-speed systems that are spread over a large area. In order to alleviate this limitation, shift registers commonly utilize master-slave flip-flops.

The most widespread application of the shift register is probably as a memory unit. The chain of flip-flops can be made long—shift registers consisting of 1000 flip-flops are not uncommon—and a large amount of data can be stored. In the simplest case as in Figure 7.16, all data are entered into the first flip-flop, shifted along the chain, and retrieved at the output of the last flip-flop ("serial-in serial-out" shift register).

When the chain is not too long and inputs to every flip-flop can be made available, data can also be entered in parallel as shown in the circuit of Figure 7.17. When control input $SERIAL = \overline{PARALLEL} = 1$, the circuit is equivalent to that of Figure 7.16, terminals P_1, P_2, and P_3 are inactive, and the circuit performs as a "serial-in serial-out" shift register. When control input $SERIAL = \overline{PARALLEL} = 0$, data can be entered into the

---

* In many circuits direct set and direct reset inputs are also provided.

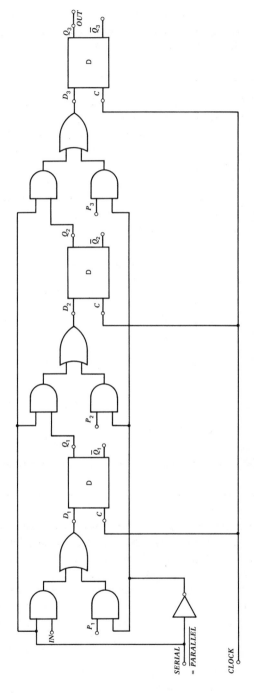

FIGURE 7.17. A 3-stage shift-register with parallel inputs.

flip-flops via parallel inputs P_1, P_2, and P_3; the data thus entered can be read out serially on output OUT after returning the control input to $SERIAL = 1$ ("parallel-in serial out" shift register).

EXAMPLE 7.9. The inputs and outputs in the shift register circuit of Figure 7.17 using master-slave D flip-flops is shown in the table of Figure 7.18 for a sequence of four subsequent clock pulses. Initial states of the flip-flops are unknown. During the first clock pulse, control input $SERIAL = \overline{PARALLEL} = 0$; and the flip-flops are set by parallel inputs P_1, P_2, and P_3 to 1, 0, and 0, respectively. During clock pulses 2, 3, and 4, control input $SERIAL = 1$; and data are read out serially at output OUT.

The circuit of Figure 7.18 can also be utilized as a bidirectional (shift-right shift-left) shift register (see Problem 7.10).

Clock pulse	1	2	3	4
$SERIAL =$ $\overline{PARALLEL}$	0	1	1	1
IN	x	0	0	0
P_1	1	x	x	x
P_2	0	x	x	x
P_3	0	x	x	x
Q_1	?	1	0	0
Q_2	?	0	1	0
$Q_3 = OUT$	?	0	0	1

FIGURE 7.18.   Input and output sequence in Example 7.9.

When connections are made to the outputs of each flip-flop in a shift register, the stored data can also be read out in parallel. Hence, there are four basic readin-readout methods: serial-in serial-out, serial-in parallel-out, parallel-in serial-out, and parallel-in parallel-out. Because of this flexibility, the shift register is very useful for conversion between serial bit-by-bit data such as present on a transmission line, and parallel data such as the contents of a computer word. In principle a single shift register of the type shown in Figure 7.17 can be used in all four applications. In practice, however, the number of available pin connections provide a limitation, particularly in the case of multi-stage shift registers contained in one IC package.

7.5 Counters

Counters are special purpose digital systems designed to count the number of transitions at an input of the system. Thus, for example, the transition diagram of Figure 7.19 describes a synchronous count-by-3 circuit that counts the number of clock pulses that occur while the input is 1. When the system is started in state 00, the number of 1 inputs is stored up to two (binary 10) in binary code; and a "carry" output is produced for three (binary 11).

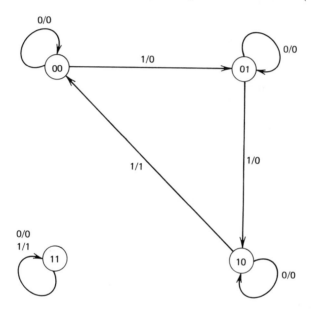

FIGURE 7.19. A count-by-3 circuit.

Thus, if the initial state was 00 and seven input 1's took place, the final state will be 01 and two carry output pulses will have occured. A transition diagram with unspecified state assignment describing a count-by-4 circuit was shown in Figure 7.7; the states can be assigned as in Figure 7.9a resulting in a *natural binary code* (8-4-2-1), or as in Figure 7.10a resulting in what is known as a *Gray code* (see Chapter 9), or as in Figure 7.11a resulting in a *down counter* (see page 208). Circuits for a larger number of counts can be designed in a similar way.

EXAMPLE 7.10.   Design a synchronous count-by-8-counter that counts the number of clock pulses using the natural binary code.

The transition diagram is shown in Figure 7.20a, a state table describing a realization with T flip-flops in Figure 7.20b.* Karnaugh maps for T_0, T_1, T_2, and OUT are shown in Figure 7.20c, the resulting circuit in Figure 7.20d. The number of clock pulses that took place can be obtained as the sum $2^0 Q_0 + 2^1 Q_1 + 2^2 Q_2 + C_n$, where C_n is the number of carry pulses that occured on carry output OUT.

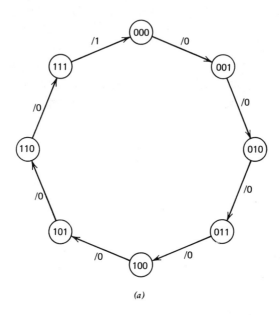

(a)

FIGURE 7.20.   Count-by-8 synchronous counter of Example 7.10: (a) transition diagram.

---

* The counter in this and in the subsequent example is realized using T flip-flops. With present technology, however, the use of J-K or D flip-flops may be preferable. A J-K flip-flop can be substituted for a T flip-flop by connecting the J and K inputs together as a T input (see Figure 6.23). The T flip-flop can be also replaced by a D flip-flop and logic gates.

Present states			Next states			Control inputs			Carry output
Q_2	Q_1	Q_0	Q_2	Q_1	Q_0	T_2	T_1	T_0	OUT
0	0	0	0	0	1	0	0	1	0
0	0	1	0	1	0	0	1	1	0
0	1	0	0	1	1	0	0	1	0
0	1	1	1	0	0	1	1	1	0
1	0	0	1	0	1	0	0	1	0
1	0	1	1	1	0	0	1	1	0
1	1	0	1	1	1	0	0	1	0
1	1	1	0	0	0	1	1	1	1

(b)

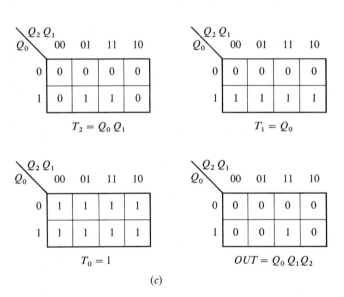

(c)

FIGURE 7.20. (*b*) **state table,** (*c*) **Karnaugh maps for** T_2, T_1, T_0, **and** OUT. (*See next page for* (*d*).)

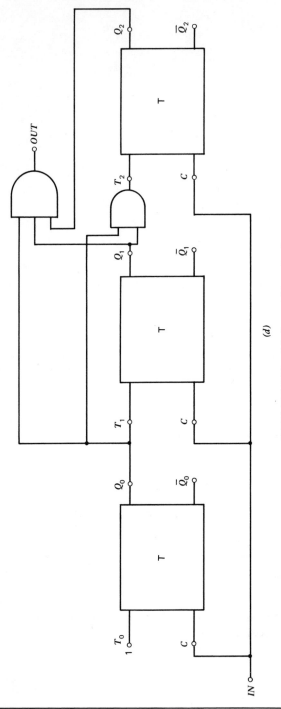

FIGURE 7.20. (*d*) the resulting circuit.

As a rule, a *self-starting* counter is preferable: such a counter has no isolated or persistent states, nor a group of such states, outside the main counting loop of the transition diagram. If this is not the case, the system may get into one of these states by some means not accounted for in the design (e.g., power turn-on) and then can be returned to the main counting loop of the transition diagram only by the application of direct set or reset inputs, if these are provided. Thus, the counter of Figure 7.12g is self-starting and is preferable to that of Figure 7.19 which is not self-starting. This problem need not arise in the case of count-by-2^m circuits when m is integer; it arises, however, in cases when the number of counts is not 2^m.

EXAMPLE 7.11. Design a synchronous count-by-10 counter that counts clock pulses in natural binary code.

The required transitions and control inputs for T flip-flops are shown in the first ten lines of the state table in Figure 7.21a; contents of the last six lines will be developed later. The first ten lines can be translated to the Karnaugh maps of Figure 7.21b; for all states $T_0 = 1$, and hence it is not shown. The circuit realization is not unique because of the don't care conditions in the last six states. The logic functions shown for T_1, T_2, T_3, and OUT represent reasonable but arbitrary choices.

Next we should see what happened to the remaining six states as a result of the logic functions chosen. In order to achieve this, six additional lines are now added to the state table of Figure 7.21a and the next states are entered into them based on the particular choice of realizations for T_1, T_2, T_3, and OUT and on the characteristic table of the T flip-flop. As the last step, the resulting state table is translated to the transition diagram shown in Figure 7.21c. This shows that don't care states 10 through 15 all funnel into the main counting loop consisting of states 0 through 9; hence, the counter is self-starting. The resulting circuit is shown in Figure 7.21d.

Figures 7.21a-d
are shown on
pages 204, 205, 206, 207.

Present states					Next states					Control inputs				Carry output
Decimal	Q_3	Q_2	Q_1	Q_0	Decimal	Q_3	Q_2	Q_1	Q_0	T_3	T_2	T_1	T_0	OUT
0	0	0	0	0	1	0	0	0	1	0	0	0	1	0
1	0	0	0	1	2	0	0	1	0	0	0	1	1	0
2	0	0	1	0	3	0	0	1	1	0	0	0	1	0
3	0	0	1	1	4	0	1	0	0	0	1	1	1	0
4	0	1	0	0	5	0	1	0	1	0	0	0	1	0
5	0	1	0	1	6	0	1	1	0	0	0	1	1	0
6	0	1	1	0	7	0	1	1	1	0	0	0	1	0
7	0	1	1	1	8	1	0	0	0	1	1	1	1	0
8	1	0	0	0	9	1	0	0	1	0	0	0	1	0
9	1	0	0	1	0	0	0	0	0	1	0	0	1	1
10	1	0	1	0	11	1	0	1	1	0	0	0	1	0
11	1	0	1	1	6	0	1	1	0	1	1	0	1	1
12	1	1	0	0	13	1	1	0	1	0	0	0	1	0
13	1	1	0	1	4	0	1	0	0	1	0	0	1	1
14	1	1	1	0	15	1	1	1	1	0	0	0	1	0
15	1	1	1	1	2	0	0	1	0	1	1	0	1	1

(a)

FIGURE 7.21.   Count-by-10 synchronous counter circuit of Example 7.11:
(a) state table.

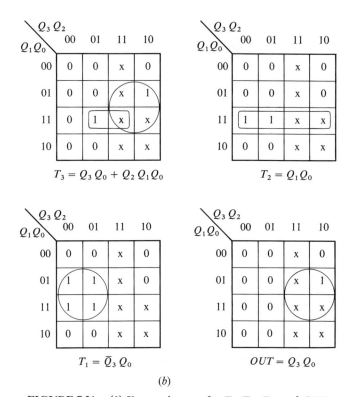

FIGURE 7.21. (*b*) Karnaugh maps for T_1, T_2, T_3, and OUT.

(*b*)

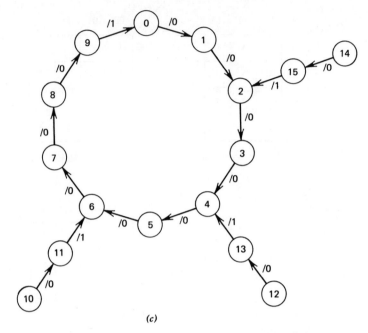

FIGURE 7.21. (*c*) complete transition diagram including all sixteen states.

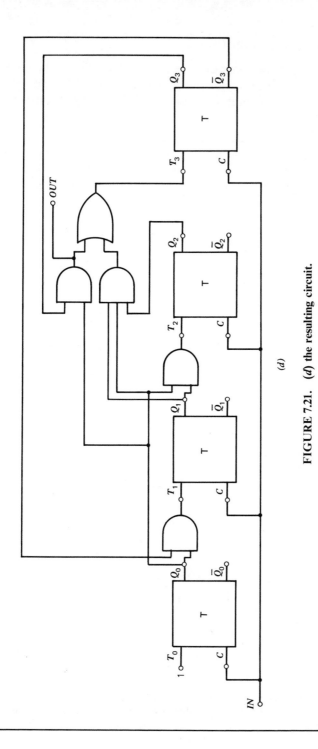

FIGURE 7.21. (d) the resulting circuit.

The counters in the preceding two examples counted the clock pulses sequentially. It is also possible to design *bidirectional* (up-down) counters that count the clock pulses up or down depending on the state of a control input.

EXAMPLE 7.12.   A count-by-4 synchronous up-down counter is described in Figure 7.22. When control input *IN* is 0, the counter counts up (increments); when it is 1, the counter counts down (decrements).

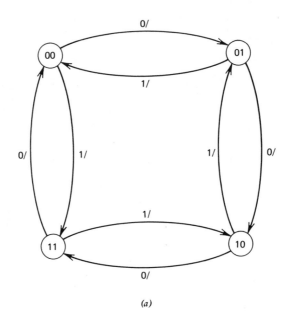

(a)

Q_n	Q_{n+1}	J_n	K_n
0	0	0	x
0	1	1	x
1	0	x	1
1	1	x	0

(b)

FIGURE 7.22. Up-down count-by-4 synchronous counter circuit of Example 7.12: (*a*) **transition diagram,** (*b*) **excitation table of the J-K flip-flop.**

Present states		IN$_n$ = 0						IN$_n$ = 1					
Q_{1_n}	Q_{0_n}	$Q_{1_{n+1}}$	$Q_{0_{n+1}}$	J_{1_n}	K_{1_n}	J_{0_n}	K_{0_n}	$Q_{1_{n+1}}$	$Q_{0_{n+1}}$	J_{1_n}	K_{1_n}	J_{0_n}	K_{0_n}
0	0	0	1	0	x	1	x	1	1	1	x	1	x
0	1	1	0	1	x	x	1	0	0	0	x	x	1
1	0	1	1	x	0	1	x	0	1	x	1	1	x
1	1	0	0	x	1	x	1	1	0	x	0	x	1

(c)

FIGURE 7.22. (c) **state table.** (*See next page for (d) and (e).*)

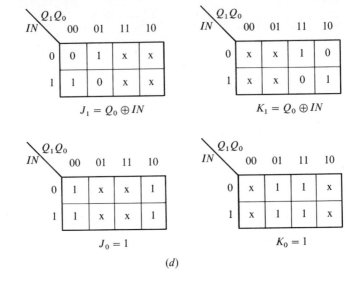

$$J_1 = Q_0 \oplus IN$$

$$K_1 = Q_0 \oplus IN$$

$$J_0 = 1$$

$$K_0 = 1$$

(d)

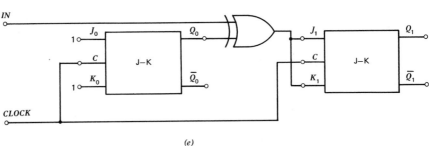

(e)

FIGURE 7.22. (*d*) **Karnaugh maps of control inputs** J_1, K_1, J_0, **and** K_0, (*e*) **the resulting circuit.**

Decoding. In addition to storing the number of counts, it is frequently required to provide *complete decoding* of a count-by-*p* circuit onto *p* lines.

EXAMPLE 7.13. Outputs Q_0 and Q_1 of the count-by-4 circuit of Figure 7.9 have to be decoded in order to generate signals S_1, S_2, S_3, and S_4 for the circuit of Figure 6.6. This is attained by the following logic functions: $S_1 = \overline{Q_1}\,\overline{Q_0}$, $S_2 = \overline{Q_1}\,Q_0$, $S_3 = Q_1\overline{Q_0}$, and $S_4 = Q_1\,Q_0$. Thus, in this case, complete decoding can be achieved by the use of four 2-input AND gates.

Complete decoding of a count-by-2^m counter with integer m can similarly be realized with 2^m gates with m inputs each. For large m, however, such a *one-stage decoding circuit* could become quite cumbersome. An alternate approach is the use of *multi-stage decoding*, illustrated for two stages in the example that follows.*

EXAMPLE 7.14. A count-by-16 counter using the natural binary code has to be fully decoded onto sixteen lines. A 1-stage decoding circuit would consist of sixteen gates with four inputs each. When using a 2-stage decoder circuit, first separate decoding of Q_0, Q_1, and of Q_2, Q_3, are performed as $L_0 = \overline{Q_1}\,\overline{Q_0}$, $L_1 = \overline{Q_1}\,Q_0$, $L_2 = Q_1\,\overline{Q_0}$, $L_3 = Q_1\,Q_0$, and $M_0 = \overline{Q_4}\,\overline{Q_3}$, $M_1 = \overline{Q_4}\,Q_3$, $M_2 = Q_4\,\overline{Q_3}$, $M_3 = Q_4\,Q_3$. From these, the required decoded outputs can be generated by sixteen gates with two inputs each (see also Problem 7.18). The resulting decoder circuit consists of 24 gates with two inputs each. This circuit, even though it has longer propagation delays, is in some cases preferable to the 1-stage decoder circuit consisting of sixteen gates with four inputs each.

Shift Register Counters. We have seen previously how a count-by-2^m counter can be realized by using m flip-flops. For large m, however, the logic circuits for the control inputs, and the decoding circuits, could become quite complex. An alternate approach is the use of shift register ring counters: This uses 2^m flip-flops, but the logic circuits for the control inputs are simple and no decoding circuitry is required. Such a ring counter can be constructed simply by connecting the last output OUT of a shift register to the first D input IN. This is illustrated in Figure 7.23a for the case of a count-by-3; it can be shown that the resulting transition diagram is that of Figure 7.23b. By presetting the circuit of Figure 7.23a to its 001 ($Q_1 = 0, Q_2 = 0, Q_3 = 1$) state, the decoded outputs are directly available on Q_1, Q_2, and Q_3. A significant disadvantage can also be seen in the transition diagram of Figure 7.23b: the counter is not self-starting, and for correct operation it has to be initialized into its 001-100-010 loop by means of direct set and reset inputs that are not shown. This disadvantage can be overcome, however, by the use of more complex feedback.

---

* Decoders are discussed in detail in Chapter 10.

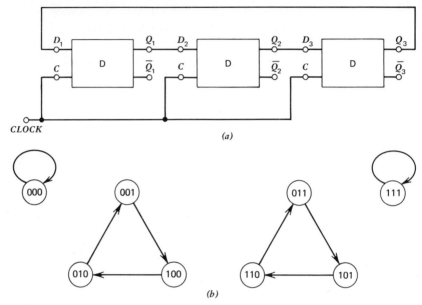

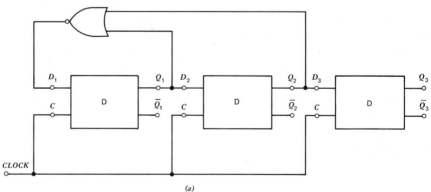

FIGURE 7.23. Three-stage shift-register ring counter: (*a*) circuit diagram, (*b*) complete transition diagram including all eight states.

EXAMPLE 7.15. A self-starting 3-stage shift register ring counter is shown in Figure 7.24*a*. Instead of the feedback $D_1 = Q_3$ of the non-self-starting counter, a $D_1 = \overline{Q_1 + Q_2}$ is used. The resulting state table is shown in Figure 7.24*b* and the resulting transition diagram in Figure 7.24*c*. These show that all states are funneled into the 001-100-010 loop; hence the counter is self-starting.

FIGURE 7.24. Self-starting 3-stage shift-register ring counter: (*a*) circuit diagram.

Present states			Next states		
Q_1	Q_2	Q_3	Q_1	Q_2	Q_3
0	0	0	1	0	0
0	0	1	1	0	0
0	1	0	0	0	1
0	1	1	0	0	1
1	0	0	0	1	0
1	0	1	0	1	0
1	1	0	0	1	1
1	1	1	0	1	1

(b)

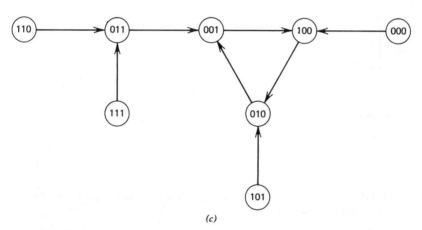

(c)

FIGURE 7.24. (b) state table, (c) complete transition diagram including all eight states.

When the D input of the first stage in a shift register is connected to the *complement* output of the last stage, a *Johnson counter* results. One such counter with three stages is shown in Figure 7.25a; it can be shown that the transition diagram is as shown in Figure 7.25b. The 3-stage circuit is a count-by-6 counter; in general, a Johnson counter consisting of m stages is a count-by-$2m$ counter. An important application of the Johnson counter is as a decade (count-by-10) counter consisting of five stages.

We can see that the Johnson counter of Figure 7.25 is not self-starting. As was the case for the ring counter, however, here too this disadvantage can be removed by use of more complex feedback.

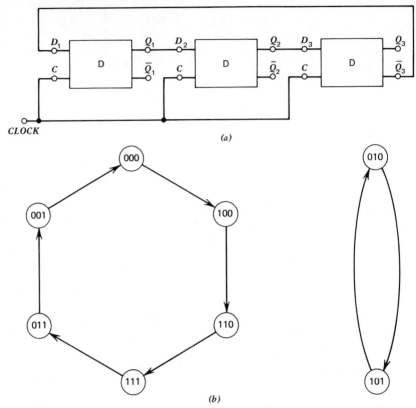

FIGURE 7.25.   Three-stage count-by-6 Johnson counter: (a) circuit diagram, (b) transition diagram.

Ripple Counters.   An additional counter configuration is the ripple counter illustrated in Figure 7.26a for a count-by-8 circuit using master-slave T flip-flops. Note that, unlike the previous counters, this counter is *not* synchronous; that is, transitions of Q_1 and Q_2 are not occurring *at* the trailing edge of *IN*, but are subsequent to it. It is, however, a simple circuit since there is no logic between the stages, and this holds for counters of any lengths. The timing diagram shown in Figure 7.26b illustrates the price we pay: the propagation delays of the flip-flops add up as the transition "ripples" through the chain of flip-flops. Another disadvantage is also evident in the decoding, as illustrated by the decoded $\overline{Q}_0 \, \overline{Q}_1 \, \overline{Q}_2$ signal showing the presence of spurious signals. These "spikes" could be eliminated by AND-ing input *IN* with the decoded signals, provided that the total propagation delay through the chain does not exceed the duration of the 0's of input *IN*, and this is just barely the case in Figure 7.25b. Thus simplicity is attained at the expense of operating speed.

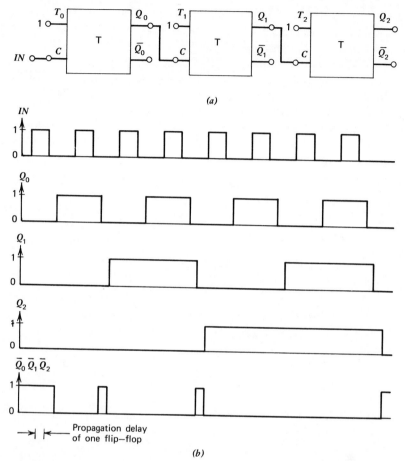

(a)

(b)

FIGURE 7.26. Count-by-8 ripple counter using master-slave T flip-flops: (a) circuit diagram, (b) timing diagram.

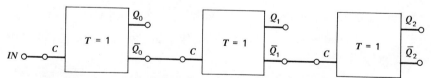

FIGURE 7.27. Count-by-8 ripple counter using edge-triggered T flip-flops with $T = 1$.

Further simplification of the circuit can be achieved by use of edge-triggered flip-flops and also by connecting the T inputs permanently to 1, that is, by using $T = 1$ flip-flops. The resulting circuit is shown in Figure 7.27.

When a count-by-p counter is required with $p \neq 2^m$ (m is integer), the direct reset inputs of the flip-flops may be utilized to reset the counter after the accumulation of a desired number of counts.

EXAMPLE 7.16. A count-by-5 circuit using three $T = 1$ edge-triggered flip-flops is shown in Figure 7.28. The counter counts up from 0 to 4 (binary 100) in the same manner as the binary counter of Figure 7.27. The next input, however, sets the counter to binary 000 via the direct reset (R_D) inputs of the first and last flip-flops.*

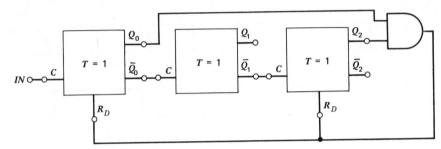

FIGURE 7.28. Count-by-5 ripple counter using edge-triggered T flip-flops with $T = 1$.

PROBLEMS

7.1 Draw a transition diagram for the J-K flip-flop by utilizing the state table of Figure 7.1c.

7.2 Draw a transition diagram for the R-S flip-flop of Figure 6.22.

7.3 Draw a state table and a transition diagram for the D-E flip-flop of Figure 6.30.

7.4 Describe isolated, transient, and persistent states by use of the terms reflecting branch, diverging branch, and converging branch.

7.5 Realize the state table of Figure 7.2b by means of a J-K flip-flop. Compare the resulting circuit with that of Figure 7.2a.

---

* An alternate way of designing a count-by-p counter is by use of a decoder (see Problem 9.32).

7.6 Change the state assignment in the D flip-flop realization of the state table of Figure 7.5a. Draw the resulting transition diagram and circuit, and compare the latter with that of Figure 7.5e.

7.7 Show that the circuit of Figure 7.14b is indeed a realization of the transition diagram of Figure 7.14a.

7.8 Draw the transition diagram of a digital system that consists of two states and is equivalent to that described by the transition diagram of Figure 7.14a. Prove equivalence with the help of the two theorems given in the text.

7.9 Prepare a state table and a complete transition diagram for the shift register shown in Figure 7.16.

7.10 In the shift register of Figure 7.17, the parallel inputs are connected as follows: $P_1 = Q_2$, $P_2 = Q_3$, $P_3 = Q_1$. Show that data from the shift register can be shifted out to the right via output OUT when control input $SERIAL = 1$ and it can be shifted out to the left via Q_1 when control input $SERIAL = 0$ ("shift-right shift-left" shift-register). How can data be entered into the shift register?

7.11 Because of long wires in the shift register of Figure 7.16, the clock pulse at the C input of the last flip-flop is delayed by a time that exceeds the propagation delay of each flip-flop. Discuss the consequences and illustrate by an example.

7.12 Realize the shift register of Figure 7.16 by using three R-S flip-flops and an inverter.

7.13 Design a synchronous count-by-8 natural binary counter using three J-K flip-flops and logic gates. Provide the logic functions for the control inputs, a circuit diagram, and a timing diagram.

7.14 In the decimal counter of Example 7.11 the resolution of the don't care conditions resulted in the transition diagram of Figure 7.21c which had no isolated or persistent states, nor groups of such states. Hence the counter was self-starting. In order to illustrate an inferior design, change the realization so that there is no return to the main counting loop of Figure 7.21c from states 10 through 15. Provide Karnaugh maps for T_1, T_2, and T_3, a circuit diagram, and a complete transition diagram.

7.15 Design a count-by-16 synchronous natural binary counter using four master-slave J-K flip-flops and logic gates. Draw the resulting circuit.

7.16 Design a self-starting synchronous count-by-10 natural binary counter using four J-K flip-flops and logic gates. Provide the logic functions for the control inputs, a complete transition diagram, and a circuit diagram.

7.17 Design a synchronous up-down count-by-8 circuit using three master-slave J-K flip-flops and logic gates. Provide the logic functions for the control inputs, and a circuit diagram.

7.18 Write the logic functions in Example 7.14 for the sixteen outputs S_0 through S_{15} utilizing L_0 through L_3 and M_0 through M_3.

7.19 Prepare a state table for the shift register of Figure 7.23a and derive from it the transition diagram of Figure 7.23b.

7.20 Design a self-starting 4-stage shift-register ring counter. Provide a circuit diagram, a state table, and a complete transition diagram.

7.21 Derive the transition diagram of Figure 7.25b from the circuit of Figure 7.25a.

7.22 Design a 5-stage count-by-10 Johnson counter. Prepare a state table, a complete transition diagram, and a circuit diagram.

7.23 Compare the properties of the counters shown in Figure 7.12, Figure 7.23, and Figure 7.24.

7.24 Sketch the signals of all eight decoded outputs in the count-by-8 ripple counter of Figure 7.26.

7.25 Draw a timing diagram for the count-by-8 ripple counter of Figure 7.27 showing signals of IN, Q_0, Q_1, Q_2, and of all eight decoded outputs.

7.26 Draw a timing diagram for the count-by-5 ripple counter of Figure 7.28.

7.27 A compromise between the synchronous and the ripple counter is shown in Figure 7.29 for a count-by-2^{16}. Each count-by-16 circuit is a synchronous natural binary counter described in Problem 7.15. The resulting *hybrid counter* is simpler than a synchronous count-by-2^{16} counter, and its propagation delay is shorter than that of a count-by-2^{16} ripple counter. What is the propagation delay of the hybrid counter from input *IN* to output *OUT*, if each flip-flop in the count-by-16 circuits has a clock propagation delay of 20 ns?

7.28 Add the requisite gating to make the 16-stage counter of Figure 7.29 fully synchronous.

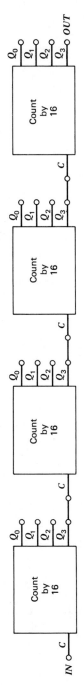

FIGURE 7.29.

CHAPTER 8

Arithmetic Circuits

Arithmetic circuits are of prime importance in digital computers and in many other digital systems. While other circuits, such as memories and input-output devices, are used for storing and transferring data, operations on the data are performed by arithmetic circuits. Some operations are executed on one number at a time, for example, complementation, multiplication by 2^m (m integer) by means of shifting, truncation, and other bit manipulations.

This chapter concentrates on arithmetic circuits performing operations that require the participation of at least two numbers, i.e., on magnitude comparators, adders, subtractors, multipliers, and dividers. In all these operations, accuracy, speed, and circuit complexity are principal considerations; in many cases low power consumption is also of interest. When the accuracy is specified by the number of digits to be processed, there is, in general, a tradeoff between the speed of operation and the complexity of the circuit, and both of these are limited by the available technology.

Thus, although it seems straightforward to set forth general principles, the "best" solution to a given problem depends on many economical and technical limitations. For these reasons, a general "optimization" of the circuits will not be attempted; but the effects of various tradeoffs will be illustrated.

8.1 Digital Comparators

The simplest arithmetic circuit operating on two numbers is the *digital comparator*, also known as *magnitude comparator*, or simply *comparator*. The comparators to be described here can be utilized for comparison of two numbers in any binary code; for simplicity, however, it will be assumed that the numbers are in natural binary code, and also that they are not negative.

When two numbers A and B are compared, the magnitude relations of interest are $A > B$, $A \geq B$, $A = B$, $A \leq B$, and $A < B$. It can be easily shown

220

that the following logic statements are true:

$$(A > B) = \overline{(A \leq B)}, \tag{8.1}$$

$$(A \geq B) = \overline{(A < B)}, \tag{8.2}$$

$$(A = B) = \overline{(A > B)} \cdot \overline{(A < B)}, \tag{8.3}$$

$$(A \leq B) = \overline{(A > B)}, \tag{8.4}$$

$$(A < B) = \overline{(A \geq B)}. \tag{8.5}$$

Thus, for example, from eq. (8.1), if $A > B$ is true, then $A \leq B$ is false, and vice versa.

In the simplest case when A and B are 1-bit numbers and can take on the values of 0 or 1 only, the magnitude relations can be translated to logic statements as shown in Figure 8.1. Thus, for $A > B$ to be true, it is

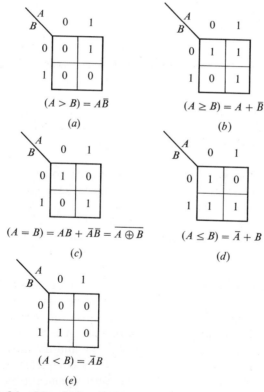

FIGURE 8.1. Magnitude relations and the corresponding Karnaugh maps and logic functions for 1-bit comparators.

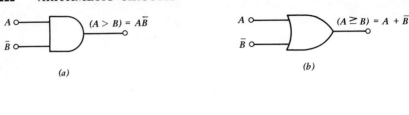

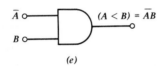

FIGURE 8.2.   Logic circuits realizing the logic functions of Figure 8.1.

necessary that $A = 1$ and $B = 0$; that is $A\bar{B} = 1$. Logic circuits realizing the functions of Figure 8.1 are shown in Figure 8.2. (Note the new EXCLUSIVE-NOR symbol introduced in Figure 8.2c.)

$A = B$ **for** n-**bit Numbers.**   Two binary or binary coded numbers are equal *iff* all corresponding bits of the two numbers have equal binary values. Therefore, based on Figure 8.1c, numbers $A = A_{n-1} A_{n-2} \cdots A_i \cdots A_0$ and $B = B_{n-1} B_{n-2} \cdots B_i \cdots B_0$ are equal *iff* $\overline{A_{n-1} \oplus B_{n-1}} \cdot \overline{A_{n-2} \oplus B_{n-2}} \cdots \overline{A_i \oplus B_i} \cdots \overline{A_0 \oplus B_0} = 1$. A circuit realizing this function for $n = 4$ is shown in Figure 8.3.

$A > B$ **for** n-**bit Numbers.**   The determination of $A > B$ for two n-bit numbers A and B may be obtained by visual inspection. First we look at the most significant bits of A and B. If the most significant bit of A is 1 and the most significant bit of B is 0, then $A > B$ and the process is completed. If the most significant bit of A is 0 and the most significant bit of B is 1, then $A \not> B$ and the process again is completed. If, however, the most significant bit of A is equal to the most significant bit of B, then we have to inspect the next to most significant bits in a similar manner; and we have to proceed until $A > B$ or $A \not> B$ has been established.

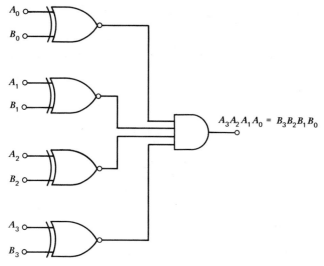

FIGURE 8.3. Logic circuit realizing $A_3 A_2 A_1 A_0 = B_3 B_2 B_1 B_0$.

EXAMPLE 8.1. Is the statement $A > B$ true if $A = 1110$ and $B = 1101$? The most significant (leftmost) bits are equal; thus we have to look at the next most significant (second from the left) bits. These are equal, too. However, the third from the left bits show that $A > B$; hence the process is terminated, and there is no need to inspect the rightmost bits.

Consider two 3-bit numbers, A and B, expressed in positional notation $A = A_2 A_1 A_0$ and $B = B_2 B_1 B_0$. The logic equation for $A > B$ may be written as

$$(A_2 A_1 A_0 > B_2 B_1 B_0)$$
$$= (A_2 > B_2) + (A_2 = B_2)[(A_1 > B_1) + (A_1 = B_1)(A_0 > B_0)] \quad (8.6)$$

or as

$$(A_2 A_1 A_0 > B_2 B_1 B_0)$$
$$= (A_2 > B_2) + (A_2 = B_2)(A_1 > B_1) + (A_2 = B_2)(A_1 = B_1)(A_0 > B_0). \quad (8.7)$$

By use of the logic functions of Figure 8.1, eq. (8.6) may be expressed as

$$(A_2 A_1 A_0 > B_2 B_1 B_0) = A_2 \overline{B_2} + \overline{A_2 \oplus B_2}[A_1 \overline{B_1} + \overline{A_1 \oplus B_1} A_0 \overline{B_0}], \quad (8.8)$$

while eq. (8.7) may be written as

$$(A_2 A_1 A_0 > B_2 B_1 B_0)$$
$$= A_2 \overline{B_2} + \overline{A_2 \oplus B_2} A_1 \overline{B_1} + \overline{A_2 \oplus B_2} \ \overline{A_1 \oplus B_1} A_0 \overline{B_0}. \quad (8.9)$$

which after considerable simplification reduces to

$$(A_2 A_1 A_0 > B_2 B_1 B_0)$$
$$= A_2 \overline{B_2} + A_2 B_2 A_1 \overline{B_1} + A_2 B_2 A_1 A_0 \overline{B_0} + A_2 B_2 A_0 B_1 \overline{B_0}$$
$$+ \overline{A_2}\,\overline{B_2} A_1 \overline{B_1} + \overline{A_2}\,\overline{B_2} A_1 A_0 \overline{B_0} + \overline{A_2}\,\overline{B_2} A_0 B_1 \overline{B_0}. \quad (8.10)$$

Thus, the logic function for $A_2 A_1 A_0 > B_2 B_1 B_0$ can be written in several forms. The first form, eq. (8.8), results in the *ripple comparator* of Figure 8.4. This is the slowest circuit: to reach the output, B_0 has to propagate

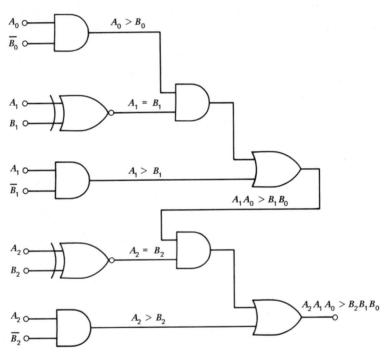

FIGURE 8.4.   A 3-bit ripple comparator.

through five gates, and also through an inverter if $\overline{B_0}$ is not available. Another realization, the *simultaneous carry comparator*, or *look-ahead carry comparator*, is based on eq. (8.9) and is shown in Figure 8.5. Here the longest propagation delay is that of four gates (with the EXCLUSIVE-NOR gate counted as two) and an inverter. The third realization is the *two-stage comparator* based on eq. (8.10) and shown in Figure 8.6; this is the fastest circuit since each input has to propagate through only two gates and an inverter.

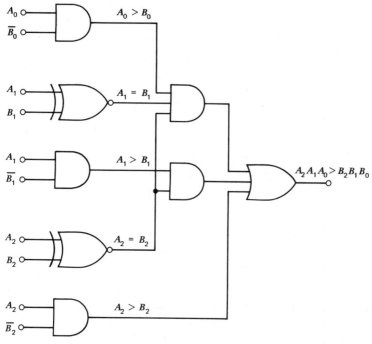

FIGURE 8.5.   A 3-bit simultaneous carry comparator.

In the case of two n-bit numbers, it can be shown that $A > B$ may be expressed as

$$(A_{n-1}A_{n-2}\cdots A_i \cdots A_0 > B_{n-1}B_{n-2}\cdots B_i \cdots B_0)$$
$$= A_{n-1}\overline{B_{n-1}} + \sum_{i=1}^{n-1} [A_{n-1-i}\overline{B_{n-1-i}} \prod_{j=1}^{i} (\overline{A_{n-j} \oplus B_{n-j}})]. \quad (8.11)$$

For example, for two 4-bit numbers, eq. (8.11) can be written as

$$(A_3 A_2 A_1 A_0 > B_3 B_2 B_1 B_0)$$
$$= A_3 \overline{B_3} + \overline{A_3 \oplus B_3}[A_2 \overline{B_2} + \overline{A_2 \oplus B_2}(A_1 \overline{B_1} + \overline{A_1 \oplus B_1} A_0 \overline{B_0})], \quad (8.12)$$

or as

$$(A_3 A_2 A_1 A_0 > B_3 B_2 B_1 B_0)$$
$$= A_3 \overline{B_3} + \overline{A_3 \oplus B_3}\, A_2 \overline{B_2} + \overline{A_3 \oplus B_3}\, \overline{A_2 \oplus B_2}\, A_1 \overline{B_1}$$
$$+ \overline{A_3 \oplus B_3}\, \overline{A_2 \oplus B_2}\, \overline{A_1 \oplus B_1}\, A_0 \overline{B_0}, \quad (8.13)$$

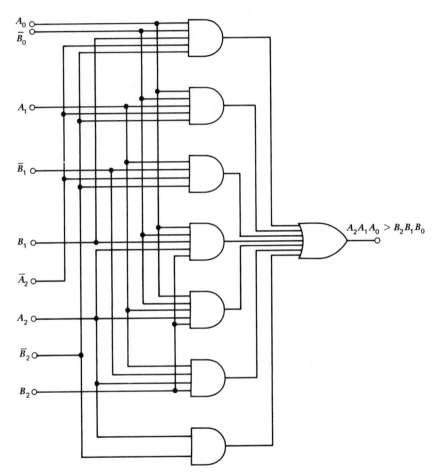

FIGURE 8.6. A 3-bit 2-stage comparator.

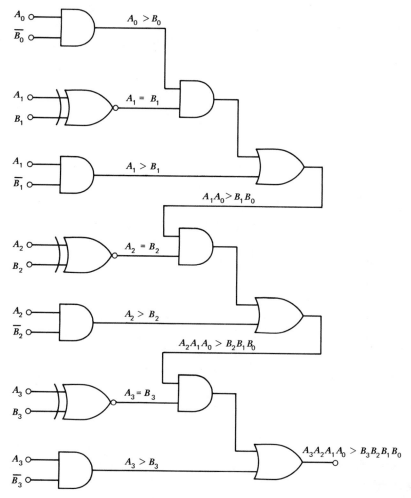

FIGURE 8.7. A 4-bit ripple comparator.

or, after simplification, as

$$(A_3 A_2 A_1 A_0 > B_3 B_2 B_1 B_0)$$
$$= A_3 \overline{B_3} + (A_3 B_3 + \overline{A_3}\, \overline{B_3})$$
$$\cdot [A_2 \overline{B_2} + (A_2 B_2 + \overline{A_2}\, \overline{B_2})(A_1 \overline{B_1} + A_1 A_0 \overline{B_0} + A_0 \overline{B_1}\, \overline{B_0})]. \qquad (8.14)$$

Thus, similarly to the case of $n = 3$, the logic equation for $A_3 A_2 A_1 A_0 >$ $B_3 B_2 B_1 B_0$ can be written in several different forms. Equation (8.12) results in the ripple comparator shown in Figure 8.7. Here the propagation delay of B_0 to the output is that of seven gates and an inverter; for an n-bit comparator it is that of $2n - 1$ gates and an inverter. Realization

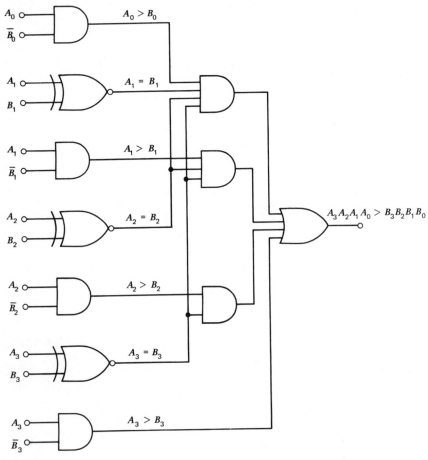

FIGURE 8.8.   A 4-bit simultaneous carry comparator.

of eq. (8.13) results in the simultaneous carry comparator shown in Figure 8.8 (in reality, NAND gates are used instead of AND and OR gates). Here the longest propagation delay is that of four gates (counting the EXCLUSIVE–NOR gate as two), plus that of an inverter. This delay, furthermore, does not increase for larger n. The shortest propagation delays are attainable by a 2-state realization of eq. (8.14); the resulting 2-stage logic circuit is quite complex, but it can be realized by use of LSI circuits discussed in Section 11.9.

Hybrid Comparators. In principle, the techniques discussed above can be applied to any n. In reality, however, some limitations will arise for large n. The extension of the $A = B$ circuit of Figure 8.3 to large n will require an AND gate with many inputs or its replacement by several gates. The propagation delay through the ripple comparators of Figure 8.4 and Figure 8.7, ignoring the inverter, increases as $2n - 1$, although this delay is tolerable in many cases. The complexity of the simultaneous (look-ahead) carry comparators of Figure 8.5 and Figure 8.8 increase rapidly for large n; the 2-stage comparator requires even more circuitry.

For large n, the *hybrid comparator* represents a reasonable compromise between speed and complexity. It is attained through cascading several simultaneous (look-ahead) carry comparators or two-stage comparators.

EXAMPLE 8.2. A symbol for a 4-bit magnitude comparator is shown in Figure 8.9. The box represents the circuit of Figure 8.8 with inverters included to generate the complements of the input variables. (Because of economy and reliability considerations, pin limitations are dominant; hence, such MSI circuits invariably generate complements of input variables internally.) Three such comparators are cascaded in Figure 8.10 to provide a 10-bit hybrid comparator.

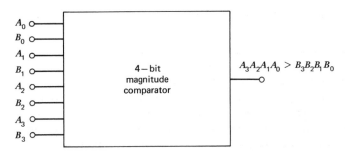

FIGURE 8.9. A symbol for the 4-bit magnitude comparator.

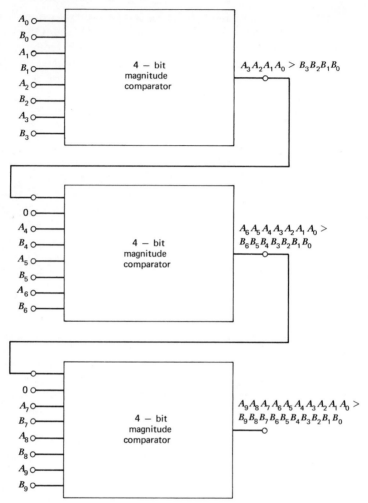

FIGURE 8.10. A 10-bit hybrid comparator utilizing three 4-bit comparators.

A 4-bit simultaneous (look-ahead) carry comparator with an added input for the partial $A_c > B_c$ "carry" is shown in Figure 8.11. It is, in fact, an extension of the realization of Figure 8.8 to $n = 5$, but with the AND gate omitted for the least significant input bits. The nomenclature used in Figure 8.11 is somewhat more general than that of Figure 8.8 and is intended to facilitate the design of comparators with $n > 4$.

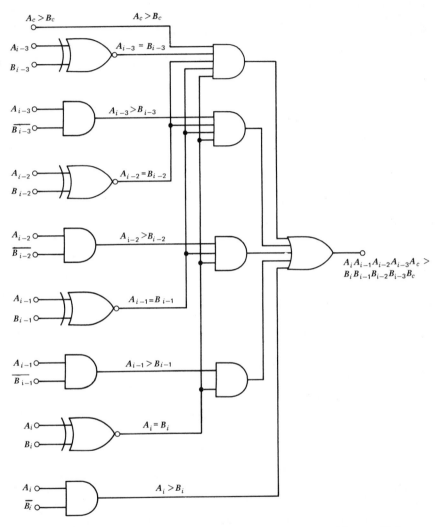

FIGURE 8.11. A 4-bit simultaneous carry comparator with an added $A_c > B_c$ input for partial carry. Subscript c refers to the "carry" from the next less significant bits.

EXAMPLE 8.3. A symbol for the 4-bit comparator with partial carry input is shown in Figure 8.12. The box represents the circuit of Figure 8.11, including inverters to generate the complements of the inputs if they are not

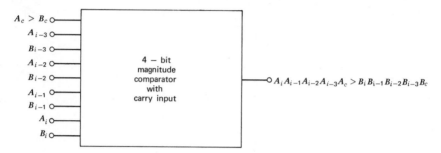

FIGURE 8.12. Symbol for the 4-bit comparator with carry input of Figure 8.11.

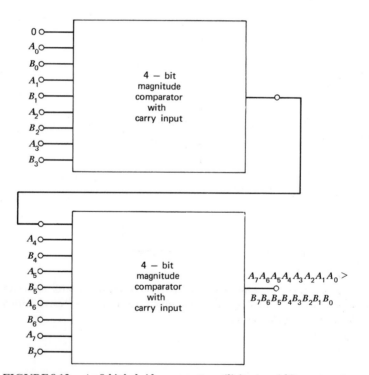

FIGURE 8.13. An 8-bit hybrid comparator utilizing two 4-bit comparators with carry input.

available. Two such comparators are cascaded in Figure 8.13 to provide an 8-bit hybrid comparator. A 16-bit hybrid comparator can be assembled in a similar manner from four 4-bit comparators with carry inputs.

Sequential Comparators. The time required to perform the comparison is immaterial in many applications. In such cases, particularly if the comparator is part of a sequential digital system, the use of a *sequential comparator* may be considered which compares two numbers A and B bit-by-bit sequentially. Thus, on the first clock pulse the most significant bits are compared, on the next clock pulse the next to most significant bits, etc., in a manner similar to that described earlier on page 222. The process is terminated when two bits compared are not equal or when all bits have been compared.

EXAMPLE 8.4. The magnitudes of two n-bit numbers A and B are to be compared to determine whether $A > B$. The two numbers are stored in two shift registers, from where they are shifted out serially with the *most significant bits first*. We also use two J-K flip-flops designated as $A > B$ and as END. Initial state of flip-flop $A > B$ is not known; flip-flop END is reset to its 0 state to start the comparison process.

Present states		Inputs		Next states		Control inputs			
$A > B$	END	A_i	B_i	$A > B$	END	$J_{A>B}$	$K_{A>B}$	J_{END}	K_{END}
0	0	0	0	0	0	0	x	0	x
0	0	0	1	0	1	0	x	1	x
0	0	1	0	1	1	1	x	1	x
0	0	1	1	0	0	0	x	0	x
0	1	0	0	0	1	0	x	x	0
0	1	0	1	0	1	0	x	x	0
0	1	1	0	0	1	0	x	x	0
0	1	1	1	0	1	0	x	x	0
1	0	0	0	0	0	x	1	0	x
1	0	0	1	0	1	x	1	1	x
1	0	1	0	1	1	x	0	1	x
1	0	1	1	0	0	x	1	0	x
1	1	0	0	1	1	x	0	x	0
1	1	0	1	1	1	x	0	x	0
1	1	1	0	1	1	x	0	x	0
1	1	1	1	1	1	x	0	x	0

FIGURE 8.14. State table for the sequential comparator of Example 8.4.

When the state of flip-flop *END* is 0, each clock pulse sets flip-flop $A > B$ to 1 if $A_i > B_i$ for the i-th bit under inspection, to 0 otherwise. Furthermore, flip-flop *END* is set to its 1 state if $A_i \neq B_i$, and the states of flip-flops $A > B$ and *END* are preserved if the state of flip-flop *END* is 1. The result of the comparison is available at the output of the $A > B$ flip-flop after the nth clock pulse or when the state of flip-flop *END* becomes 1, whichever occurs first.

From above specifications, the state table shown in Figure 8.14 can be prepared. The Karnaugh maps for the control inputs are shown in Figure 8.15. The resulting circuit is shown in Figure 8.16.

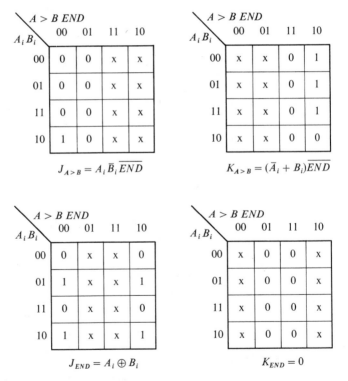

FIGURE 8.15. Karnaugh maps for the flip-flop control inputs of the sequential comparator of Example 8.4.

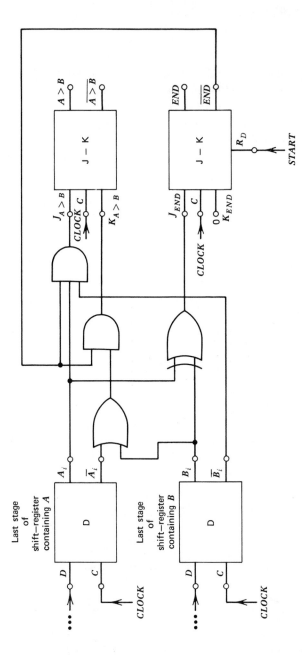

FIGURE 8.16. Sequential comparator circuit of Example 8.4.

8.2 Adders

Addition is one of the most important operations in digital arithmetic since it is also used in other arithmetic operations such as subtraction, multiplication, division, square-root extraction, etc. Thus, a large effort has been invested in the development of adder circuits; this section provides an outline of these developments.

When two n-bit binary numbers A and B are added, each bit of the sum S is a function of the corresponding bits of A and B and of the carry from the next less significant bits (see Chapter 2). A truth table for the ith sum bit S_i and for the ith carry bit C_i is shown in Figure 8.17a, the corresponding Karnaugh maps in Figure 8.17b and c. Two realizations for C_i are given: the first one uses an OR function and results in a faster circuit (circuit not shown); the second one utilizes the available EXCLUSIVE-OR function of S_i resulting in the *full adder* circuit shown in Figure 8.17d.* A symbol for the full adder is shown in Figure 8.18.

A_i	B_i	C_{i-1}	S_i	C_i
0	0	0	0	0
0	1	0	1	0
1	0	0	1	0
1	1	0	0	1
0	0	1	1	0
0	1	1	0	1
1	0	1	0	1
1	1	1	1	1

(a)

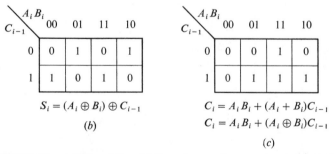

$$S_i = (A_i \oplus B_i) \oplus C_{i-1}$$

(b)

$$C_i = A_i B_i + (A_i + B_i)C_{i-1}$$
$$C_i = A_i B_i + (A_i \oplus B_i)C_{i-1}$$

(c)

FIGURE 8.17. Addition of two 1-bit numbers and a carry input: (a) truth table, (b) Karnaugh map for the sum bit S_i, (c) Karnaugh map for the carry bit C_i.

* Auxiliary outputs G_i and P_i in the circuit of Figure 8.17d will be utilized later.

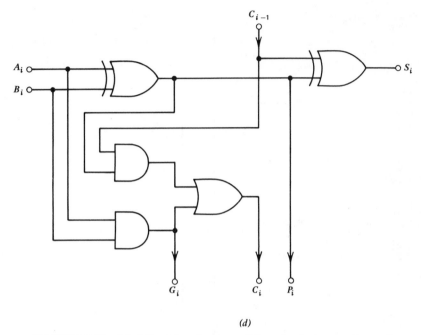

(d)

FIGURE 8.17. *(d)* **full adder circuit realizing the logic equations of** *b* **and** *c*.

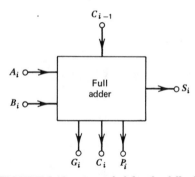

FIGURE 8.18. A symbol for the full adder.

Ripple Adders. An adder for two n-bit numbers can be realized by cascading n full adders in a *ripple adder* configuration.

EXAMPLE 8.5. Four full adders of Figure 8.18 are connected as a 4-bit ripple adder in Figure 8.19. Note that the G and P outputs are not utilized in a ripple adder.

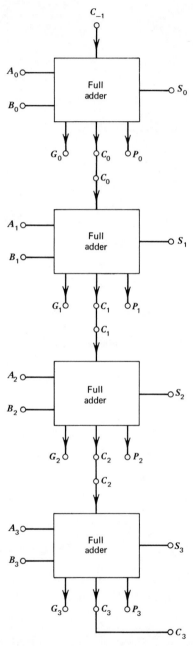

FIGURE 8.19. A 4-bit ripple adder using four full adders.

Two-Stage Adders. In the ripple adder, the least significant bits of two n-bit numbers A and B have to propagate through n carry circuits in order to reach the output; this results in long propagation delays for large n. Shorter propagation delays can be attained by use of 2-stage logic circuits for the generation of the carry output and of the sum bits. In what follows, a 2-stage logic circuit for the carry output of a 2-bit adder will be developed.*

For a 2-bit adder, the carry output can be realized (see Figure 8.17c) as

$$C_{i+1} = A_{i+1}B_{i+1} + (A_{i+1} + B_{i+1})C_i$$
$$= A_{i+1}B_{i+1} + (A_{i+1} + B_{i+1})[A_i B_i + (A_i + B_i)C_{i-1}],$$

which can also be written as

$$C_{i+1} = A_{i+1}B_{i+1} + A_{i+1}A_i B_i + A_{i+1}A_i C_{i-1} + A_{i+1}B_i C_{i-1}$$
$$+ B_{i+1}A_i B_i + B_{i+1}A_i C_{i-1} + B_{i+1}B_i C_{i-1}. \tag{8.15}$$

The resulting adder circuit with a 2-stage carry is shown in Figure 8.20. In principle, eq. (8.15) and the circuit of Figure 8.20 could also be extended to larger n; the complexity of the circuit, however, increases rapidly for large n.

Look-ahead Carry Circuits. Because of the long propagation times of the ripple adder and of the increasing complexity of 2-stage adders for large n, various other implementations of the carry have been investigated. One of these is the *look-ahead carry*, or *simultaneous carry*, circuit.

Consider the logic equations for the carry output of a 4-bit adder with no carry input $(C_{-1} = 0)$. From Figure 8.17c we can write the following equations:

$$C_0 = A_0 B_0. \tag{8.16}$$
$$C_1 = A_1 B_1 + (A_1 \oplus B_1)C_0,$$

which, by substituting eq. (8.16), can also be written as

$$C_1 = A_1 B_1 + (A_1 \oplus B_1)A_0 B_0. \tag{8.17}$$

Similarly, $\qquad C_2 = A_2 B_2 + (A_2 \oplus B_2)C_1,$

or, by utilizing eq. (8.17),

$$C_2 = A_2 B_2 + (A_2 \oplus B_2)A_1 B_1 + (A_2 \oplus B_2)(A_1 \oplus B_1)A_0 B_0.$$

Further, $\quad C_3 = A_3 B_3 + (A_3 \oplus B_3)C_2, \tag{8.18}$

---

* Realization of a 2-stage circuit to generate the sum bit is left to the reader as an exercise (see Problem 8.7).

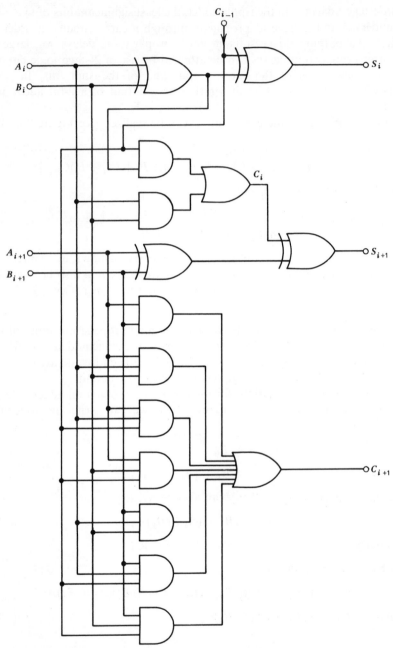

FIGURE 8.20. **A 2-bit adder circuit using 2-stage logic for the generation of the carry output.**

which, with eq. (8.18), becomes

$$C_3 = A_3 B_3 + (A_3 \oplus B_3)A_2 B_2 + (A_3 \oplus B_3)(A_2 \oplus B_2)A_1 B_1$$
$$+ (A_3 \oplus B_3)(A_2 \oplus B_2)(A_1 \oplus B_1)A_0 B_0. \quad (8.19)$$

In general,

$$C_m = A_m B_m + \sum_{i=1}^{m} \left[A_{m-i} B_{m-i} \prod_{j=1}^{i} (A_{m+1-j} \oplus B_{m+1-j}) \right]. \quad (8.20)$$

Alternatively, if a variable $p \equiv m + 1$ is introduced, eq. (8.20) can be written as

$$C_{p-1} = A_{p-1} B_{p-1} + \sum_{i=1}^{p-1} \left[A_{p-1-i} B_{p-1-i} \prod_{j=1}^{i} (A_{p-j} \oplus B_{p-j}) \right].* \quad (8.21)$$

Next we define for any integer k

$$P_k \equiv A_k \oplus B_k \quad (8.22)$$

and

$$G_k \equiv A_k B_k. \quad (8.23)$$

With these, eq. (8.20) and eq. (8.21) can be also written as

$$C_m = G_m + \sum_{i=1}^{m} G_{m-i} \prod_{j=1}^{i} P_{m+i-j} \quad (8.24)$$

and

$$C_{p-1} = G_{p-1} + \sum_{i=1}^{p-1} G_{p-1-i} \prod_{j=1}^{i} P_{p-j}. \quad (8.25)$$

In the case of a 4-bit adder with no carry input eqs. (8.16) through eq. (8.19) become

$$C_0 = G_0 \quad (8.26)$$

$$C_1 = G_1 + G_0 P_1 \quad (8.27)$$

$$C_2 = G_2 + G_1 P_2 + G_0 P_1 P_2 \quad (8.28)$$

$$C_3 = G_3 + G_2 P_3 + G_1 P_2 P_3 + G_0 P_1 P_2 P_3. \quad (8.29)$$

Equations (8.26) through (8.29) will be implemented here by AND and OR gates, although the usual realization in MSI circuits is by NAND gates. The resulting circuit, shown in Figure 8.21, is a 4-bit adder with look-ahead

---

* Note the similarity between the form of this equation and that of eq. (8.11).

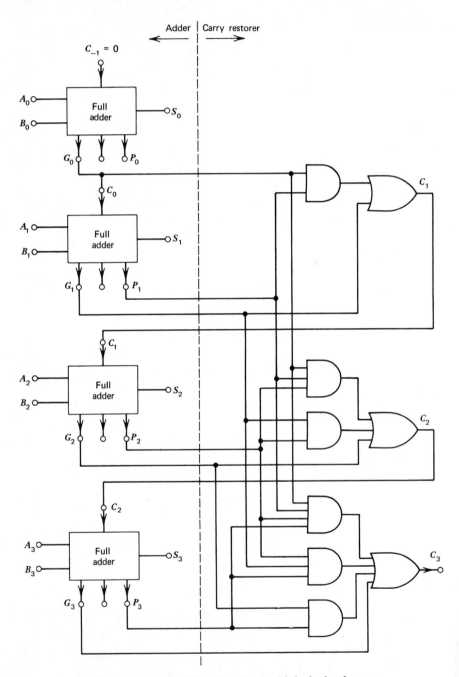

FIGURE 8.21.   A 4-bit adder circuit with look-ahead carry.

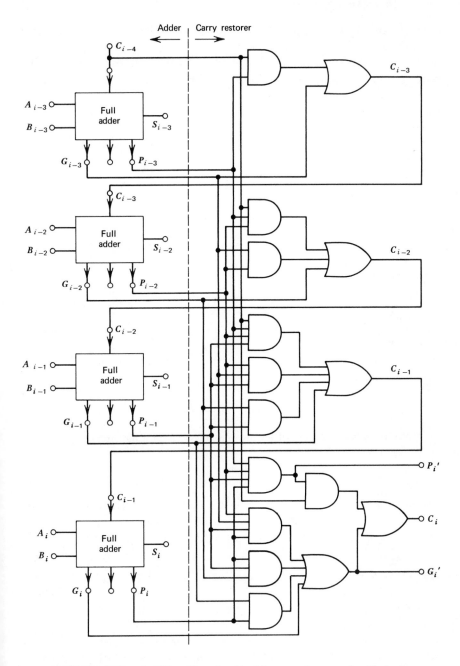

FIGURE 8.22. A 4-bit adder circuit with carry input and with look-ahead carry.

carry (or simultaneous carry); it consists of an adder circuit composed of full adders and of a *carry restorer circuit*. Note that the propagation delay of the carry in this scheme is not a function of n; it is the sum of the propagation delays of an EXCLUSIVE–OR, an AND, and an OR gate. This is longer than the propagation delay of a 2-stage adder, but it is shorter than that of the ripple adder.

To allow for the possibility of a carry input into the adder, the circuit of Figure 8.21 is extended to a 5-bit adder with the full adder for the least significant bits removed. The resulting circuit is shown in Figure 8.22, and a symbol of the circuit in Figure 8.23. Note that the subscript notation is

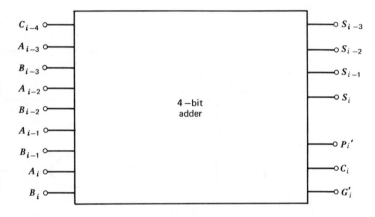

FIGURE 8.23.   A symbol for the 4-bit adder circuit with carry input of Figure 8.22.

somewhat more general than used before to facilitate further applications. In principle, the circuit of Figure 8.22 could be also extended to larger n; the complexity of the resulting circuit, however, becomes formidable for large n.

Hybrid Adders.   One way of extending the use of the 4-bit adder with look-ahead carry to $n > 4$ is to connect several of them in a ripple configuration. Such a circuit is designated a *hybrid adder*.

EXAMPLE 8.6.   Two 4-bit adders of Figure 8.23 are connected as an 8-bit hybrid adder in Figure 8.24. By utilizing carry output C_7, the circuit can be also extended to $n > 8$.

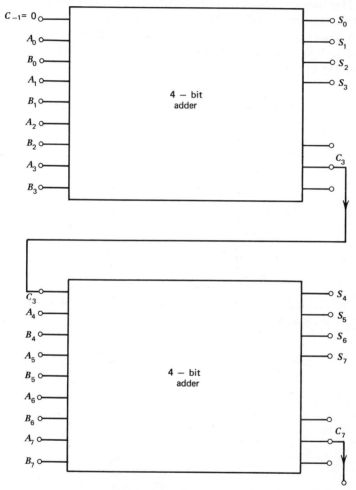

FIGURE 8.24. Block diagram of the hybrid adder using two 4-bit adders with carry input of Figure 8.23.

Multi-level Look-ahead Carry Circuits.   For large n, several 4-bit adders may also be connected in a configuration that is faster than a hybrid adder provided additional carry restorer circuits are used. Figure 8.25 shows a symbol for the carry restorer circuit depicted in the right half of Figure 8.22. This restorer circuit is utilized in the 16-bit adder circuit shown in Figure 8.26. The propagation delay of the resulting circuit is significantly shorter than that of a hybrid adder. The look-ahead carry principle can also be extended to $n > 16$.

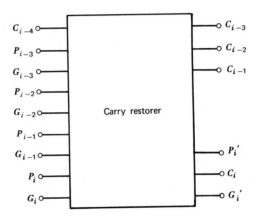

FIGURE 8.25.   Symbol for the carry restorer circuit shown in the right half of Figure 8.22.

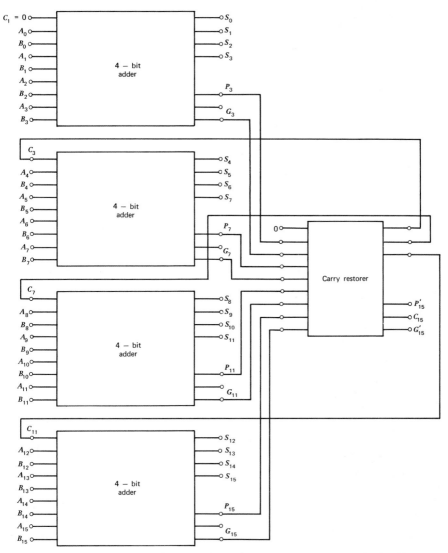

FIGURE 8.26. Block diagram of a 16-bit adder with 2-level look-ahead carry.

EXAMPLE 8.7. A 64-bit adder with a 3-level look-ahead carry circuit is outlined in Figure 8.27. It utilizes sixteen 4-bit adders of Figure 8.23 and five carry restorers of Figure 8.25. Interconnections within each of the four 16-bit adder portions are identical to those of Figure 8.26 and are not shown.

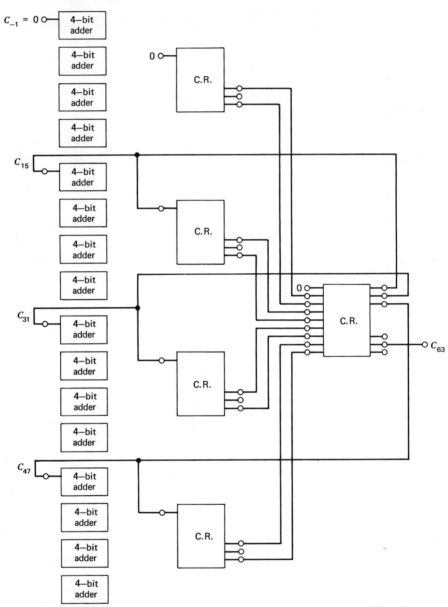

FIGURE 8.27. Block diagram of a 64-bit adder with 3-level look-ahead carry. "C.R." designates the carry restorer of Figure 8.25; "4-BIT ADDER" is that of Figure 8.23. Refer to Figure 8.26 for additional interconnections.

Sequential Adders. When the speed of operation is of no importance, the use of sequential adders performing the addition bit-by-bit sequentially may be considered. One such adder circuit is shown in Figure 8.28. The two n-bit numbers A and B to be added are stored in two shift registers from where they are shifted out *least significant bits first.* Sum S_i and carry C_i are generated by logic circuits in accordance with Figure 8.17b and c. The carry is stored in a flip-flop and is added to the next higher significant bits of A and B at the time of the subsequent clock pulse. The sum bits are stored in a shift register. The addition is completed in n clock pulses.

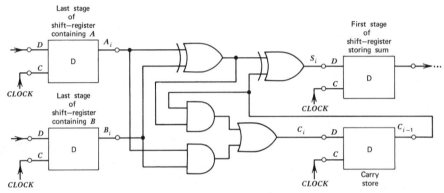

FIGURE 8.28. A sequential adder circuit.

BCD Adders. In many digital systems the inefficiencies of BCD codes are outweighed by their convenience. A BCD adder operating on decimal digits presented in an 8-4-2-1 code is shown in Figure 8.29. The left half of the circuit is a 4-bit binary adder; a ripple adder is shown in the figure, although a look-ahead carry circuit could be used equally well. The binary adder provides four binary sum bits S_0, S_1, S_2, S_3, and a carry bit C_3. These are translated to four BCD sum bits D_0, D_1, D_2, D_3, and a decimal carry bit C_{OUT} by a binary-to-BCD converter which is realized by the circuit shown in Figure 8.30; the operation of the circuit is as follows:

The "1" bit of the 8-4-2-1 BCD output, D_0, is always equal to the least significant bit of the binary sum, S_0. BCD bits D_1, D_2, D_3 are equal to binary bits S_1, S_2, S_3, respectively, only if the sum is below ten. Decimal carry out C_{OUT} is 1 if the binary carry $C_3 = 1$, or if the "8" bit S_3 and the "4" bit S_2 are 1, or if the "8" bit S_3 and the "2" bit S_1 are 1. Also, when $C_{OUT} = 1$, six is added to the resulting BCD sum $D_3 D_2 D_1 D_0$ by a 3-bit adder, and sixteen is subtracted by not using the carry output of this adder;

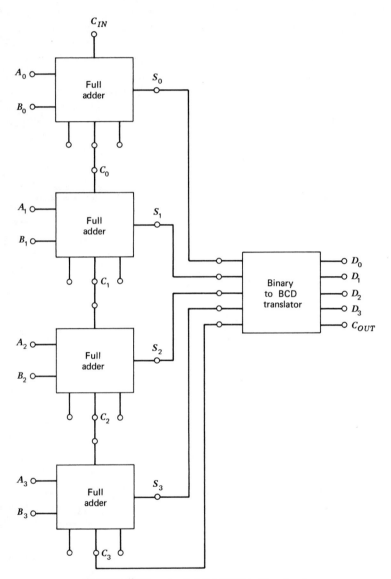

FIGURE 8.29. An 8-4-2-1 BCD adder circuit.

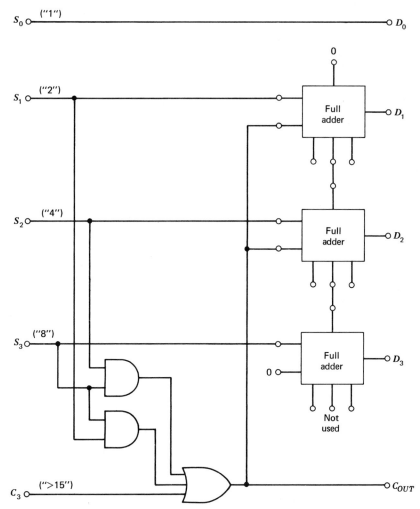

FIGURE 8.30.   A binary-to-BCD converter for the BCD adder of Figure 8.29.

the result is a subtraction of ten when $C_{OUT} = 1$, which is the desired correction since C_{OUT} represents a carry of ten.*

While the circuit of Figure 8.30 is straightforward, it does not necessarily yield the simplest configuration. Indeed, significant simplifications are possible when the don't care conditions are taken into consideration (see Problem 8.15).

* Binary-to-BCD conversion is discussed in detail in Chapter 9.

8.3   Subtractors

When a binary number B is to be subtracted from a binary number A, the result may be positive, negative, or zero, depending on the values of minuend A and subtrahend B. A subtractor has to provide the correct magnitude and sign of the difference $A - B$. The realization depends on the particular representation of negative numbers.

Sign-and Magnitude Subtractors.   When A and B are in sign-and-magnitude representation (see Chapter 2), first the signs and the relative magnitudes of minuend A and subtrahend B have to be inspected and the correct sign of $A - B$ established. Then the smaller magnitude is subtracted from the larger magnitude; this can be performed by a *full subtractor*. Assuming that $|A| > |B|$, the subtraction can be executed by a process based on the pencil-and-paper method as shown in the truth table of Figure 8.31a. This gives the *difference* bit D_i and the *subtraction carry*, or *borrow*, bit C_i; the corresponding Karnaugh maps are shown in Figure 8.31b and c. As was the

A_i	B_i	C_{i-1}	D_i	C_i
0	0	0	0	0
0	1	0	1	1
1	0	0	1	0
1	1	0	0	0
0	0	1	1	1
0	1	1	0	1
1	0	1	0	0
1	1	1	1	1

(a)

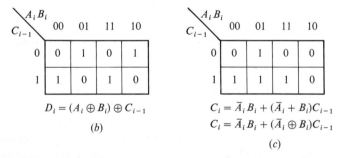

$$D_i = (A_i \oplus B_i) \oplus C_{i-1}$$

(b)

$$C_i = \bar{A}_i B_i + (\bar{A}_i + B_i)C_{i-1}$$
$$C_i = \bar{A}_i B_i + (\bar{A}_i \oplus B_i)C_{i-1}$$

(c)

FIGURE 8.31.   Subtraction of two 1-bit numbers and a carry input: (a) truth table, (b) Karnaugh map for the difference bit, (c) Karnaugh map for the carry bit.

case for the full adder, here too several realizations are possible, and full subtractors can be connected in a ripple subtractor or look-ahead-carry subtractor configuration. The realization of the full subtractor is left to the reader as an exercise.

1's-Complement Subtractors.   In this representation, the difference $A - B$ is generated by taking the 1's-complement of B [see eq. (2.16)] and adding it to A, and by providing an end-around carry connection. Such an arrangement, however, represents 0 as a negative number, and in some cases this is objectionable (this is the case, for example, in the switching circuit divider of Figure 8.36 discussed later in this chapter). Thus, a preferable implementation, shown in Figure 8.32a, realizes the subtraction $A - B$ as $-(B - A)$, and provides an end-around carry output when $A - B$ is positive *or* zero. A symbol of the circuit without the end-around carry connection is shown in Figure 8.32b.

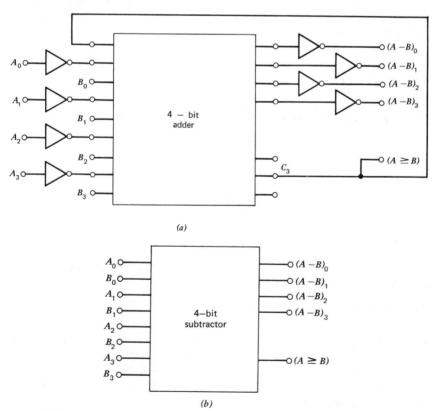

(a)

(b)

FIGURE 8.32.   A 4-bit 1's-complement subtractor utilizing the 4-bit adder of Figure 8.23: (*a*) circuit, (*b*) symbol.

2's-Complement Subtractors. In this representation, subtraction is performed by first taking the 2's-complement of the subtrahend, and then adding it to the minuend [see discussion following eq. (2.21)]. Thus, the following operations are performed: inversion of all bits of the subtrahend (by inverters), adding 1 to the result (by an adder), and adding the minuend (by another adder). Hence, a subtractor in this representation involves a substantial amount of circuitry. Because of the absence of the end-around carry, however, it is in many cases preferable to a 1's-complement subtractor.

8.4 Multipliers

Multiplication of two binary numbers M and N can follow the pencil-and-paper multiplication algorithm of two decimal numbers: the multiplicand is multiplied by successive digits of the multiplier and the appropriately positioned resulting partial products are added. In binary multiplication, however, each digit can be either 1 or 0; hence each partial product is either equal to the multiplicand or it is zero. In general, the multiplication of an m-bit multiplicand M by an n-bit multiplier N results in an $(m + n)$-bit product P.

EXAMPLE 8.8. Multiply $M = M_3 M_2 M_1 M_0 = 1101$ by $N = N_2 N_1 N_0$ $= 101$.

$$
\begin{array}{r}
1101 \times 101 \\
\hline
1101 \\
0000 \\
1101 \\
\hline
100001
\end{array}
$$

Multiplication, as was the case for addition and subtraction as well, can be realized by combinational or sequential circuits.

Switching-Circuit Multipliers. It is feasible to build multiplier circuits consisting of combinational logic circuits only: such circuits are designated *switching-circuit multipliers* or *switching-network multipliers*. A 4-bit by 3-bit multiplier following the principle of multiplication on paper (see Example 8.8 above) is shown in Figure 8.33. Multiplicand $M_3 M_2 M_1 M_0$ is first multiplied by N_0 by means of four AND gates. The result is added by the first 4-bit adder to the appropriately positioned partial product $M_3 M_2 M_1 M_0 \times N_1$, and the resulting sum is added by the second 4-bit adder to the appropriately positioned $M_3 M_2 M_1 M_0 \times N_2$. The circuit can

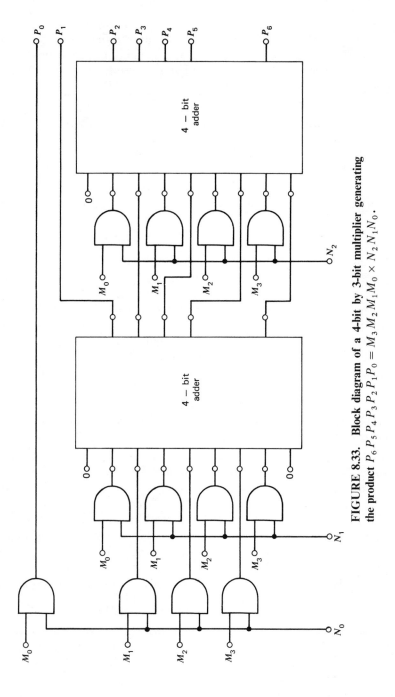

FIGURE 8.33. Block diagram of a 4-bit by 3-bit multiplier generating the product $P_6 P_5 P_4 P_3 P_2 P_1 P_0 = M_3 M_2 M_1 M_0 \times N_2 N_1 N_0$.

be extended to longer multiplicands and multipliers; the complexity of the resulting circuit, however, usually increases roughly in proportion to the *product* of the lengths of the multiplicand and the multiplier.

EXAMPLE 8.9.   The circuit of Figure 8.33 can be extended to accommodate an 8-bit multiplicand resulting in an 11-bit product by utilizing the carry outputs of the two 4-bit adders as carry inputs to two additional 4-bit adders. The resulting circuit consists of four 4-bit adders and of 24 2-input AND gates.

Thus, although combinational logic circuit multipliers are available for large n (64-bit by 64-bit circuits are not uncommon), such circuits are usually quite complex.

Sequential Multipliers.   Whenever constraints on the multiplication time are not severe, the use of sequential circuits, or combinations of combinational circuits and sequential circuits, should be considered. In principle, the complete multiplication operation could be performed on a bit-by-bit basis. With the availability of multi-bit adder circuits, however, several attractive compromise solutions are available; one of these is illustrated in Figure 8.34. Multiplier $N_2 N_1 N_0$ consists of $n = 3$ bits and is initially entered into a 3-stage shift register. Multiplicand $M_2 M_1 M_0$ consists of $m = 3$ bits and is initially entered into m bits of an $(m + n - 1 = 5)$-stage shift register with the bit sequence reversed. Partial products $B_4 B_3 B_2 B_1 B_0$ are added repetitively whenever the multiplier bit is 1; this is accomplished by a circuit designated as an *accumulator* to be discussed later.

EXAMPLE 8.10.   In the sequential multiplier circuit of Figure 8.34, $M_2 M_1 M_0 = 111$ $(= 7_{10})$ and $N_2 N_1 N_0 = 101$ $(= 5_{10})$. Thus initially $B_4 B_3 B_2 B_1 B_0 = 00111$.

The operation of the accumulator is controlled by the *accumulator gate* N_0. When N_0 is 0, the contents of the accumulator remain unchanged; when it is 1, $B_4 B_3 B_2 B_1 B_0$ is added to the contents. Since the accumulator gate is initially $N_0 = 1$, on the first clock pulse $P_5 P_4 P_3 P_2 P_1 P_0 = 000111$ is stored in the accumulator. The first clock pulse also shifts right $N_2 N_1 N_0$, resulting in an accumulator gate of $N_1 = 0$; and it also shifts right the 5-stage shift register, resulting in $B_4 B_3 B_2 B_1 B_0 = 01110$.

At the second clock pulse, the accumulator gate is $N_1 = 0$ and the contents of the accumulator remain unchanged. The second clock pulse, however, changes the accumulator gate to $N_2 = 1$ and shifts the 5-stage shift register, resulting in $B_4 B_3 B_2 B_1 B_0 = 11100$

The third clock pulse adds $B_4 B_3 B_2 B_1 B_0 = 11100$ to the contents of the accumulator which is 000111. The result is 100011 $(= 35_{10})$.

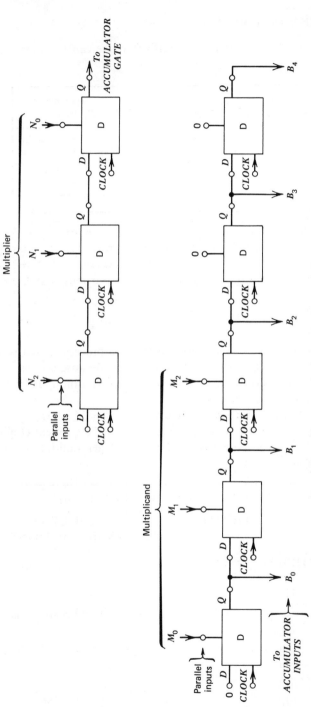

FIGURE 8.34. A 3-bit by 3-bit sequential multiplier using the accumulator shown in Figure 8.35.

The sequence of the operations described above can be summarized as follows:

Clock pulse	N_2 N_1 N_0	B_4 B_3 B_2 B_1 B_0	Accumulator P_5 P_4 P_3 P_2 P_1 P_0
0	1 0 1	0 0 1 1 1	0 0 0 0 0 0
1	? 1 0	0 1 1 1 0	0 0 0 1 1 1
2	? ? 1	1 1 1 0 0	0 0 0 1 1 1
3	? ? ?	1 1 0 0 0	1 0 0 0 1 1

Details of the accumulator circuit are shown in Figure 8.35. Six D flip-flops store the accumulated sum of partial products. The latest partial sum is added to this by the adder; the new sum is stored in the flip-flops when the accumulator gate is 1, or the previous sum is preserved when it is 0.

8.5   Dividers

The binary division of two numbers can follow a procedure that is similar to the division of two decimal numbers on paper. Because a binary "digit" can be only 0 or 1, however, each step is simpler, although the division operation will require more steps.

In general, when an integer m-bit dividend $M = M_{m-1}M_{m-2} \cdots M_0$ is divided by an n-bit divisor $N = N_{n-1}N_{n-2} \cdots N_0$, the quotient Q will have $q = m - n + 1$ digits provided that M is an integer multiple of N or Q is truncated to an integer.

EXAMPLE 8.11.   Division of 1010 $(= 10_{10})$ by 11 $(= 3_{10})$.

	QUOTIENT	RESULT OF SUBTRACTION
$11\overline{)1010} =$	???	
10		
-11		
$\overline{<0}$	0??	Not utilized because it is < 0
Remainder: 10		
101		
$-\ 11$		
$\overline{010}$	01?	Utilized
Remainder: 010		

$$\begin{array}{r} 0100 \\ -11 \\ \hline 0001 \end{array}$$ 011   Utilized

Remainder: 0001

Thus, ten is divided by three in three steps. The result is a quotient of 011 $(= 3_{10})$ and a remainder of 1.

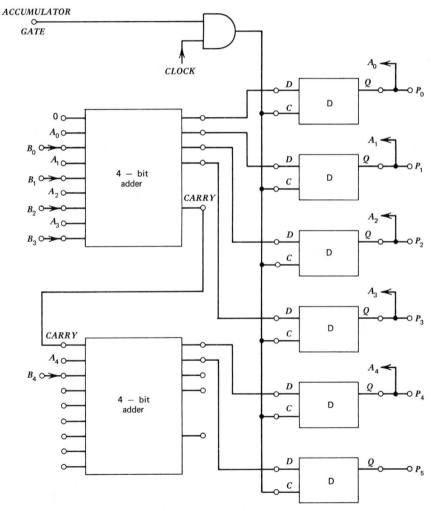

FIGURE 8.35. An accumulator for the multiplier of Figure 8.34. Connections of A_0, A_1, A_2, A_3, and A_4 are not drawn.

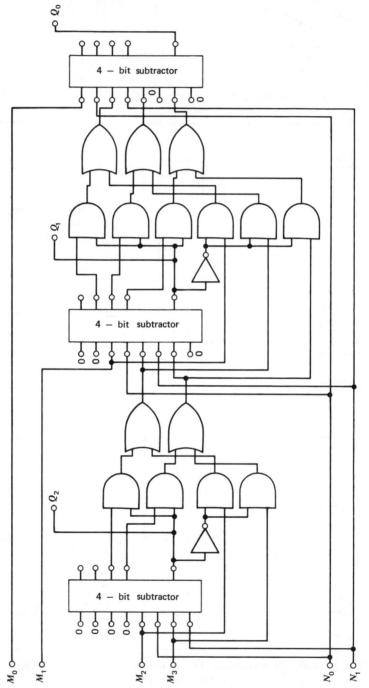

FIGURE 8.36. A 4-bit by 2-bit switching circuit divider.

Switching-Circuit Dividers. For small m and n, dividers consisting only of combinational logic circuits are practical. Such circuits are designated *switching-circuit dividers* or *switching-network dividers*. Figure 8.36 shows such a divider that uses the 4-bit subtractor of Figure 8.32b and that follows closely the procedure of the division on paper illustrated in Example 8.11 above. The divisor is subtracted from the remainder augmented by the appropriate bit of the dividend (in the first step the divisor is subtracted from the first n bits of the dividend). The resulting difference is carried as a remainder when it is not negative; otherwise the subtraction is not performed.

Sequential Dividers. As was the case for the switching circuit multiplier, the complexity of the switching circuit divider increases roughly as $m \times n$. Thus, for large m and n the use of sequential circuits, or combinations of sequential circuits and switching circuits, may be desirable. Many such circuits have been built, some of them based on further improvements of the pencil and paper method (see reference 1). In as much as the choice of a particular circuit is strongly influenced by the detailed requirements and by the available technology, these will not be described here; they are discussed in the references listed at the end of the book.

PROBLEMS

8.1 Write in the logic levels (0 or 1) at each point in Figure 8.4 if $A_2 A_1 A_0 = 111$ and $B_2 B_1 B_0 = 110$.

8.2 Modify the comparator circuit of Figure 8.4 such that its output is 1 when $A_2 A_1 A_0 \leq B_2 B_1 B_0$.

8.3 Realize the comparator circuit of Figure 8.11 using NAND gates and inverters only.

8.4 Extend the comparator circuit of Figure 8.11 such that, in addition to its $A > B$ output, it also provides an $A = B$ and an $A < B$ output.

8.5 Find the propagation delay from A_0 to the output in the hybrid comparator circuit of Figure 8.13 if each AND and OR gate has a propagation delay of 10 ns, and if the propagation delays of the inverters are negligible.

8.6 Demonstrate that the comparator circuits described in this chapter can be used with *any* binary code if the bits are connected such that for any j the weight of A_{j+1} is greater than or equal to that of A_j and the weight of B_{j+1} is greater than or equal to that of B_j.

8.7   Design a 2-stage logic circuit that generates the sum bit S_i of Figure 8.17*a*.

8.8   When the carry input of a full adder is permanently 0, such as in the top full-adder of Figure 8.21, simplifications of the circuit become possible; the simplified circuit is designated a *half-adder*. Draw a circuit diagram of the half-adder and show that a full-adder can be realized by using two half-adders and an OR gate.

8.9   The ripple adder of Figure 8.19 is synthesized with AND gates, OR gates, and inverters only. Find the propagation delay from carry input C_{-1} to carry output C_3, if the propagation delay of each AND gate and of each OR gate is 10 ns.

8.10   Find the propagation delay from the carry input C_{i-4} to carry output C_i in the adder circuit of Figure 8.22, if each AND gate and each OR gate has a propagation delay of 10 ns.

8.11   A 16-bit hybrid adder circuit utilizes four 4-bit adders of Figure 8.23. Each 4-bit adder is built by utilizing logic gates and inverters that have a propagation delay of 10 ns each. Find the propagation delay from input A_0 to carry output C_{15}.

8.12   The 16-bit adder circuit of Figure 8.26 is built by using logic gates and inverters that have a propagation delay of 10 ns each. Find the propagation delay from input A_0 to carry output C_{15}.

8.13   The 64-bit adder circuit of Figure 8.27 is built by using logic gates and inverters that have a propagation delay of 10 ns each. Find the propagation delay from input A_0 to carry output C_{63}.

8.14   Draw the block diagram of a 32-bit adder with 3-level look-ahead carry. Show all interconnections.

8.15   Utilize the don't care conditions and demonstrate that in the BCD adder circuit of Figure 8.30, D_1, D_2, and D_3 can also be implemented as $D_1 = S_1 \oplus C_{OUT}$, $D_2 = S_2 \overline{C_{OUT}} + S_1 S_2 + \overline{S}_1 C_3$, and $D_3 = (A_3 + B_3 + C_2)\overline{C_{OUT}} + S_1 C_3$, provided A_3, B_3, and C_2 are made available from the circuit of Figure 8.29.

8.16   Design a combinational logic circuit that realizes the full subtractor described by Figure 8.31*b* and *c*. Use EXCLUSIVE–OR, AND, and OR gates.

8.17   Draw a circuit diagram for the multiplier described in Example 8.9.

8.18   Design an 8-bit by 4-bit switching circuit multiplier. Provide a block diagram showing all interconnections.

8.19 Design an 8-bit by 4-bit sequential multiplier.

8.20 Extend the divider circuit of Figure 8.36 such that, in addition to the quotient, it also provides the remainder.

8.21 Design an *arithmetic logic unit* (ALU) that performs the following operations for 4-bit numbers A and B: $A + B$, $A - B$, $B - A$, $A > B$, $A \geq B$, $A = B$, $A \leq B$, and $A < B$. Aim at a design that uses a minimum amount of circuitry, but do not use sequential circuits. Use 1's-complement representation for the subtraction operations.

CHAPTER 9

Coding,
Code Conversion,
Error Detection
and Correction

Coding, as applied in this chapter, is a means of expressing in binary form a given quantity or a symbol. Decoding is the reverse process, i.e., it is the reconstitution of a number, a character of the alphabet, or a symbol from a string of binary digits. Code converters are combinational or sequential circuits translating characters that are presented in one coded form into a different code. This chapter discusses a variety of codes, their operational properties, code conversion techniques, and codes suitable for error detection and correction.

9.1 Applications of Codes

Human communication with a computer represents a typical example for the need of encoding and decoding. For arithmetic operations and for internal decision-making the computer uses devices having two stable states. The choice of such a binary system is based on the ready availability of reliable switching devices exhibiting two stable states. However, it takes more binary than decimal digits to express a given number, N. Assume that i is the number of binary digits and j the number of decimal digits required to express N. Then

$$N = 2^i = 10^j; \text{ hence } i \log_{10} 2 = j, \text{ and } i = 3.32 j. \tag{9.1}$$

Thus a binary number will have roughly 3.32 times as many digits as the same quantity expressed in base 10. The representation of alphabetic characters and other common symbols requires as many as six bits, 8-bit codes being also employed. The large number of digits required to express an information word in binary representation cannot be handled with convenience by human operators. Input devices to many computers are alpha-numeric requiring coding in a binary, or binary-coded-decimal form. The outputs from computers also require decoding to transform the binary to an alpha-numeric representation so that the results may be more conveniently interpreted by the operator.

When the input to a computer consists of numbers only, we can reduce somewhat the inconvenience associated with the many bits of the binary representation by use of a number base that is a power of 2, commonly octal (base $= 2^3$) or hexadecimal (base $= 2^4$). The number of digits is thus considerably reduced, as was shown in Figure 2.2.

A requirement for code conversion exists also within a computer. Some computers carry out their arithmetic operation in a binary-coded-decimal (BCD) form. Many BCD codes have been devised, each providing certain advantages in efficiency as far as arithmetic operations are concerned, depending on the circuit details and the algorithms used. Instructions, entered into a computer in a program language, also have to be translated into binary representation.

Communication between two pieces of digital equipment that use different codes requires code conversion. Redundancy, i.e., additional bits, may be included with the transmitted binary word to check that the information was not distorted due to noise in the transmission channel. Also, with more than one redundant bit, it is possible to detect the faulty bit and take corrective measures.

9.2 Binary-Coded-Decimal (BCD) Numbers

BCD codes are used more often than any other binary coded form, and there is a wide choice of such codes. A minimum of four bits is required to represent the ten decimal digits. Since four bits can represent 2^4 different states, there is a redundancy of six states in a 4-bit BCD code. The maximum number of BCD combinations is therefore

$$N_{BCD} = \frac{16!}{(16-10)!} \approx 2.8 \times 10^{10}. \tag{9.2}$$

Not all of these are unique codes. It has been shown that there are about 7.6×10^7 codes that could, in principle, represent the ten decimal digits.

However, most would be inconvenient to work with; and only a few have merited practical use.

8-4-2-1 BCD Code. Consider the binary number 1001 expressed in the positional notation:

$$N_2 = 2^3 \times 1 + 2^2 \times 0 + 2^1 \times 0 + 2^0 \times 1 = 1001_2 = 9_{10}.$$

The BCD representation in this form is called the naturally weighted, or 8-4-2-1, BCD since the weights are increasing powers of the base 2. The ten decimal digits coded in the 8-4-2-1 code and the six redundant states are shown in the second column of Figure 9.1. Each group of four bits representing a decimal digit can be used to express a multi-digit decimal number in a BCD weighting scheme, as shown in Figure 9.2.

Decimal	Binary 8-4-2-1	Excess-3	
0	0 0 0 0	0 0 1 1	
1	0 0 0 1	0 1 0 0	
2	0 0 1 0	0 1 0 1	
3	0 0 1 1	0 1 1 0	
4	0 1 0 0	0 1 1 1	
5	0 1 0 1	1 0 0 0	
6	0 1 1 0	1 0 0 1	
7	0 1 1 1	1 0 1 0	
8	1 0 0 0	1 0 1 1	
9	1 0 0 1	1 1 0 0	
10	1 0 1 0	1 1 0 1	Redundant states
11	1 0 1 1	1 1 1 0	
12	1 1 0 0	1 1 1 1	
13	1 1 0 1	0 0 0 0	
14	1 1 1 0	0 0 0 1	
15	1 1 1 1	0 0 1 0	

FIGURE 9.1.   8-4-2-1 natural BCD and excess-3 (XS-3) codes.

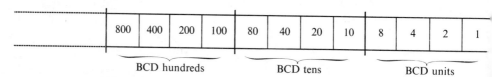

| 800 | 400 | 200 | 100 | 80 | 40 | 20 | 10 | 8 | 4 | 2 | 1 |

      BCD hundreds   BCD tens   BCD units

FIGURE 9.2.   Decimal weights of 8-4-2-1 natural BCD digits in a 3-digit decimal number.

EXAMPLE 9.1. Express the decimal number 759 in BCD notation.

Decimal:	7	5	9
BCD:	$\overparen{0111}$	$\overparen{0101}$	$\overparen{1001}$
Decimal weights:	Hundreds	Tens	Units

The codes discussed in this section are fixed-weight BCD codes, i.e., the position of each bit within a number has a definite weight, or multiplication factor, attached to it. All fixed-weight BCD codes have to satisfy the following rules

(a) The least significant weight must be 1 otherwise the decimal number 1 could not be coded.

(b) The next higher weight must be 1 or 2 to enable coding of decimal 2.

(c) The sum of the values of all weights must be at least nine and cannot be greater than 15 in a 4-bit code.

The Excess-3 Code. This code can be derived from the natural binary 8-4-2-1 code by adding 3 to each coded number, as is shown in the last column of Figure 9.1; hence it is also referred to as a *biased weighted code*. Note that the decimal digits 0 to 9 are represented by the middle ten of the sixteen entries of the 8-4-2-1 code. The desirable property of this code is that nine's-complementation can be performed by interchanging 1's and 0's. The arithmetic operation of subtraction thus reduces to complementation of the subtrahend followed by addition of the minuend. A simple inverter can perform the nine's-complementation of the excess-3 coded number.

The 4-2-2-1 Code. The nine's-complement of this code is also obtained from the inversion of the bits constituting the BCD number, as can be seen in the

Decimal	BCD 4-2-2-1	BCD 4-2-2-1	
0	0 0 0 0	0 0 0 0	
1	0 0 0 1	0 0 0 1	
2	0 0 1 0	0 1 0 0	
3	0 0 1 1	0 0 1 1	
4	1 0 0 0	1 0 0 0	← Axis of symmetry
5	0 1 1 1	0 1 1 1	
6	1 1 0 0	1 1 0 0	
7	1 1 0 1	1 0 1 1	
8	1 1 1 0	1 1 1 0	
9	1 1 1 1	1 1 1 1	

FIGURE 9.3. Two 4-2-2-1 BCD codes. Note the axis of symmetry for 9's-complementation.

two codes of Figure 9.3. Note that the second *and* the third bits have a weight of 2. The BCD code for 7 could be either 1101 or 1011 since the sum of the weights of the 1's yields 7 in either case, and a similar situation pertains to the code for decimal 2. A BCD code whose sum of the weights is 9 and which is symmetrically situated with respect to the axis as shown in Figure 9.3, will yield a nine's-complement through the inversion of the bits.

Although the 4-2-2-1 code is both weighted and nine's-complementing, it is not used as much in digital arithmetic as is the excess-3 code.

9.3 Unit-Distance Codes

Unit-distance codes derive their name from their property that only one bit changes between representations of two consecutive numbers. Most, but not

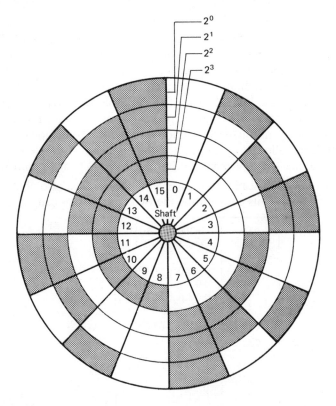

FIGURE 9.4. An encoder pattern for measurements of quantized angular displacements using the natural binary code which is prone to large reading errors.

all, unit-distance codes are unweighted. The numerical value of an unweighted-code number can be found only from a table showing the state assignments. Unit-distance codes are used in instrumentation, and in particular in quantized transducers such as shaft-encoders to measure angular displacement, or in linear encoders for measurement of linear displacement.

An example of a 4-bit binary coded pattern of an encoder of angular position is shown in Figure 9.4. Each of the four concentric rings, or zones, represents a binary weight, while the dark and white patterns represent 1's and 0's, respectively. The outer zone represents the lsb. Four transducers, mechanical or optical, are located along the radius to detect the pattern configuration corresponding to a binary number. Additional concentric zones would increase the resolution of the device. Consider, however, the ambiguity arising from the finite size of the detecting element traversing a position where more than one state is changing at the same time, e.g., between 7 and 8. If, for instance, the most significant bit is detected while the signals from the lesser bits are still in the "1" state, the binary number obtained will be $1111_2 = 15$ instead of $1000_2 = 8$. Thus the natural binary pattern shown in Figure 9.4 can lead to serious reading errors.

With the introduction of integrated circuits, and especially MSI's and LSI's, use of *unit-distance codes* in high-resolution applications has become attractive in providing codes without ambiguity. The coded quantities can be readily translated either into other, computer-compatible codes or into decimal numbers for display. The reflected-binary, or Gray code, shown in Figure 9.5a, is commonly used with such transducers; the resulting encoder pattern is shown in Figure 9.5b.

From Figure 9.5a we may also note that the Gray code is not suited as a BCD unit-distance code since a change of three bits occurs at the transition between 9 and 0. The unit-distance property is, however, preserved by another code derived from the middle ten entries of the Gray code, as shown in the third column of Figure 9.5a. The code in that column has been essentially shifted up three numbers, and hence it is known as the excess-3 (or XS-3) Gray code. Other unit-distance codes can be devised by simply traversing a Karnaugh map in unit steps between adjacencies.

9.4 Codes for Error Detection

Although transmission of information in digital form is much more immune to noise than transmission of analog information, the need for error checking does exist. Such errors may occur due to marginally operating

Decimal	Gray	XS-3 Gray
0	0 0 0 0	0 0 1 0
1	0 0 0 1	0 1 1 0
2	0 0 1 1	0 1 1 1
3	0 0 1 0	0 1 0 1
4	0 1 1 0	0 1 0 0
5	0 1 1 1	1 1 0 0
6	0 1 0 1	1 1 0 1
7	0 1 0 0	1 1 1 1
8	1 1 0 0	1 1 1 0
9	1 1 0 1	1 0 1 0
10	1 1 1 1	1 0 1 1
11	1 1 1 0	1 0 0 1
12	1 0 1 0	1 0 0 0
13	1 0 1 1	0 0 0 0
14	1 0 0 1	0 0 0 1
15	1 0 0 0	0 0 1 1

(a)

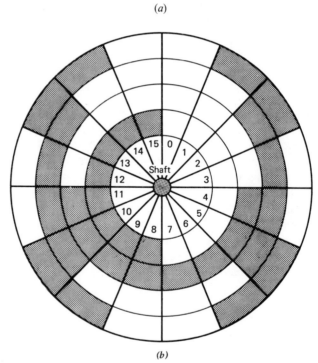

(b)

FIGURE 9.5. (a) The Gray code and the excess-3 Gray code. (b) An encoder pattern using the Gray code and producing a maximum reading error of $\pm\frac{1}{2}$ lsb.

equipment, occasional power-line transients, extraneous noise signals of high energy, intermittent faults in the transmitting or receiving equipment, or due to fading in the transmission.

When n bits of a code are utilized to represent 2^n units of information (*symbols*), no redundancy exists in such a code. A change of one bit in any symbol, due to electrical noise, for example, will result in another valid symbol that cannot be recognized as being faulty. A code in which symbols are separated by one bit only is called a *distance-1 code*. (Consider the differences between distance-1 codes and unit-distance codes).

An error check can be continuously executed provided there is sufficient redundancy in the coded message. An 8-4-2-1 BCD code has limited redundancy and thus can check only those errors that result in a binary number greater than 9, i.e., in states that are not valid in the code under discussion. The addition of a single bit to any code will make any two code symbols differ by at least two bits. Such an error detecting code is therefore called a *distance-2 code*.

Odd and Even Parity Error Check. Two distance-2 codes are common in digital applications for single bit error checking: the *odd parity* and the *even parity* code.

For the odd parity check, an additional bit is chosen such that the sum of all 1's in the transmitted word, including the check bit, is odd. This also can be expressed in the form of a modulo-2 sum as:

$$X_n \oplus X_{n-1} \oplus X_{n-2} \oplus \cdots \oplus X_1 \oplus X_0 \oplus p = 1, \qquad (9.3)$$

where p is the parity bit. Similarly, the even parity check results in an even sum i.e., the modulo-2 sum is 0:

$$X_n \oplus X_{n-1} \oplus X_{n-2} \oplus \cdots \oplus X_1 \oplus X_0 \oplus p = 0. \qquad (9.4)$$

Since $p \oplus p = 0$ and $1 \oplus 1 = 0$, it can be shown that

$$p(\text{odd}) = X_n \oplus X_{n-1} \oplus \cdots \oplus X_0 \oplus 1 \qquad (9.5)$$

and

$$p(\text{even}) = X_n \oplus X_{n-1} \oplus \cdots \oplus X_0. \qquad (9.6)$$

EXAMPLE 9.2. A message consisting of eight different symbols is to be transmitted in distance-1 code and in two distance-2 codes. Show the eight possible combinations for each code.

Symbol	Distance-1 code $X_2\ X_1\ X_0$	Distance-2 codes $X_2\ X_1\ X_0$	Distance-2 codes $X_2\ X_1\ X_0$
0	0 0 0	0 0 0 1	0 0 0 0
1	0 0 1	0 0 1 0	0 0 1 1
2	0 1 0	0 1 0 0	0 1 0 1
3	0 1 1	0 1 1 1	0 1 1 0
4	1 0 0	1 0 0 0	1 0 0 1
5	1 0 1	1 0 1 1	1 0 1 0
6	1 1 0	1 1 0 1	1 1 0 0
7	1 1 1	1 1 1 0	1 1 1 1
		odd parity bit	even parity bit

FIGURE 9.6. **Eight symbols represented in distance-1 and in distance-2 codes.**

Three bits are required for a distance-1 code to represent eight different symbols, as shown in column 1 of Figure 9.6. Columns 2 and 3 show distance-2 codes in which odd and even parity check bits, respectively, have been added.

Equations (9.3) through (9.6) are valid for any number of bits. Circuit implementation is easy since EXCLUSIVE-OR gates can serve as the basic circuit elements.

EXAMPLE 9.3. A transmission channel of eight information bits and one parity bit is shown in Figure 9.7. Two identical 9-bit parity circuits (MSI) are used, one to generate the parity bit at the transmitting end and the other to check the parity at the receiving end. A faulty symbol is recognized at the receiving end which may request its retransmission.

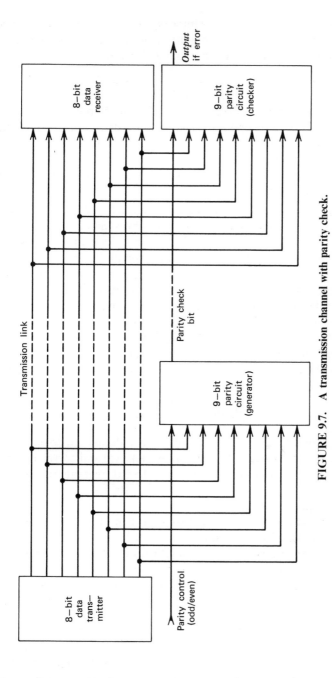

FIGURE 9.7. A transmission channel with parity check.

Odd parity in codes is more commonly used than even parity since the latter would not recognize a fault condition in which all zeros are transmitted. Properties of parity checks are summarized below:

(a)   The parity check holds when a number is shifted in a shift-register provided no new 1's are added.

(b)   It is independent of the position of a binary point.

(c)   Its error detecting power is increased if a detection circuit is added to recognize unallowed code combinations in n-bit codes that utilize less than 2^n symbols.

(d)   It will not detect an error in which an even number of bits has been changed.

EXAMPLE 9.4.   A message consisting of 8-bit symbols is transmitted at a rate of 10^5 symbols/sec via a circuit similar to that of Figure 9.7. It has been established through tests that the probability, P_1, of a single bit error occuring in a symbol is $P_1 = 2 \times 10^{-8}$. Determine (a) the number of errors received per second, and (b) the improvement in the error rate after a parity bit has been added to the transmission channel.

(a)   The number of errors received per second is $P_1 \times$ (number of symbols/sec) $= 2 \times 10^{-3}$/sec, i.e., 7.2 errors/hr. (b) After addition of a parity check, the message consists of 9-bit symbols. The probability, P_2, of a double error in a 9-bit symbol is equal to $(9/8 \times P_1)^2$ times the number of possible combinations of 2 bits in a 9-bit symbol. Thus

$$P_2 = P_1{}^2 (9/8)^2 \frac{9!}{2!\,7!} = 1.8 \times 10^{-14}.$$

The number of error symbols per second due to double-bit errors only has to be considered since single-bit errors can be detected and corrective measures can be taken (e.g., retransmission). The undetected error rate is thus $P_2 \times$ rate of transmission $= 1.8 \times 10^{-14} \times 10^5 = 1.8 \times 10^{-9}$/sec, which is equivalent to one error in ≈ 6000 days.

Thus the addition of the ninth (parity) bit in the transmission channel, as shown in Figure 9.7, reduced the undetected error rate by a factor of 10^6. An odd number of error bits in a symbol is detected by a parity check, as can be seen from the last two columns of Figure 9.6. The probabilities of even numbers of error bits having more than two errors in a symbol is very small compared with double-bit errors and can be neglected for all practical purposes.

Other Error Detecting Codes.   The *two-out-of-seven code*, Figure 9.8a, derives its name from the two 1's in a 7-bit code. The redundancies offer easy error detection: specifically we note two groups of binary numbers.

Decimal	Weights 5 0	4 3 2 1 0
0	0 1	0 0 0 0 1
1	0 1	0 0 0 1 0
2	0 1	0 0 1 0 0
3	0 1	0 1 0 0 0
4	0 1	1 0 0 0 0
5	1 0	0 0 0 0 1
6	1 0	0 0 0 1 0
7	1 0	0 0 1 0 0
8	1 0	0 1 0 0 0
9	1 0	1 0 0 0 0

(a)

Decimal	Weights 7 4 2 1	Added bit
0	1 1 0 0	0
1	0 0 0 1	1
2	0 0 1 0	1
3	0 0 1 1	0
4	0 1 0 0	1
5	0 1 0 1	0
6	0 1 1 0	0
7	1 0 0 0	1
8	1 0 0 1	0
9	1 0 1 0	0

(b)

FIGURE 9.8. (a) 2-out-of-7 error detecting code. (b) 2-out-of-5 error detecting code.

The first 2-bit group (msb) has weights 0 or 5; the position of the 1 in the 5-bit group carries weights between 0 and 4. Each group contains exactly one " 1." Note the axis of symmetry between numbers 4 and 5.

Although this code is inefficient in bit utilization it is easily applicable for arithmetic operations. Note that the " 1 " in the second group shifts to the left as the number increases, which simplifies algorithms for arithmetic operations.

The *two-out-of-five code*, shown in Figure 9.8b, is a 7-4-2-1 weighted BCD code with a parity bit added to obtain exactly two 1's and three 0's in each symbol. Note that numeral zero is represented by the code 1100 to achieve the required count of 1's and 0's. All single errors are detected in this code; in addition those double errors are detected that generate the six symbols which are not utilized in this BCD code. This additional redundancy enables detection of 40% of double errors.

9.5 Codes for Error Correction

One method of preventing errors in messages utilizes the *triple modular redundancy* (TMR) in which all messages are transmitted in three parallel channels. At the receiving end the messages are compared bit-by-bit in a voter or majority circuit (see Section 4.3), and the majority decision is assumed to yield the correct information. Such systems, although quite expensive, protect the data from successive errors in one of the channels.

Should an identical error occur *simultaneously* in two channels, e.g., due to transmission fading, it will pass undetected.

Block Parity Codes. The parity check of Section 9.4 was discussed in the context of error checks performed on each individual symbol. It is a powerful tool in error detection as long as the probability is negligibly low for two corrupted bits to occur in any symbol. However, transmission of messages experience bursts of noise which generate double-bit error rates that are much higher than those calculated from measurements of single-bit error rates. Such errors can be more efficiently detected with *block parity (geometric) codes.*

A geometric code using *horizontal and vertical parity* bits is shown in Figure 9.9. The message to be transmitted is arranged in an array of n rows

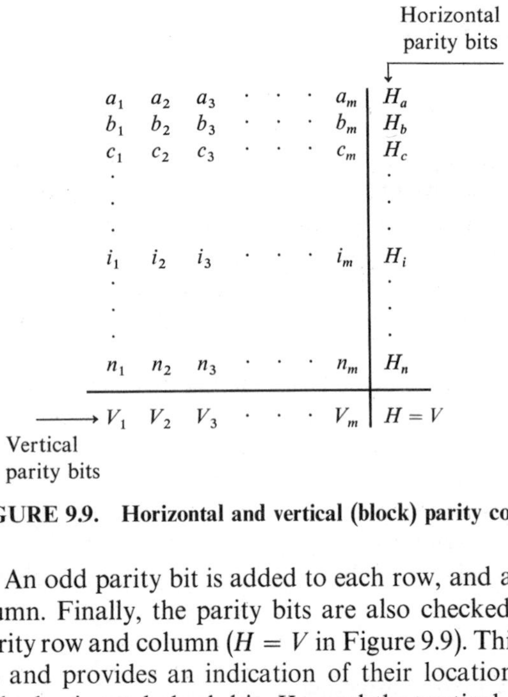

FIGURE 9.9. Horizontal and vertical (block) parity code.

and m columns. An odd parity bit is added to each row, and an even parity bit to each column. Finally, the parity bits are also checked at the intersection of the parity row and column ($H = V$ in Figure 9.9). This code detects all single errors and provides an indication of their location through the intersection of the horizontal check bit, H_h, and the vertical check bit, V_v, where the subscripts h and v refer to the row and column, respectively, in which the error has occurred. A correction of the corrupted bit can thus be executed once its location has been isolated.

Note that an even number of errors that are located symmetrically along two coordinates will remain undetected. For example, an error in b_1 and b_3 occurring simultaneously with an error in i_1 and i_3 will not be detected by the parity checks since the modulo-2 sum of the rows and columns has not changed.

The code is inexpensive to implement and is widely used in magnetic tapes and other computer peripheral equipment.

Distance-3 Codes. We have seen in Section 9.4 that one additional bit, the parity check bit, was required to detect an error in a symbol. We said that this type of code has a distance $d = 2$; in other words, d is the number of elements in which the code symbols differ. Correction of errors is not possible with distance-2 codes; but a distance of at least 3, $d \geq 3$, is required to allow correction of one erroneous bit in a code symbol. For correction of two errors $d = 5$, for three errors $d = 7$, etc., are required.

The *Hamming code*, named after its inventor, is a distance-3 code capable of detecting and correcting one error in a word. The code is constructed by incorporating several check bits, C, with the information bits, I. The check bits within the composite word occupy positions $1, 2, 4, \ldots 2^i$, as shown below:

Position	7	6	5	4	3	2	1
BIT	I_7	I_6	I_5	C_4	I_3	C_2	C_1

The number of check bits required for a given word length is shown in Figure 9.10a. If an error should occur in the word during transmission, its location will be detected at the receiving end through the decimal equivalent of a binary number that is composed of the *check bits* only. The error may be either in a C-bit or in an I-bit. The truth table of Figure 9.10b shows the check bit pattern and the respective position of the bit that contains an error for three check bits. C_1 will be 1 if there is an error in position 1, 3, 5, or 7. C_2 will identify errors in positions 2, 3, 6, or 7. When C_4 is 1, an error in

Number of check bits C	Number of information bits I	Total word length $N = C + I$
2	1	3
3	4	7
4	11	15
5	26	31
6	57	63
7	120	127

(a)

FIGURE 9.10. Hamming code: (a) check bits required for a given word length. (*See next page for* (b).)

Bit position containing error	Number represented by check bits $C_4\ C_2\ C_1$
No error	0 0 0
1	0 0 1
2	0 1 0
3	0 1 1
4	1 0 0
5	1 0 1
6	1 1 0
7	1 1 1

(b)

FIGURE 9.10. (b) **truth table for identification of error position.**

position 4, 5, 6, or 7 is indicated. If the check bits are 000, the composite word is error-free. Thus C bits can identify 2^C states of which one state represents no error, while $2^C - 1$ states identify the error bit position through the decimal equivalent of the word composed of the C bits only with proper binary weights attached.

EXAMPLE 9.5.   Calculate the number of check bits, C, required to correct all possible single error bits in a word containing I information bits.

The total word length to be checked is $(C + I)$ bits. In the previous discussion we have seen that $2^C - 1$ error locations can be identified by C check bits. Thus

$$2^C - 1 \geq C + I. \tag{9.7}$$

Equation (9.7) is represented in Figure 9.10a for I up to 120 bits.

EXAMPLE 9.6.   Construct a Hamming code for the 4-bit word 1101, using even parity check bits.

Three check bits are required. Taking C_4 first, we note from Figure 9.10b that this bit checks bits 4, 5, 6, and 7. For even parity we assign to C_4 a value such that the modulo-2 sum

$$C_4 \oplus I_5 \oplus I_6 \oplus I_7 = 0.$$

This can be manipulated algebraically to obtain C_4, by a mod-2 addition of C_4 to both sides of the equation. The reader can easily verify that $C_4 \oplus C_4 = 0$. Thus, the check bit C_4 for the word 1101 is

$$C_4 = I_5 \oplus I_6 \oplus I_7 = 1 \oplus 1 \oplus 0 = 0.$$

Similarly, it can be shown that

$$C_2 = I_3 \oplus I_6 \oplus I_7 = 1$$

and

$$C_1 = I_3 \oplus I_5 \oplus I_7 = 0.$$

The above step-by-step procedure is shown in Figure 9.11.

Position $\rightarrow$	7	6	5	4	3	2	1
Bit $\rightarrow$	I_7	I_6	I_5	C_4	I_3	C_2	C_1
Word	1	1	0		1		
Insert $C_4 = I_5 \oplus I_6 \oplus I_7$	1	1	0	0	1		
Insert $C_2 = I_3 \oplus I_6 \oplus I_7$	1	1	0	0	1	1	
Insert $C_1 = I_3 \oplus I_5 \oplus I_7$	1	1	0	0	1	1	0

FIGURE 9.11. Constructing a Hamming code for the information word 1101.

At the receiving end the coded word undergoes an even parity check. If the message has been transmitted error free, then

$$C_4 \oplus I_5 \oplus I_6 \oplus I_7 = 0, \tag{9.8}$$

$$C_2 \oplus I_3 \oplus I_6 \oplus I_7 = 0, \tag{9.9}$$

$$C_1 \oplus I_3 \oplus I_5 \oplus I_7 = 0. \tag{9.10}$$

If, however, a different bit pattern has been received, then the bit position where the error has been introduced is found from the decimal number formed from the check bits. Assume that the Hamming code constructed above has been received to read 1110110. Is the message correct? If it is incorrect, what is the position of the erroneous bit?

Equations (9.8) through (9.10) yield $C_4 = 1, C_2 = 0, C_1 = 1$; and the error is thus in bit position $101 = 5_{10}$. To correct the error, bit 5 has to be complemented. Removal of the check bits yields now the original message.

Hamming Code Transmitter and Receiver. The implementation of a Hamming code *transmitter* with even parity is shown in Figure 9.12. The information and check bits are stored in a 7 bit shift-register.* The information bits are loaded via four AND gates; the check bits are generated by

---

* See also the error correcting code generator, shown in Problem 9.29, using a 4-bit shift-register. Generation of Hamming codes using LSI read-only memories (ROM's) is discussed in Section 11.8.

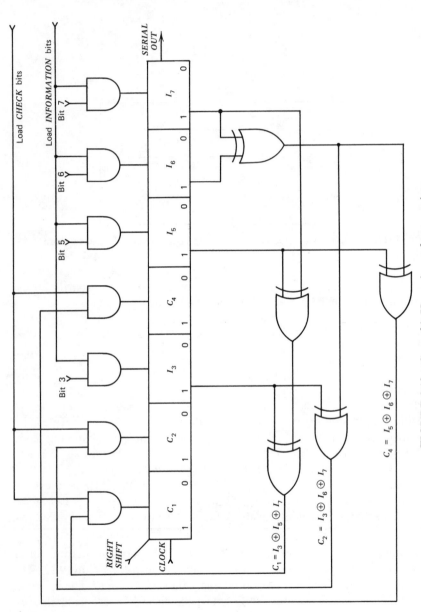

FIGURE 9.12. Seven-bit Hamming code transmitter.

five EXCLUSIVE–OR gates (see left column of Figure 9.11) and loaded in the appropriate bit positions via three additional AND gates.

The components required for a Hamming code *receiver* are: (*i*) a 7-bit register; (*ii*) EXCLUSIVE–OR gates for checking the message; (*iii*) a complementing circuit to correct an error bit when required; and (*iv*) control circuitry to generate the appropriate sequence of operations. The design of the circuit is left to the reader as an exercise.

The distance-3 Hamming code has the limitation that one error only can be corrected; a correction executed in case of a double error may introduce additional errors.

9.6   Code Converters

Code converters are required at the interface of man-machine communication and for transmission of data between various digital subsystems that do not employ the same code. As an example, Figure 9.13 illustrates a control

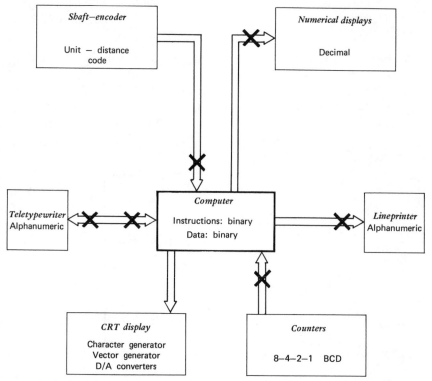

FIGURE 9.13.   A computer system with peripheral equipment employing various codes.

computer and its peripheral equipment. Assume that the internal operation of the computer uses the natural binary code while the peripheral equipment uses, for good technical reasons, a variety of codes as shown in the drawing. Communication of data and/or instructions in a system such as Figure 9.13 would be chaotic if code converters were not interposed in the transmission lines; locations of these converters are marked by × 's. Note that a bi-directional transmission channel requires two converters.

In this section we demonstrate conversion of codes by combinational methods. The procedure is applicable for conversion between any codes. In practice it is limited to 6-variable codes, since the Karnaugh maps employed in the process become unwieldy for larger number of variables.

EXAMPLE 9.7.   Convert the excess-3 Gray code to the natural BCD code. The procedure is as follows:

1. Draw truth tables for both codes (Figure 9.14). Mark the variables of the binary code B_3, B_2, B_1, B_0 in descending order of binary weights. The corresponding variables of the excess-3 Gray code mark G_3, G_2, G_1, G_0.

2. For each variable B_i draw a Karnaugh map (Figure 9.15). Look up those B_i terms that have a binary value 1 in the table. Enter into the map the corresponding minterms from the excess-3 Gray code table. For example, in the last term of the BCD code B_3 is 1; and the corresponding excess-3 Gray term in 1010, i.e., minterm m_{10}. Enter this term in the map as 1. In the BCD code for 8_{10}, B_3 is

	XS-3 Gray	8-4-2-1 BCD
Decimal	$G_3\ G_2\ G_1\ G_0$	$B_3\ B_2\ B_1\ B_0$
0	0 0 1 0	0 0 0 0
1	0 1 1 0	0 0 0 1
2	0 1 1 1	0 0 1 0
3	0 1 0 1	0 0 1 1
4	0 1 0 0	0 1 0 0
5	1 1 0 0	0 1 0 1
6	1 1 0 1	0 1 1 0
7	1 1 1 1	0 1 1 1
8	1 1 1 0	1 0 0 0
9	1 0 1 0	1 0 0 1

FIGURE 9.14.   Truth table for the excess-3 Gray and the natural BCD codes.

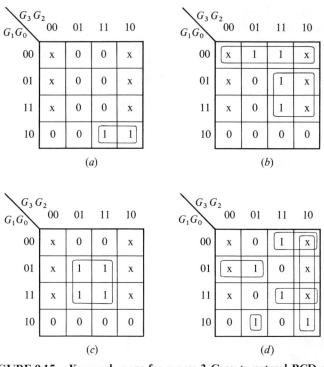

FIGURE 9.15. Karnaugh maps for excess-3 Gray-to-natural BCD code conversion: (*a*) map for B_3, (*b*) map for B_2, (*c*) map for B_1, (*d*) map for B_0.

also 1; and its corresponding excess-3 Gray term is 1110, i.e., minterm m_{14}. Enter this minterm in the appropriate location of the map.

3. Mark the six redundant terms of the excess-3 code as *don't care conditions* since these terms are not available as valid inputs to the combinational network. The don't care conditions are:

$$\Sigma d(0,1,3,8,9,11).$$

4. Obtain minimized terms from the maps of the respective B_i's.

$$B_3 = G_3 G_1 \bar{G}_0,$$

$$B_2 = G_3 G_0 + \bar{G}_1 \bar{G}_0,$$

$$B_1 = G_2 G_0,$$

$$B_0 = G_3 \bar{G}_2 + G_3 \bar{G}_1 \bar{G}_0 + \bar{G}_3 \bar{G}_1 G_0 + G_3 G_1 G_0 + \bar{G}_3 G_2 G_1 \bar{G}_0.$$

An alternative solution for B_0 can be derived noting that two minterm pairs, namely m_5, m_{12} and m_6, m_{15} are diagonally located with respect to each other. This topology suggests an implementation with EXCLUSIVE–OR gates, as was shown in Section 4.11. Thus, ignoring don't care terms m_2, m_8, and m_{11} we obtain, after some algebraic manipulation,

$$B_0 = G_3 \bar{G}_2 + G_2(G_3 \oplus G_1 \oplus G_0).$$

Binary-to-Gray Code Converters. Conversion between the two codes shown in Figure 9.16 can be done by the map method, as shown in Example 9.7.

Decimal	Binary $B_3\ B_2\ B_1\ B_0$	Gray $G_3\ G_2\ G_1\ G_0$	
0	0 0 0 0	0 0 0 0	
1	0 0 0 1	0 0 0 1	
2	0 0 1 0	0 0 1 1	
3	0 0 1 1	0 0 1 0	
4	0 1 0 0	0 1 1 0	
5	0 1 0 1	0 1 1 1	
6	0 1 1 0	0 1 0 1	
7	0 1 1 1	0 1 0 0	← Axis of
8	1 0 0 0	1 1 0 0	symmetry
9	1 0 0 1	1 1 0 1	
10	1 0 1 0	1 1 1 1	
11	1 0 1 1	1 1 1 0	
12	1 1 0 0	1 0 1 0	
13	1 1 0 1	1 0 1 1	
14	1 1 1 0	1 0 0 1	
15	1 1 1 1	1 0 0 0	

FIGURE 9.16. Truth table for the binary and Gray codes.

However, there are certain symmetries in the binary and the (reflected) Gray code that result in a set of simple rules applicable to any number of bits in the codes.

It can be verified by inspection that

$$G_n = B_n, \tag{9.11}$$

$$G_{n-1} = B_n \oplus B_{n-1},$$

$$G_i = B_{i+1} \oplus B_i, \tag{9.12}$$

$$G_0 = B_1 \oplus B_0.$$

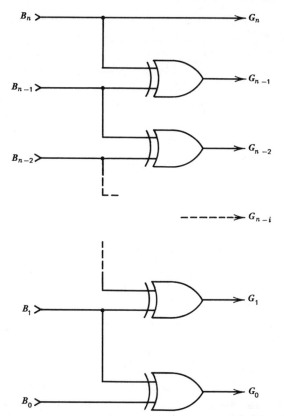

FIGURE 9.17.   Implementation of a binary-to-Gray-code converter using EXCLUSIVE–OR gates.

Thus, to convert an n-bit character, $(n - 1)$ EXCLUSIVE–OR gates are required. A binary-to-Gray converter realization based on the above equations is shown in Figure 9.17.

Gray-to-Binary Code Converters.   The symmetrical properties of these two codes result in the following conversion equations:

$$B_n = G_n,$$

$$B_{n-1} = G_n \oplus G_{n-1} = B_n \oplus G_{n-1},$$

$$B_{n-2} = G_n \oplus G_{n-1} \oplus G_{n-2} = B_{n-1} \oplus G_{n-2},$$

$$B_k = G_n \oplus G_{n-1} \oplus \cdots \oplus G_k = B_{k+1} \oplus G_k = \sum_{i=k}^{n} G_i(\bmod 2), \qquad (9.13)$$

$$B_0 = G_n \oplus G_{n-1} \oplus \cdots \oplus G_0 = B_1 \oplus G_0.$$

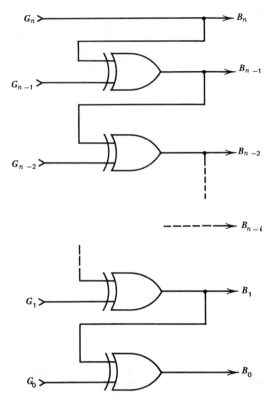

FIGURE 9.18. Implementation of a Gray-code-to-binary converter using EXCLUSIVE–OR gates.

A converter for n bits using $(n - 1)$ EXCLUSIVE–OR gates is shown in Figure 9.18.

Note that the number of EXCLUSIVE–OR gates is the same for either binary-to-Gray or Gray-to-binary conversion of a given number of bits. It is practical, therefore, to design a gating network with a control circuit to utilize the same EXCLUSIVE–OR gates for conversion from binary-to-Gray code, and vice versa. Specifications for such a converter are given in Problem 9.16.

Serial Gray to Natural Binary Code Converters. Suppose that the Gray code, available in serial form with msb first, is to be converted to the natural binary code. The nature of the requisite circuit may be deduced from the set of eq. (9.13) as follows: 1. The msb's of both codes have the same binary values i.e., $B_n = G_n$. 2. Having obtained B_n, examine the next serial bit of the Gray code and generate the EXCLUSIVE–OR

$B_{n-1} = B_n \oplus G_{n-1}$. In general $B_k = B_{k+1} \oplus G_k$; in other words, if $B_{k+1} = 1$, then the next lower significant bit $B_k = \overline{G}_k$, otherwise $B_k = G_k$. The natural binary code is thus constituted serially bit-by-bit starting with the msb.

EXAMPLE 9.8.   Convert the Gray code number 1 1 0 1 1 0 0 1 0 1 0 0 1 to a natural binary number.

Gray code: 1 1 0 1 1 0 0 1 0 1 0 0 1

Natural binary code: 1 0 0 1 0 0 0 1 1 0 0 0 1

It can be seen that a toggle (T) flip-flop executes the conversion shown in this example. The initial state of the flip-flop must be 0, and the Gray code must be applied msb first. The latter condition is contrary to the sequence used in most computer applications, which limits the applicability of this type of serial converter.

9.7   Natural-Binary-to-BCD Converters

The extensive use of the decimal and binary number systems in digital equipment calls for efficient conversion methods between the two number representations. The Karnaugh map method of code conversion, discussed in Section 9.6 is practical for a relatively small number of digits. In this and the following two sections we shall discuss several conversion systems, some of which will also demonstrate the power of MSI and LSI devices in these applications.

Serial Code Converters Using Counters.   This method is applicable when a relatively long time is available to execute the conversion. The circuit, shown in Figure 9.19, functions as follows: the binary number to be converted is loaded into a binary "down-counter" via parallel inputs. A gated clock is applied simultaneously to the inputs of the binary and the BCD counters. At each clock pulse the binary counter is *decremented* by 1, while the BCD counter is *incremented* by 1. When the "down-counter" has reached zero, the output of the AND gate stops the clock. The contents of the BCD counter at this time are the BCD equivalent of the binary number previously loaded into the binary "down-counter."

A slight modification of the conversion method described above utilizes a binary "up-counter." This counter is parallel loaded at its input with the *2's-complement* of the binary number that is to be converted to the BCD form. The remainder of the logic network is as shown in Figure 9.19.

Note that this serial conversion method is applicable for any pair of weighted codes. It is merely necessary to code the counter sequences with the appropriate weights. Thus, for example, a similar system can be used for

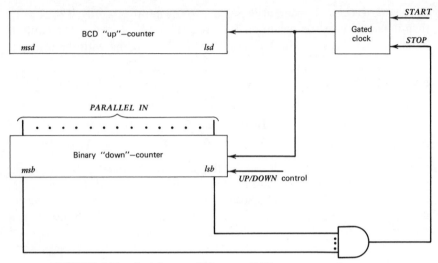

FIGURE 9.19.   Serial natural-binary-to-BCD converter using counters.

BCD-to-binary conversion with the functions of the two counters in Figure 9.19 interchanged.

Serial code converters are practical either for a small number of bits, or for a large number of bits if conversion time is not a restricting parameter.

EXAMPLE 9.9.   Calculate the time, t, required to convert a 16-bit binary number to BCD form using counters with a clock frequency $f = 50$ MHz.

The largest number expressed by n bits is $2^n - 1$. Therefore, $t = (2^n - 1)/f = (6.5536 \times 10^4 - 1)/(50 \times 10^6) = 1,310.7 \ \mu\text{sec}$.

One Clock-Period per Bit Serial Converters.   These conversion methods are derived from the algorithms for conversion of numbers of different bases as discussed in Section 2.4. Let N_B be an integer binary number of n bits, expressed in positional notation

$$N_B = b_{n-1}b_{n-2}\cdots b_1 b_0.$$

The decimal value of N_B is given by the sum

$$N_D = \sum_{i=0}^{n-1} b_i 2^i = b_{n-1}2^{n-1} + b_{n-2}2^{n-2} + \cdots + b_1 2^1 + b_0.$$

This can be expressed in a different form through repeated multiplication by 2,

$$N_D = (\{[(b_{n-1} \cdot 2 + b_{n-2}) \cdot 2 + b_{n-3}] \cdot 2 + \cdots\} + b_1) \cdot 2 + b_0. \quad (9.14)$$

The form of eq. (9.14) suggests the shift-register as a convenient element for number conversion, since each shift to the left multiplies the shifted number by a factor of 2.

To achieve binary-to-BCD conversion, a shift-register of length $4n$ is divided into n groups of four bits each as shown in Figure 9.20. Each group of four bits represents a decade, and the weight assignments in the decade are 8-4-2-1. (Any BCD assignment is permitted.) The position of each decade within the $4n$-bit shift-register determines its decimal weight, e.g., 10^0 for the least significant decade, 10^1, 10^2 for the next decades, etc.

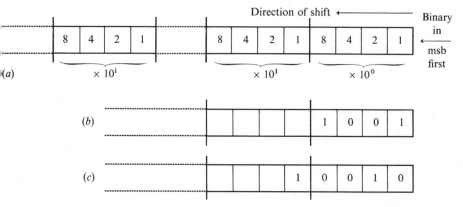

FIGURE 9.20. Shift-register for binary-to-BCD conversion: (*a*) shift-register partition, (*b*) state of shift-register after four shifts left, (*c*) state of shift-register after five shifts left.

Suppose now that we start shifting a binary number into the shift-register of Figure 9.20. The direction of shifting is to the left, the most significant bit, msb, of the binary number being entered into the lsb of the 10^0th decade first. The allowed states in the 4-bit register are the decimal equivalents of numbers from 0 to 9. When the number 1001 is shifted to the left, a carry is shifted from the 10^0-decade to the 10^1-decade, the BCD value of the decimal number increasing from 9_{10} to 12_{10} as shown in Figure 9.20*b* and *c*. Let us disregard for a moment the partitioning of the shift-register into decades, and let us read the contents of the register after shifting as a natural binary number. In this way we obtain the equivalent 18_{10}. The value of 6, lost due to the decade partitioning, has to be added to make the binary and BCD numbers correspond. An additional correction has to be applied whenever the contents of the BCD decade *before* shifting contain numbers 5, 6, or 7. A subsequent shift to the left, though not crossing the boundary between BCD digits, will produce $\geq 10_{10}$, which is not an allowed state in the BCD code. This can be corrected by subtracting 10_{10} from the contents of the 4-bit register containing a number > 9 and adding 1 to the lsb of the next higher decade. We will show below that both conditions requiring a correction can be satisfied in one circuit.

Assume that at time t_i the converter is in state σ_i; then after one shift to the left the state of the converter will be

$$\sigma_{i+1} = 2\sigma_i + b_i, \tag{9.15}$$

where b_i is the last bit shifted in, and may be 1 or 0.

A diagram of the transitions is shown in Figure 9.21a. For convenience we have chosen to start with a shift register in the zero state. Also we have assigned numbers to the new states to correspond to the number that would have been obtained from the input and from the shifting. Note, for example, that in state 0 with a 0 input the register remains in state 0 which coincides with the decimal number represented by the state of the register. A 1 applied to the register in the zero state, followed by a string of zeros, will take the register through states 2, 4 ... 16. Other transitions can be easily recognized from the diagram. The numbers along the arrows represent the inputs at the present state to the lsb of the register.

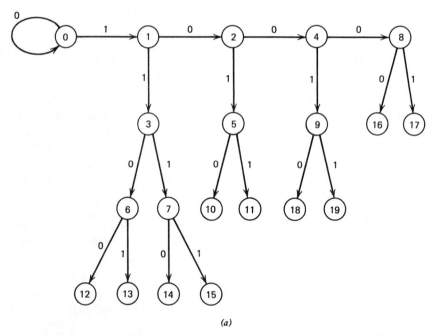

(a)

FIGURE 9.21. Shift-register transition diagram: (a) no corrections applied.

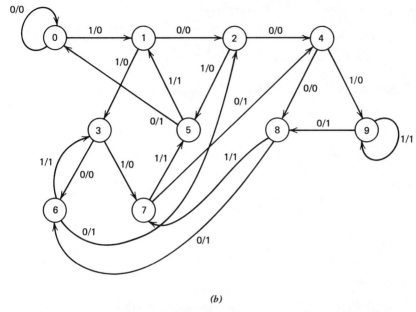

(b)

FIGURE 9.21. (*b*) corrections applied for non-valid BCD combinations.

We now have to apply a correction to ensure that after each shift every decade of the shift register will be in a state representing a valid and correct BCD combination. Thus, in accordance with the earlier discussion, when the register decade has reached a value 5 or greater, we shift and then subtract 10_{10} from the resultant number, and add 1 to the next higher decade. The former state transitions can now be redefined

$$\sigma_i(\geq 5) = 2\sigma_i(\geq 5) + b_i - 10_{10}, \qquad (9.16)$$

where b_i is the bit shifted in last.

Equation (9.16) results in a transition diagram as shown in Figure 9.21*b*. The 1 after the slash on the lines showing state transitions signifies that a 1 has to be loaded into the lsb of the next higher decade during the clock period in which the particular transition has taken place.

EXAMPLE 9.10. An 8-bit shift register is divided into two BCD decades as shown in Figure 9.22. Successive rows in the figure show the sequences in a conversion of binary 110110 to BCD 54. Corrections are applied in rows six and nine in form of a carry that is propagated from the units decade to the tens decade whenever the unit decade contains a numerical value >9.

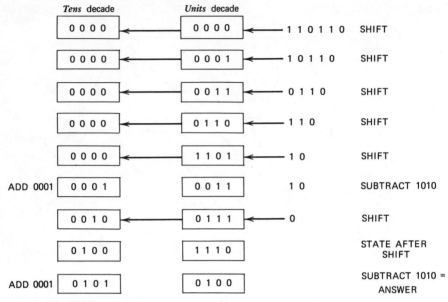

FIGURE 9.22. Shift-register states including corrections for a binary-to-BCD conversion of $110110 = 54_{10}$.

The nature of the combinational network required to carry out the corrections is deduced as follows. The information of the transition diagram, Figure 9.21b, is transferred to the state table, Figure 9.23, which shows the present state, σ_i, of a 4-bit register and the next state, σ_{i+1}, as a function of the shifting operation. Two columns for the next state are shown, one for the serial input $C_i = 0$, the other for $C_i = 1$. The last column of the table shows the output C_{i+1} (carry) which is the result of a shifting operation followed by the correction.

Four flip-flops, FF_3 through FF_0, are required to implement this sequential circuit. The output, C_{i+1}, derived from the state table is

$$C_{i+1} = Q_3 + Q_2 Q_1 + Q_2 Q_0, \qquad (9.17)$$

where Q_3, Q_2, etc., are the outputs of the respective flip-flops FF_3, FF_2, etc. Next we assume that D-type flip-flops are used as the memory elements.*

---

* Other bistable elements can be also used in the implementation of the converter.

| Present state | Next state | | Output to |
| 8 4 2 1 | $C_i = 0$: | $C_i = 1$: | next decade |
Q_3 Q_2 Q_1 Q_0	8 4 2 1 Q_3 Q_2 Q_1 Q_0	8 4 2 1 Q_3 Q_2 Q_1 Q_0	C_{i+1}
0 0 0 0	0 0 0 0	0 0 0 1	0
0 0 0 1	0 0 1 0	0 0 1 1	0
0 0 1 0	0 1 0 0	0 1 0 1	0
0 0 1 1	0 1 1 0	0 1 1 1	0
0 1 0 0	1 0 0 0	1 0 0 1	0
0 1 0 1	0 0 0 0	0 0 0 1	1
0 1 1 0	0 0 1 0	0 0 1 1	1
0 1 1 1	0 1 0 0	0 1 0 1	1
1 0 0 0	0 1 1 0	0 1 1 1	1
1 0 0 1	1 0 0 0	1 0 0 1	1

FIGURE 9.23.   State table for transition diagram of Figure 9.21b.

Note that under the column "next state," the states of the corresponding flip-flops differ only in Q_0: when $C_i = 0$, $Q_0 = 0$; when $C_i = 1$, $Q_0 = 1$. Thus

$$D_0 = C_i. \tag{9.18}$$

The input equations to the other flip-flops are derived from the truth table of the D flip-flops and the state table of Figure 9.23.

$$D_1 = Q_3\bar{Q}_0 + \bar{Q}_3\bar{Q}_2 Q_0 + Q_2 Q_1 \bar{Q}_0, \tag{9.19}$$

$$D_2 = Q_3\bar{Q}_0 + Q_1 Q_0 + \bar{Q}_2 Q_1, \tag{9.20}$$

$$D_3 = Q_3 Q_0 + Q_2 \bar{Q}_1\bar{Q}_0. \tag{9.21}$$

From the above equations we draw the circuit required for each decade of the converter. Figure 9.24 shows the resulting MSI circuit containing four D-type edge-triggered flip-flops with a *CLEAR* terminal. Clearing of the shift register is required before each conversion cycle. In addition, the circuit requires six 3-input positive NAND gates and seven 2-input positive NAND gates.

A block diagram showing interconnections of the decades and the signal flow is given in Figure 9.25. Conversion of an *n*-bit binary number into BCD requires *n* clock periods. The ratio of time required for conversion with

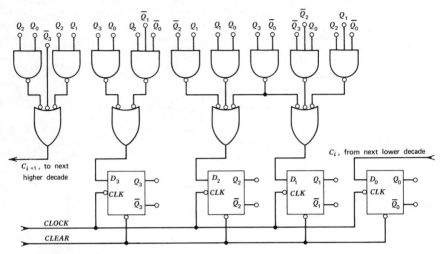

FIGURE 9.24. One decade of a "one clock-period per bit" binary-to-BCD converter using D flip-flops.

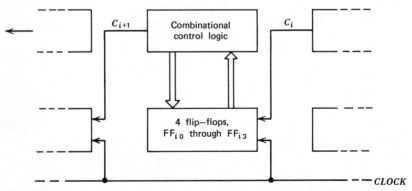

FIGURE 9.25. Block diagram and signal flow of the "one clock-period per bit" binary-to-BCD converter.

a shift-register converter such as shown in Figure 9.25 as compared with the counter method of Figure 9.19 operating at the same frequency is $n/(2^n - 1)$. For conversion of a 16-bit number this ratio is $16/65,535 = 2.44 \times 10^{-4}$, an improvement of better than three orders of magnitude in speed.

Combinational Binary-to-BCD Converters.   In the discussion of corrections to be applied in the shift-register method, the following rules were established: (a) when the number contained in the shift-register is 4 or less, no correction is required; (b) for numbers between 5 and 9, subtract 10_{10} from the resultant number *after* shifting, and add 1 to the contents of the next higher decade to correct for the effective loss of 6 whenever a carry has been propagated to the next higher decade. This latter step is equal to adding 16 and subtracting 10_{10}, i.e., an addition of 6, as required.

Rather than effectively adding 6 *after* shifting, we can achieve the same result through addition of 3 *before* shifting. The shift-left operation, being equivalent to a multiplication by 2, will then result in a correction having a numerical value 6. Thus when the correction is applied *before* shifting, rule (b) has to be modified as follows: (b') add 3 prior to shifting if the contents of the decade are in the range of 5 to 9.

This algorithm can be implemented with shift-registers in an analogous way to the previous example. In the combinational method, the shifting operation is achieved through suitably interconnecting combinational networks that perform the addition of 3 whenever the input conditions demand it.

We shall illustrate in Example 9.11 the operation of the combinational converter using an 8-bit binary input. After that we shall develop general equations for the logic gates, since the gating structure required for binary-to-BCD conversion can be made modular and hence suitable for any number of applied bits. This fact results in availability of MSI and LSI conversion circuits from commerical sources, greatly facilitating the implementation of a converter.

EXAMPLE 9.11.   Figure 9.26 consists of nine identical combinational networks. Each network examines the binary pattern and adds 3 for states corresponding to decimal numbers 5 through 9. The outputs of each stage are "hardwire"-shifted by one bit to the next combinational circuit. The only restriction on the converter is that the input to the first circuit to which the most significant bits are applied shall not exceed 9. All unused terminals are connected to ground representing the logic 0 state. The decimal numbers next to the circuits show the input and output states, respectively. Note that for inputs 5 through 9, the corresponding outputs are 8 through 12. The binary input 11101011 appears correctly at the output as BCD 235.

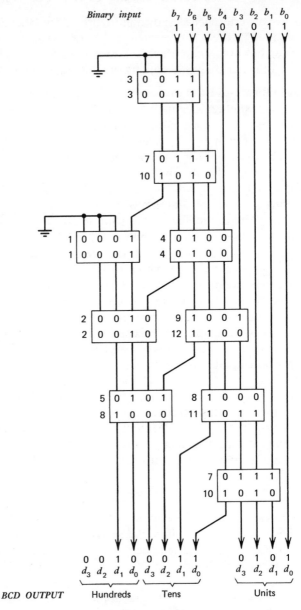

FIGURE 9.26. Combinational binary-to-BCD converter for an 8-bit binary number. States of the combinational networks illustrate conversion of 11101011 to BCD 235.

We shall now derive the gating structure of each combinational circuit from the following conditions: (*a*) for each input in the range of 0 through 4 the output shall remain unchanged; (*b*) for each input in the range 5 through 9 the corresponding outputs shall be 8 through 12. These conditions are shown in the truth table of Figure 9.27. Since we imposed the restriction

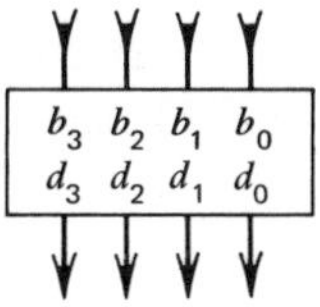

	Input	Output
Decimal	$b_3\ b_2\ b_1\ b_0$	$d_3\ d_2\ d_1\ d_0$
0	0 0 0 0	0 0 0 0
1	0 0 0 1	0 0 0 1
2	0 0 1 0	0 0 1 0
3	0 0 1 1	0 0 1 1
4	0 1 0 0	0 1 0 0
5	0 1 0 1	1 0 0 0
6	0 1 1 0	1 0 0 1
7	0 1 1 1	1 0 1 0
8	1 0 0 0	1 0 1 1
9	1 0 0 1	1 1 0 0

(*b*)

FIGURE 9.27. (*a*) Block diagram, (*b*) Truth table. Combinational networks used in Figure 9.26.

that inputs greater than 9 are not allowed, we have six redundant states that can be useful in simplification of the term describing the converter gating. Denoting the binary inputs b_3 (msb), b_2, etc., and the corresponding BCD outputs d_3 (msb), d_2, etc., we deduce from the truth table the following minimum gating network applicable to any decade:

$$\left.\begin{aligned}
d_3 &= b_3 + b_2 b_1 + b_2 b_0, \\
d_2 &= b_3 b_0 + b_2 \bar{b}_1 \bar{b}_0, \\
d_1 &= \bar{b}_2 b_1 + b_3 \bar{b}_0 + b_1 b_0, \\
d_0 &= b_3 \bar{b}_0 + \bar{b}_3 \bar{b}_2 b_0 + b_2 b_1 \bar{b}_0.
\end{aligned}\right\} \qquad (9.22)$$

The method illustrated above can be used for any weighted code, such as the 4-2-2-1 BCD, two-out-of-five code, etc., to yield gating structures that can be used as building blocks for the required converters.

The combinational method has an advantage of higher speed, and has no need for parallel-to-serial conversion when the binary number is presented in parallel form. The disadvantages are that the hardware required for the combinational method increases approximately as $0.15\ n^2$ (for $n \geq 8$), where n is the number of binary bits, whereas in the clocked method the hardware increases linearly with n.

9.8 BCD-to-Natural-Binary Converters

The techniques shown in Section 9.7 for binary-to-BCD conversion can be applied, with suitable modifications, in BCD-to-binary converters. These modifications are briefly discussed in this section, and additional techniques are illustrated.

Serial Code Converters Using Counters. This technique is analogous to the method shown in Figure 9.19. The only modification involves the interchange of the functions of the binary and BCD counters. Note that if a BCD "down-counter" is not conveniently available, the 10's-complement of the number to be converted is loaded into a BCD "up-counter." The conversion time is the numerical value contained in the BCD "down-counter" times the clock period.

One Clock-Period per Bit BCD-to-Binary Converters. The method described in this paragraph makes use of the repeated division-by-2 algorithm that was discussed in Section 2.4. Let D be a decimal integer; then repeated division by 2 produces

$$\frac{D}{2} = D_1 + b_0, \text{ where } b_0 \text{ is the remainder}$$

$$\frac{D_1}{2} = D_2 + b_1, \text{ and in general}$$

$$\frac{D_i}{2} = D_{i+1} + b_i, \text{ while the } n\text{-th division produces}$$

$$D_{n-1} = D_n + b_{n-1}, \text{ where } D_n = 0.$$

The equivalent binary integer in positional notation is

$$N_B = b_{n-1}b_{n-2}\cdots b_i \cdots b_1 b_0.$$

Assume a register composed of m identical decades. A division by 2, i.e., shift right, will produce a remainder only if the original number in the decade was odd. For decades higher than 10^0, e.g., in the ith decade, the remainder represents $\frac{1}{2}(10^i) = 5(10^{i-1})$. This means that a carry, C_{i-1}, of value 5 has to be *added* to the next lower decade whenever a remainder has been produced by the division. Figure 9.28 shows the resultant state table. The first column shows the states of the four flip-flops FF_3 through FF_0 at time t_k, the second and third columns show the states of the flip-flops at t_{k+1} with a "carry-in" $C_i = 0$ and $C_i = 1$, respectively, applied from the

Present state 8 4 2 1 Q_3 Q_2 Q_1 Q_0	Next state		Output to next decade C_{i-1}
	$C_i = 0$: 8 4 2 1 Q_3 Q_2 Q_1 Q_0	$C_i = 1$: 8 4 2 1 Q_3 Q_2 Q_1 Q_0	
0 0 0 0	0 0 0 0	0 1 0 1	0
0 0 0 1	0 0 0 0	0 1 0 1	1
0 0 1 0	0 0 0 1	0 1 1 0	0
0 0 1 1	0 0 0 1	0 1 1 0	1
0 1 0 0	0 0 1 0	0 1 1 1	0
0 1 0 1	0 0 1 0	0 1 1 1	1
0 1 1 0	0 0 1 1	1 0 0 0	0
0 1 1 1	0 0 1 1	1 0 0 0	1
1 0 0 0	0 1 0 0	1 0 0 1	0
1 0 0 1	0 1 0 0	1 0 0 1	1

FIGURE 9.28.   State table for transitions in a BCD-to-binary converter.

next higher decade. C_{i-1} is the carry to the next lower decade, and is 1 if the present decade contains an odd integer. The C_{i-1} signal from the 10^0-decade is collected in a shift-register and forms the desired binary number.

Assuming that J-K flip-flops are used for FF_3 through FF_0, we deduce the following input equations for the J and K terminals:

$$J_3 = Q_2 Q_1 C_i, \qquad\qquad K_3 = \overline{C}_i,$$
$$J_2 = \overline{Q}_3 C_i + Q_3 \overline{C}_i = Q_3 \oplus C_i, \quad K_2 = \overline{Q}_3 \overline{C}_i + \overline{Q}_3 Q_1 + Q_3 \overline{Q}_1 C_i,$$
$$J_1 = Q_2, \qquad\qquad K_1 = Q_2 C_i + \overline{Q}_2 \overline{C}_i = \overline{Q_2 \oplus C_i}, \qquad (9.23)$$
$$J_0 = Q_1 \overline{C}_i + \overline{Q}_1 C_i = Q_1 \oplus C_i, \quad K_0 = \overline{Q}_1 \overline{C}_i + Q_1 C_i = \overline{Q_1 \oplus C_i} = \overline{J_0},$$
$$C_{i-1} = Q_0.$$

Implementation of one decade using J-K flip-flops is shown in Figure 9.29a; other bistable elements may also be used (see Problem 9.23). A block diagram showing interconnections and signal flow in a general "one clock-period per bit" BCD-to-binary converter is given in Figure 9.29b.

Combinational BCD-to-Binary Converter I.   This method, based on an algorithm of "*nested*" *decimal multiplication*, utilizes binary full adders for conversion. A decimal number N of $n - 1$ digits can be expressed as

$$N = \{[(10 D_{n-1} + D_{n-2})10 + (10 D_{n-3} + D_{n-4})]10 + \cdots$$
$$+ (10 D_2 + D_1)\}10 + D_0, \quad (9.24)$$

where $D_i = \{0, 1, \ldots, 9\}$, and is represented in the natural 8-4-2-1 BCD code by four bits: d_3 (msb), d_2, d_1, and d_0.

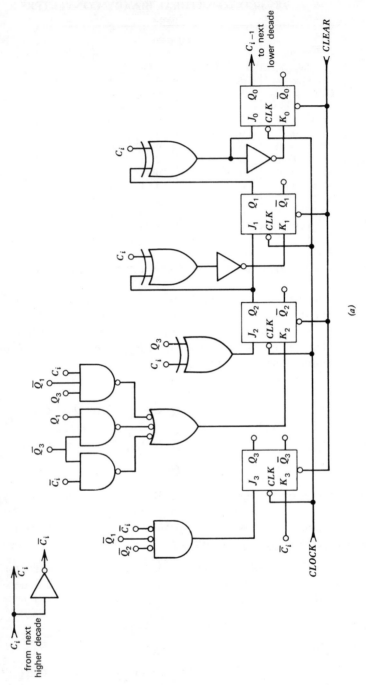

FIGURE 9.29. BCD-to-binary converter. "One clock-period per bit" method: (a) logic diagram for one decade.

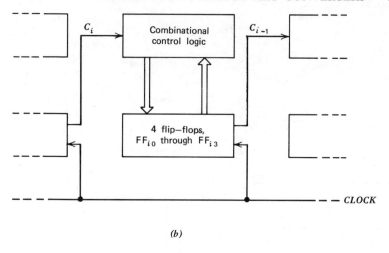

(b)

FIGURE 9.29. *(b)* **block diagram showing interconnections and signal flow.**

Multiplication by 10 can be accomplished with binary adders if we recognize that

$$10 D_i = (2^3 + 2^1)D_i = 2^3 D_i + 2^1 D_i.$$

The $2^3 D_i$ term can be obtained in a binary register through shifting the $2^1 D_i$ term two places left. The shifting operation can be eliminated using adders only with the $2^3 D_i$ term "hardwired" two places to the left of the $2^1 D_i$ term, as shown below:

$$
\begin{array}{lcccccccc}
D_i = & & & & d_3 & d_2 & d_1 & d_0 \\
2 D_i = & & & d_3 & d_2 & d_1 & d_0 & 0 \\
8 D_i = & d_3 & d_2 & d_1 & d_0 & 0 & 0 & 0 \\
\end{array}
$$

$$10 D_i = 2 D_i + 8 D_i = S_{32}\, S_{16}\, S_8\, S_4\, S_2\, S_1\, S_0$$

Note that $S_0 = 0$ since the lsb of D_i did not participate in the summation.

EXAMPLE 9.12.   Convert the BCD number 397 into natural binary representation using binary full adders.

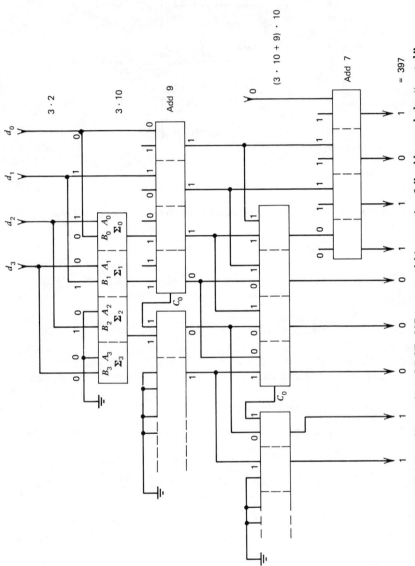

FIGURE 9.30. Conversion of BCD 397 to natural binary using full adders and the "nested" decimal multiplication algorithm.

Since $397 = [(10 \cdot 3 + 9) \cdot 10 + 7]$, by starting with the most significant digit we obtain

$$d_3 \, d_2 \, d_1 \, d_0$$

(0 0 1 1)	(3)	
0 0 1 1 0	$\left. \begin{matrix} 3 \cdot 2 \\ 3 \cdot 8 \end{matrix} \right\} +$	
0 0 1 1 0 0 0		
0 0 1 1 1 1 0	$3 \cdot 10$	
1 0 0 1	Adding 9	
0 1 0 0 1 1 1	$(3 \cdot 10 + 9)$	
0 1 0 0 1 1 1 0	$\left. \begin{matrix} (3 \cdot 10 + 9) \cdot 2 \\ (3 \cdot 10 + 9) \cdot 8 \end{matrix} \right\} +$	
1 0 0 1 1 1 0 0 0		
1 1 0 0 0 0 1 1 0	$(3 \cdot 10 + 9) \cdot 10$	
1 1 1	Adding 7	
1 1 0 0 0 1 1 0 1	Result, 397	

Implementation of the conversion process is shown in Figure 9.30, requiring six 4-bit full adders; note that more adders would be required if the 3-digit BCD number to be converted were ≥ 400.

In the general case eight 4-bit full adders are required for conversion of *any* 3-digit BCD into natural binary. Figure 9.31 shows 4-bit full adders and their interconnections for a BCD-to-binary conversion of four decimal digits. The number of 4-bit adders is shown at each stage in parentheses. Fourteen units are required for the conversion of a 4-digit decimal number. The conversion time equals the sum of the propagation delay times of the adders, which compares favorably with the "shift-add" serial conversion.

Combinational BCD-to-Binary Converter II. In this method we shall first establish an algorithm for a converter utilizing shift-registers; as a second step, we shall replace the shifting operation through hardwire interconnections of *identical* combinational networks.

Consider a long shift-register into which a BCD number has been entered through "parallel load" gates. We are given the task of designing a circuit that will produce a serial binary number as the contents of the BCD shift-register are serially shifted out to the right. Shifting in that direction represents division by 2, except at the boundary between two decades, as shown below:

0 0 0 1	0 1 1 0	BCD 16
0 0 0 0	1 0 1 1	Shift right
	1 1	Apply correction, (subtract 3)
0 0 0 0	1 0 0 0	Result, BCD 8

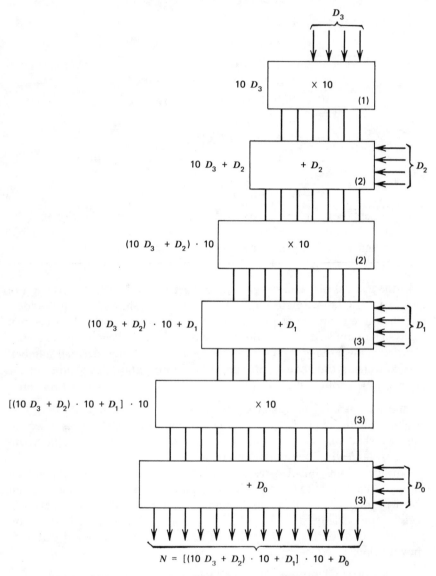

FIGURE 9.31. BCD-to-natural-binary converter for four decimal digits, using 4-bit full adders based on the nested decimal multiplication algorithm.

The BCD number 16 shifted one bit to the right results in a non-valid state of the BCD code in the unit decade. The lsb of the 10^1-decade has a value of 10 before shifting. Since a shift to the right represents a division by 2, we expect the shifted bit to carry a value of 5 into the next lower decade. However, the carry-in, being the msb of that decade, has a value of 8. A subtraction of 3 is thus required *after* shifting, to correct for the discrepancy.

This leads to the following rules for BCD-to-binary conversion utilizing shift-registers:

(a) Shift one bit right.
(b) Examine to see if any msb of a decade contains 1.
(c) If all msb's are 0, apply the next shift pulse.
(d) If an msb of any decade contains 1 *after* shifting, subtract 3 from that decade before applying the next shift pulse.

Notice that a subtraction of 3 is applied once only. Thus if the msb of any decade should contain a 1 after subtraction of 3, the contents of that decade are correct and no further correction is required.

We have shown previously that a shift-add sequence can be replaced by addition only, the shifting being effectively implemented through suitable hardwiring between the adder elements. Similarly, we can replace the shift-subtract operation by a combinational circuit that is hardwired as follows: the input to the most significant bit, X_3, of the decade is taken from the lsb of the next higher decade. The other inputs are the three msb's of the decade in which the conversion process is taking place. The combinational circuit also has to sense whether its msb is 0 or 1. If it is 0, no correction is required. If the msb is 1, the output of the circuit has to deliver a number that is equal to its input *minus 3*, in accordance with rule (d) above. Figure 9.32 shows the truth table of the combinational circuit for one decade.

Minterms m_0 through m_4 appear at the input when the lsb of the next higher decade is 0. Minterms m_8 through m_{12} are due to a 1-state of the bit carried in from the next higher decade. The output column shows that for each $X_3 = 1$ the corresponding output number has been corrected and has a value that is smaller by 3 than the input. Minterms m_5, m_6, m_7, m_{13}, m_{14}, and m_{15} do not appear as valid input combinations, and can be used as don't care conditions in finding minimized expressions for Y_3, Y_2, Y_1, and Y_0, yielding the following equations:

$$Y_3 = X_3 X_2 + X_3 X_1 X_0,$$
$$Y_2 = \overline{X}_3 X_2 + X_3 \overline{X}_2 \overline{X}_1 + X_3 \overline{X}_2 \overline{X}_0 = \overline{X}_3 X_2 + X_3 \overline{X}_2 (\overline{X}_1 + \overline{X}_0),$$
$$Y_1 = X_1 \overline{X}_0 + \overline{X}_3 X_1 + X_3 \overline{X}_1 X_0, \tag{9.25}$$
$$Y_0 = X_3 \overline{X}_0 + \overline{X}_3 X_0 = X_3 \oplus X_0.$$

Minterm m_i	Input $X_3\ X_2\ X_1\ X_0$	Output $Y_3\ Y_2\ Y_1\ Y_0$
0	0 0 0 0	0 0 0 0
1	0 0 0 1	0 0 0 1
2	0 0 1 0	0 0 1 0
3	0 0 1 1	0 0 1 1
4	0 1 0 0	0 1 0 0
8	1 0 0 0	0 1 0 1
9	1 0 0 1	0 1 1 0
10	1 0 1 0	0 1 1 1
11	1 0 1 1	1 0 0 0
12	1 1 0 0	1 0 0 1

FIGURE 9.32. Truth table for one decade of a combinational BCD-to-natural-binary converter.

The above equations describe the conversion process in any decade. Therefore we can utilize identical building blocks to convert any BCD number to its binary equivalent. The speed of the conversion process is limited only by the propagation delay times of the circuit blocks used. The number of circuits required for conversion increases rapidly with the number of BCD decades as shown in the table below.

Number of decades	Number of circuits
2	4
3	11
4	21
5	36
6	53

Figure 9.33 shows the connections of four circuits required to convert a 2-digit BCD number to its equivalent binary.

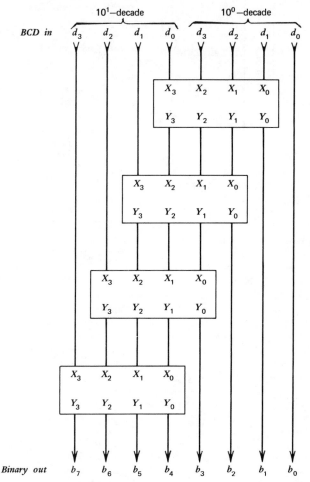

FIGURE 9.33. BCD-to-binary conversion of two decades using identical combinational networks derived from Figure 9.32.

9.9   Read-Only Memories for Code Conversion

Read-only memories (ROM's)* have the bit capacity to provide binary-to-BCD and BCD-to-binary conversion in one chip. As an example, the organization of the Motorola MC4001 ROM is shown in Figure 9.34a. The truth table of Figure 9.34b shows four inputs that can be either binary or BCD. The outputs are divided in two groups: Q_0 through Q_3 yield four binary bits when the input is BCD, while Q_4 through Q_7 provide a 4-bit

---

* Read-only memories are discussed in detail in Section 11.7.

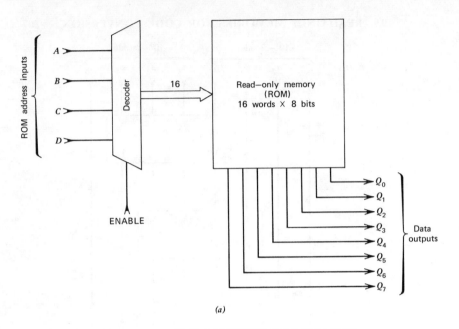

	Output	
Input	Binary to BCD	BCD to Binary
$D\ C\ B\ A$	$Q_7\ Q_6\ Q_5\ Q_4$	$Q_3\ Q_2\ Q_1\ Q_0$
0 0 0 0	0 0 0 0	0 0 0 0
0 0 0 1	0 0 0 1	0 0 0 1
0 0 1 0	0 0 1 0	0 0 1 0
0 0 1 1	0 0 1 1	0 0 1 1
0 1 0 0	0 1 0 0	0 1 0 0
0 1 0 1	1 0 0 0	0 1 0 1
0 1 1 0	1 0 0 1	0 1 1 0
0 1 1 1	1 0 1 0	0 1 1 1
1 0 0 0	1 0 1 1	0 1 0 1
1 0 0 1	1 1 0 0	0 1 1 0
1 0 1 0	1 0 0 0	0 1 1 1
1 0 1 1	1 0 0 1	1 0 0 0
1 1 0 0	1 1 1 0	1 0 0 1
1 1 0 1	1 1 1 1	0 0 1 0
1 1 1 0	1 0 1 1	0 0 1 0
1 1 1 1	0 1 0 0	0 0 0 0

(b)

FIGURE 9.34. Read-only memory (ROM) for code conversion (Motorola MC 4001): (a) organization of the ROM, (b) truth table.

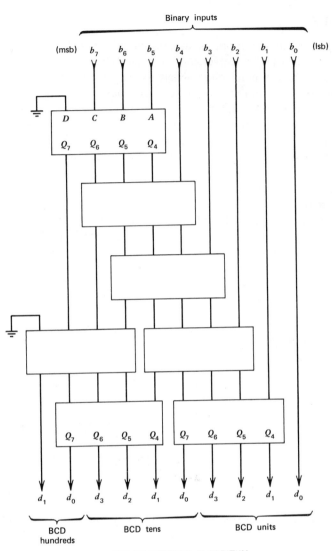

EQUIVALENT CONVERSION ALGORITHM

1. Examine three most significant bits.
If sum $\geq$ 5, add 3.
2. Shift left 1 bit.
3. Examine each BCD decade. If sum $\geq$ 5,
add 3.
4. Shift left 1 bit.
5. Is least significant bit in least significant
BCD position: If "no", loop to item 3.
If "yes", conversion is completed.

FIGURE 9.35. **Seven ROM's (Motorola MC 4001's) used in binary-to-BCD conversion.**

BCD output when the input is natural binary. Figure 9.35 shows the interconnections for the conversion of eight binary bits to three BCD digits. BCD-to-binary conversion is left to the reader as an exercise (see Problem 9.25). This combinational approach has a limitation due to the rapid increase in ROM's required as the number of input bits increases.

ROM's can be also used in sequential converters such as the "one clock-period per bit" converters discussed in Sections 9.7 and 9.8. A 24-bit binary-to-BCD sequential converter can be constructed using seven MC4001 ROM's and seven MSI quad flip-flops. Conversion speeds of less than 100 ns per bit are easily achieved.

The Texas Instruments SN54185/SN74185 binary-to-BCD converter uses a 256-bit ROM as the basic building block, and has the capacity to convert a 6-bit binary number to BCD, as shown in Figure 9.36. The units can be interconnected to convert a binary number of any bit length to BCD. For example, a 20-bit input requires 27 ROM's having a typical propagation delay of 275 ns, with a maximum specified propagation delay of 440 ns.

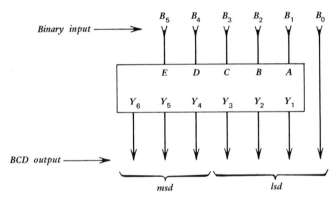

FIGURE 9.36.   Binary-to-BCD converter (Texas Instruments SN54185/ SN74185).

The Texas Instruments SN54184/SN74184 BCD-to-binary converter shown in Figure 9.37 is a 256-bit ROM with five input connections and eight outputs. Since the lsb of both codes are identical, it does not have to be applied to the converter. Conversion of six bits is thus possible with one ROM. The outputs Y_i are divided into two groups; Y_1 through Y_5 are the outputs for BCD-to-binary conversion, see Figure 9.37a; Y_6, Y_7, and Y_8 are used for conversion from the 8-4-2-1 BCD code to its 9's- or 10's-complements, as shown in Figure 9.37b and c, respectively.

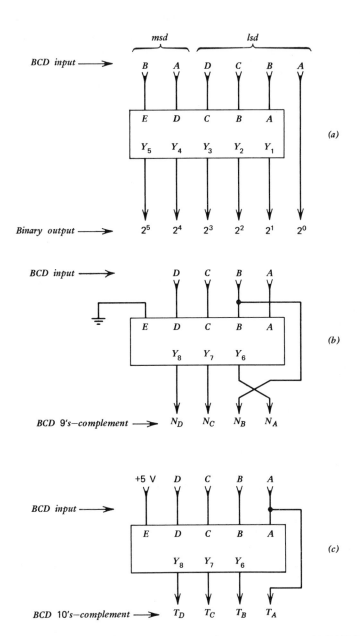

FIGURE 9.37. BCD-to-Binary converter (Texas Instruments SN54184/ SN74184): (a) 6-bit converter, (b) BCD-to-9's-complement converter, (c) BCD-to-10's-complement converter.

The ROM's can be cascaded for conversion of many decades. For example, a converter for six BCD decades requires 28 ROM's having a typical propagation delay time of 325 ns, with a specified maximum propagation delay of 520 ns.

9.10   Decoders

Decoders convert information from n coded input lines to a maximum of 2^n unique output lines. In the present section we show several examples of decoders for numerical display.

When choosing a decoder, we have to consider the driving requirements at the output. Decoders are available with several output configurations: active-low voltage, high-sink current for direct driving of indicator lamps; and active-high outputs for current sourcing applications. Output voltage ratings range from $+6$ volts to over $+100$ volts, the latter for driving cold cathode tubes.

The input may also have several auxiliary signals for "lamp test" and "zero blanking" control. Overriding blanking can be used to control lamp intensity or to inhibit the outputs. A complete display system requires storage of data, and often utilizes multiplexing of the display drivers as well as leading and trailing "zero blanking."

BCD-to-Decimal Decoder.   The truth table for the decoder is shown in Figure 9.38. The gating structure is obtained directly from the first column of the table: decimal $0 = \bar{X}_3 \bar{X}_2 \bar{X}_1 \bar{X}_0$, $1 = \bar{X}_3 \bar{X}_2 \bar{X}_1 X_0$, etc. Ten 4-input NAND gates and four inverters implement the decoder. Although there are six redundant states available as don't care conditions, they are commonly not utilized for gate minimizations. "*Full decoding*" is a term applied to decoders that uniquely specify each state, i.e., no minimization of the gate network is employed. Two advantages accrue from full decoding. First, non-valid input combinations do not yield incorrect or ambiguous outputs; the last six rows of Figure 9.38 show the false outputs that would be obtained from input combinations in the range of 10 through 15 with the minimized gating shown in the last column of the table. Second, non-valid states can be utilized for display blanking: the blanking action is achieved through setting the inputs of a decoder to non-valid states. Thus in case of full decoding, the display device will not be activated, i.e., the read-out will be blanked out.

4-Line-to-16-Line Decoder.   The gating structure of this decoder generates the 2^4 states that are possible with four input variables. Several decoders

Full decoding BCD input				Decimal output	Minimized terms			
X_3	X_2	X_1	X_0		X_3	X_2	X_1	X_0
0	0	0	0	0	0	0	0	0
0	0	0	1	1	0	0	0	1
0	0	1	0	2		0	1	0
0	0	1	1	3		0	1	1
0	1	0	0	4		1	0	0
0	1	0	1	5		1	0	1
0	1	1	0	6		1	1	0
0	1	1	1	7		1	1	1
1	0	0	0	8	1			0
1	0	0	1	9	1			1

Invalid inputs	1	0	1	0	2	False outputs if full decoding not employed
	1	0	1	1	3	
	1	1	0	0	4, 8	
	1	1	0	1	5, 9	
	1	1	1	0	6, 8	
	1	1	1	1	7, 9	

FIGURE 9.38.   Truth table for a BCD-to-decimal decoder.

can be interconnected in a decoder-tree configuration to produce a large number of mutually exclusive output signals when the input function has more than four variables. A 4-line-to-16-line decoder can be utilized to drive numerical displays from data that are derived in codes other than the 8-4-2-1 BCD code. This can be effected by merely connecting the output pins to the proper numerals represented by the respective code.

EXAMPLE 9.13.   An angular position encoder delivers a multi-digit output in the excess-3 Gray code. Design a circuit for a numerical display of the encoder output.

One solution would require executing a code conversion for each decade from excess-3 Gray to BCD. Next, BCD-to-decimal converters could be employed to drive the numerical displays. In a more economical solution we would use a 4-line-to-16-line decoder that generates sixteen unique outputs corresponding to the 2^4 possible states of a 4-variable function. Ten of these outputs are then selectively connected to drive the desired numerals. Thus, referring to Figure 9.39 we connect output "2" to drive numeral "0," output "6" to drive numeral "1," etc.

Input binary number	Outputs	Excess-3 BCD	2-4-2-1 BCD	4-2-2-1 (Berkeley)	Gray code	Excess-3 Gray	8-4-2-1 BCD
0 0 0 0	0		0	0	0		0
0 0 0 1	1		1	1	1		1
0 0 1 0	2		2	2	3	0	2
0 0 1 1	3	0	3	3	2		3
0 1 0 0	4	1	4		7	4	4
0 1 0 1	5	2	5		6	3	5
0 1 1 0	6	3	6	4	4	1	6
0 1 1 1	7	4	7	5	5	2	7
1 0 0 0	8	5					8
1 0 0 1	9	6					9
1 0 1 0	10	7				9	
1 0 1 1	11	8					
1 1 0 0	12	9		6	8	5	
1 1 0 1	13			7	9	6	
1 1 1 0	14		8	8		8	
1 1 1 1	15		9	9		7	

FIGURE 9.39.   4-line-to-16-line decoder. Output connections for six different BCD codes used in BCD-to-decimal decoding.

The same technique can be used for any 4-bit code, some of which are shown in Figure 9.39.

BCD-to-7-Segment Decoder.   The numerals shown in Figure 9.40a can be generated by selectively actuating several of seven independent segments simultaneously to produce the desired pattern. One possible assignment of variables for the seven segments is shown in Figure 9.40b. The decoder accepts a BCD code and generates seven outputs to actuate displays of numerals 0 through 9. Several alphabetic symbols can also be displayed using the seven segments. The requisite combinational circuits can be economically implemented by use of 4-input multiplexers that were described in Section 4.8 (see Problem 9.27).

(a)

FIGURE 9.40.   BCD-to-seven-segment decoder: (a) assignment of segments for display of numerals.

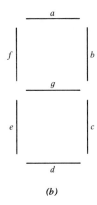

(b)

FIGURE 9.40. (*b*) assignment of variables for the seven segments.

9.11 Encoding and Decoding with Shift Registers

A shift-register with a feedback loop can generate *pseudo-random sequences* (*PRS*) which have some interesting applications, such as improvement of signal-to-noise ratio in digital communications, or in analyzing equipment failure in a random noise environment. Coding and decoding of binary messages will be used in this section as an illustration of the shift-register feedback technique.

A shift-register with n bits and a *linear* (mod-2) *feedback logic* is described by the linear recurrence

$$a_m = C_1 a_{m-1} \oplus \cdots \oplus C_i a_{m-i} \oplus \cdots \oplus C_n a_{m-n}, \qquad (9.26)$$

where a_m is the new generated state. $C_i = 1$ when bit i of the shift-register is used in the feedback loop; otherwise $C_i = 0$.

EXAMPLE 9.14. Consider the sequence produced by a 4-bit shift-register of Figure 9.41*a* and represented by the linear recurrence

$$a_m = a_{m-1} \oplus a_{m-4}.$$

For this sequence generator to be non-trivial we have to impose the condition that at no time shall all shift-register outputs be zero. This condition is imposed since the feedback path consists of an EXCLUSIVE–OR gate and thus an all-zero input to the EXCLUSIVE–OR would propagate through the shift-register yielding zero outputs, i.e., the shift-register would always remain in the zero state.

For the 4-bit shift-register in our example let us assume an initial state

$$a_0 = a_{-1} \oplus a_{-4},$$

where $\qquad a_{-1} = a_{-2} = a_{-3} = 0, \quad \text{and} \quad a_{-4} = 1.$

Figure 9.41b shows the sequences $a_0, a_1 \ \ldots \ a_m$ generated at each shift.

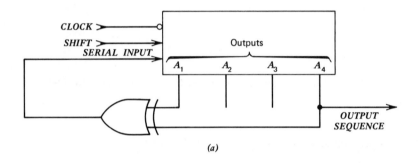

(a)

State, m	a_m	a_{m-1}	a_{m-2}	a_{m-3}	a_{m-4}
0	1	0	0	0	1
1	1	1	0	0	0
2	1	1	1	0	0
3	1	1	1	1	0
4	0	1	1	1	1
5	1	0	1	1	1
6	0	1	0	1	1
7	1	0	1	0	1
8	1	1	0	1	0
9	0	1	1	0	1
10	0	0	1	1	0
11	1	0	0	1	1
12	0	1	0	0	1
13	0	0	1	0	0
14	0	0	0	1	0
15 = 0	1	0	0	0	1

(b)

FIGURE 9.41. Pseudo-random sequence (PRS) generator: (a) 4-bit shift-register with feedback, (b) sequences produced by the PRS generator of Figure 9.41a.

Note that we have just demonstrated a maximum length sequence generator, since the table shows fifteen unique states that were achieved with a 4-bit shift-register (zero state not allowed). A maximum length sequence generator consisting of an n-bit shift-register and a linear feedback logic has a period of $p = 2^n - 1$.

A system for coding and decoding of binary characters consists of a PRS generator at the transmitting end producing a defined code sequence that is added modulo-2 to the information bits of the transmitted character. At the receiving end, another PRS generator is used having exactly the same pseudo-random sequence and phase as the transmitter. The received message is added modulo-2 to the coding sequence and yields the original information.

EXAMPLE 9.15. Using a 6-bit shift-register, design a PRS generator to code a message consisting of 4-bit characters.

It can be shown that a maximum period, $p = 63$, can be obtained in a 6-bit shift-register with a feedback logic represented by

$$a_m = a_{m-1} \oplus a_{m-6}.$$

Selecting the maximum length sequence is desirable since it produces the most powerful encoding for a given shift-register size. The coding of the message is generated by

$$c = a_m \oplus m_1,$$

where
$$c = \text{coded bits}$$

$$a_m = a_{m-1} \oplus a_{m-6} = \text{PRS coding bits}$$

$$m_1 = \text{message bits}.$$

Decoding at the receiving end is attained through mod-2 addition of the same pseudo-random sequence coding, transmitter and receiver using the PRS synchronously.

Letting $d = $ decoded bits, we have at the receiving end

$$d = c \oplus a_m = (a_m \oplus m_1) \oplus a_m.$$

But
$$a_m \oplus a_m = 0,$$

and
$$0 \oplus m_1 = m_1;$$

hence $d = m_1$, as required.

Figure 9.42a shows a circuit that is suitable at either end of the transmission link. When the input consists of message data, the output will yield the encoded message. Conversely, when the encoded message is applied to the input, the original message data will be obtained at the output. The 6-input negative NAND gate together with the inverter perform the function of "zero suppression." Thus when the register is in its zero state, the "load" terminal is actuated to load logic 1 via the input terminals D_i $(1 \le i \le 6)$.

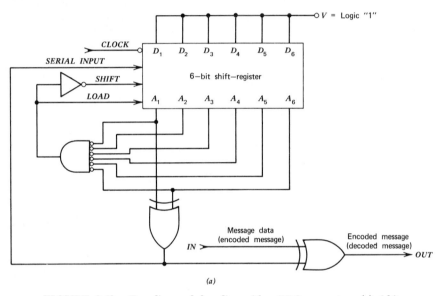

FIGURE 9.42.   Encoding and decoding with a PRS generator: (a) 6-bit shift-register with feedback loop and zero-state suppression.

Figure 9.42b shows a typical sequence of coding and encoding with a pseudo-random sequence generator. The initial state of the PRS generator shown in column 1 is a result of the operation of the "zero suppression" circuit. The other a_m states follow from the feedback loop of the circuit shown in Figure 9.42a, as can be readily verified by the reader. Message data, m_1, in the second column are identical to the decoded message, d, of the fourth column, in accordance with the equations developed in Example 9.15.

PRS state a_m	Message m_1	Encoded message $c = a_m \oplus m_1$	Decoded message $d = c \oplus a_m$
1	0 ⎫	1 ⎫	0 ⎫
1	1 ⎬ 4	0 ⎬ 11	1 ⎬ 4
1	0	1	0
1	0 ⎭	1 ⎭	0 ⎭
1	0 ⎫	1 ⎫	0 ⎫
1	0 ⎬ 2	1 ⎬ 15	0 ⎬ 2
0	1	1	1
1	0 ⎭	1 ⎭	0 ⎭
0	0 ⎫	0 ⎫	0 ⎫
1	0 ⎬ 1	1 ⎬ 4	0 ⎬ 1
0	0	0	0
1	1 ⎭	0 ⎭	1 ⎭
1	1 ⎫	0 ⎫	1 ⎫
0	0 ⎬ 8	0 ⎬ 1	0 ⎬ 8
0	0	0	0
1	0 ⎭	1 ⎭	0 ⎭
1	0 ⎫	1 ⎫	0 ⎫
0	0 ⎬ 3	0 ⎬ 8	0 ⎬ 3
1	1	0	1
1	1 ⎭	0 ⎭	1 ⎭
1	0 ⎫	1 ⎫	0 ⎫
0	1 ⎬ 5	1 ⎬ 14	1 ⎬ 5
1	0	1	0
1	1 ⎭	0 ⎭	1 ⎭
0	0 ⎫	0 ⎫	0 ⎫
1	0 ⎬ 0	1 ⎬ 4	0 ⎬ 0
0	0	0	0
0	0 ⎭	0 ⎭	0 ⎭

(b)

FIGURE 9.42. (b) encoding and decoding a message.

PROBLEMS

9.1 How many digits are required to express an i-digit decimal number in (a) binary, (b) ternary, (c) octal, and (d) hexadecimal notation.

9.2 Construct a BCD weighted code in which 9's-complementation can be achieved through logical inversion.

9.3 Using a 4-variable Karnaugh map, construct (a) a unit-distance code having sixteen states, and (b) a unit-distance BCD code.

9.4 The even parity check is not as often used as the odd parity check. Discuss the reasons.

9.5 Consider the block parity check. Can a double error be detected using this check? Discuss the limitations.

9.6 (a) Prove that the error check bit of an odd parity check is

$$p \text{ (odd)} = X_n \oplus X_{n-1} \oplus \cdots \oplus X_0 \oplus 1.$$

(b) Prove that the even parity check is

$$p \text{ (even)} = X_n \oplus X_{n-1} \oplus \cdots \oplus X_0.$$

9.7 For binary-to-Gray-code conversion verify that

$$G_i = B_{i+1} \oplus B_i,$$

using a Karnaugh map for a 4-variable code.

9.8 (a) Construct a table showing the odd parity Hamming code for the sixteen possible messages of a 4-bit information word.
(b) Repeat the exercise for an even parity Hamming code.

9.9 Find the binary values of the check bits for the information character 10011. Assume that the odd parity Hamming error correcting code is used. How many error check bits are required?

9.10 Design an odd parity Hamming code generator for transmission of 8-4-2-1 BCD coded information.

9.11 Design a Hamming code receiver for a 4-bit information word as specified on page 281.

9.12 Design a network that will receive a 7-bit character composed of four natural BCD bits and three Hamming code (even parity) check bits. This network shall (a) detect the location of the error bit and (b) correct the error.

9.13 Design an excess-3 to natural BCD converter. What are the simplest expressions for B_3(msb), B_2, B_1, and B_0 ?

9.14 Design a minimized combinational network that converts the natural BCD code to an excess-3 code.

9.15 For Gray-code-to-binary conversion verify that

$$B_i = \sum_i^n G_i \text{ (mod 2)},$$

using the Karnaugh map for a 5-variable code.

9.16 (a) Design a gating network that operates as follows. When the control signal $C = 1$, n EXCLUSIVE–OR gates shall be connected to implement a binary-to-Gray-code conversion. For $C = 0$, the same EXCLUSIVE–OR gates shall execute a Gray-code-to-binary conversion.

(b) How many bits can be converted using n EXCLUSIVE–OR gates?

9.17 Design a minimum network to convert the BCD 8-4-2-1 code to the BCD 4-2-2-1 code shown below.

8	4	2	1	4	2	2	1
X_3	X_2	X_1	X_0	Y_3	Y_2	Y_1	Y_0
0	0	0	0	0	0	0	0
0	0	0	1	0	0	0	1
0	0	1	0	0	0	1	0
0	0	1	1	0	0	1	1
0	1	0	0	0	1	1	0
0	1	0	1	1	0	0	1
0	1	1	0	1	1	0	0
0	1	1	1	1	1	0	1
1	0	0	0	1	1	1	0
1	0	0	1	1	1	1	1

9.18 Using 4-line to 16-line decoders design a network that generates 64 unique outputs.

9.19 Design a minimum combinational network that yields an output that is twice the 8-4-2-1 BCD input. The output has five variables: Y_4 through Y_0.

9.20 Refer to Figure 9.16 and write an algorithm for serial conversion of a natural-binary number to a Gray-coded number.

9.21 Design a binary-to-BCD converter having a conversion speed of one clock-period per one bit using J-K flip-flops and suitable combinational networks.

(a) Denoting the most significant flip-flop FF_3, the next FF_2, etc., write down the input equations J_3, K_3, etc.

(b) What is the equation for C_{i+1}, the carry into the next higher decade?

(c) Draw a circuit diagram for the conversion of one decade and compare the component count with the converter shown in Figure 9.24.

9.22 Verify eq. (9.23) by use of the excitation table of the J-K flip-flop and the state table of Figure 9.28.

9.23 Design an 8-4-2-1 BCD-to-binary converter using the one clock-period per bit method and D-type flip-flops. (a) write down the input equations for D_3, D_2, D_1, and D_0. (b) What is the equation for the output C_{i-1}? (c) Compare your circuit with Figure 9.29a with respect to complexity of implementation.

9.24 Prepare the truth table for the ten states (0 through 9) of a 6, 3, 1, -1 BCD code and list the redundant states.
(a) Construct the state stable for a binary to 6, 3, 1, -1 BCD conversion.
(b) Assuming that J-K flip-flops are used, what are the input equations of the J and K terminals of the four flip-flops in the ith decade? (Ref. 2) (Denote the msb as Q_3, the next bit as Q_2, etc.)

9.25 (a) Design a BCD-to-binary converter for three BCD digits using the read-only memory, type MC4001, shown in Figure 9.34.
(b) Prepare a table for the following: (i) ROM's required for binary-to-BCD conversion up to 24 bits, and (ii) ROM's required for BCD-to-binary conversion up to six BCD digits.

9.26 Using a BCD-to-decimal decoder MSI circuit, implement decoding to decimal of the codes listed below. Denoting the input variables as Y_3, Y_2, Y_1, and Y_0 (lsb), the inputs to the decoder as X_3 through X_0, and the ten outputs as A_0, A_1, ..., A_9, list the connections to be made from Y_i to X_k, and from A_i to a numerical display, to obtain correct number sequences. The codes applied to the inputs of the decoder are (see also Figure 9.39); (a) Excess-3 Gray, (b) Gray code, (c) 2-4-2-1 BCD, (d) Excess-3 BCD. (Inverters may be utilized at the inputs X_i).

9.27 Referring to Figure 9.40b
(a) Establish a truth table and derive the minimized gating structure for display of numerals 0 through 9 when the input is given in an 8-4-2-1 BCD code.
(b) Derive minimized equations for the following characters of the alphabet that can be displayed by the seven segments: A, C, E, F, H, L, P, S, U.
(c) Repeat Problem 9.27b utilizing 4-input multiplexers for implementation of the combinational logic. Assuming that two 4-input multiplexers (MSI) are available in one integrated circuit package, compare the SSI and MSI solutions in terms of package count.

9.28 Given a 4-bit binary code, design a minimum gate network that yields the natural BCD output for all valid natural BCD input combinations. For inputs corresponding to numerical values in the range of 10 to 15, the output shall generate a code that corresponds to the *unit-digit* of the input combination.

9.29 Figure 9.43 shows an error correcting code generator and the positions of the information and check bits in the coded word. SR is a parallel/serial input shift-register with parallel output. The information

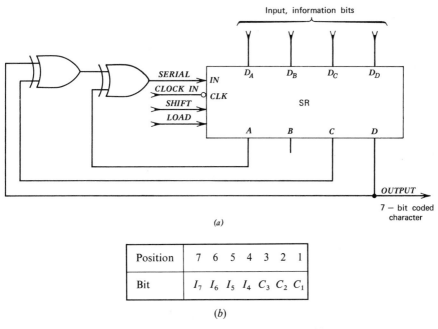

(a)

Position	7	6	5	4	3	2	1
Bit	I_7	I_6	I_5	I_4	C_3	C_2	C_1

(b)

FIGURE 9.43. (a) **Error-correcting code generator.** (b) **Bit positions:** I = information bits, C = check bits.

bits I_i are loaded in parallel via terminals D_A(lsb) through D_D(msb). After the I_i are loaded, seven clock pulses are applied to the terminal denoted *CLOCK IN*.

(a) Write down the shift-register feedback equation.
(b) What are the expressions for the check bits C_1, C_2, and C_3?
(c) Draw a state table showing the 7-bit output coded word resulting from seven shifts for each of the sixteen input combinations.
(d) Write down the mod-2 sums necessary to check any errors in the received message.

(e) Construct a truth table to identify any erroneous bit in terms of the check bits.

9.30 Using 5-bit shift-registers draw the transition diagrams for PRS generators having the following feedback functions: (a) $a_2 \oplus a_3$; (b) $a_3 \oplus a_4$; (c) $a_3 \oplus a_5$.

9.31 A 4-stage shift-register with a feedback connection $a_2 \oplus a_4$ is utilized to generate a non-maximum length of linear sequences.
(a) Draw the resultant state table and the transition diagram. Assume an initial state 0001.
(b) Repeat the problem for different initial states and comment on the results.

9.32 Design a count-by-13 circuit using a 4-bit binary counter and a 4-line-to-16-line decoder described in Section 9.10.

CHAPTER 10

Digital-to-Analog and Analog-to-Digital Converters

This chapter deals with the interdomain of continuous and discrete quantities and their conversion from one form to another. Digital-to-analog conversion is used whenever discrete information, for example, from a computer, has to be translated into a physical quantity such as voltage, current, angular motion, positioning of an electron beam on the face of a cathode ray tube, and waveform generation. Analog-to-digital conversion translates continuous physical quantities into a digital form with an approximation dependent on, among other things, the number of digits chosen for the conversion. Consider a scientific data acquisition system which measures continuous physical quantities such as temperature, pressure, optical absorption, etc., that have to be recorded at fixed time intervals, or on demand. The measuring elements, or transducers, convert the physical quantities into continuous electrical signals of proportional magnitude. A conversion from the continuous (or analog) to the discrete (or digital) is required to make these signals suitable as inputs to a digital data acquisition system. In process control systems analog-to-digital and digital-to-analog conversions are frequently executed and their results are used as feedback signals in the system.

In this chapter we present a variety of conversion methods, analyze limits of their accuracies, and discuss the properties of circuit components utilized in various converters. Although in the digital domain we deal predominantly with natural binary and 8-4-2-1 BCD codes, the general treatment of the material will enable the reader to extend the techniques to other coded forms.

10.1   Digital-to-Analog Conversion. General Considerations

We introduce the *digital-to-analog converter*, DAC, first because it is also utilized as a feedback element in some analog-to-digital converters. The DAC accepts digital information and delivers a current or a voltage that is proportional to the numerical value applied to its input. The information may be in any coded form and may represent positive or negative quantities, or both. In Figure 10.1, for example, we show the digital input versus analog output relationship of a 3-bit DAC in which the negative quantities are in sign-and-magnitude representation.

One of the advantages of using DAC's in analog applications is the ability to keep the information in digital form making it therefore independent of drift with time and temperature, and more immune to noise than analog information.

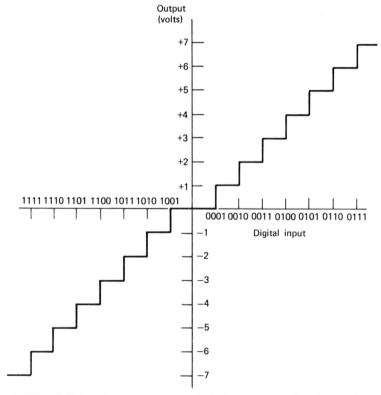

FIGURE 10.1.   Voltage output vs. digital input in a digital-to-analog converter.

The *resolution* of a DAC depends on the number of bits chosen to represent the analog output quantity, usually not exceeding sixteen bits, i.e., a resolution of 1 part in $2^{16} - 1$. The *accuracy* of a DAC is governed by the precision and stability of all the components used in the converter, and by the electrical noise and leakage in the converter circuit.

EXAMPLE 10.1. Given a DAC with a maximum output of 10 V, determine the permitted upper limit of combined errors due to non-ideal components and due to noise to achieve a conversion accuracy of (*a*) 12 bits, and (*b*) 16 bits.

Number of bits	Voltage corresponding to 1 lsb	1 lsb in terms of % of full scale
12	2.4 mV	.024%
16	152 μV	.00152%

It is evident from the above example that a DAC of 16-bit resolution requires the most stringent control and matching of all its components, a thorough shielding from electrical noise, and elimination of leakage currents, if a conversion accuracy comparable to the least significant bit (lsb) is required.

10.2 Binary Weighted Resistance Digital-to-Analog Converter

A binary weighted resistance DAC consists of the following four major components as shown in Figure 10.2: (*a*) *n* switches, one for each bit applied to the input; (*b*) a weighted resistive network; (*c*) a reference voltage V_R; and (*d*) a summing element that adds the currents flowing in the

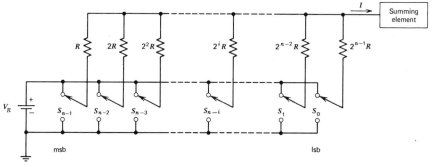

FIGURE 10.2. A binary weighted digital-to-analog converter.

resistive network to develop a signal that is proportional to the digital input.

A register, which is assumed external to the DAC, delivers a binary number, N, of n bits

$$N = a_{n-1} 2^{n-1} + a_{n-2} 2^{n-2} + \cdots + a_1 2^1 + a_0 2^0 = \sum_{i=0}^{n-1} a_i 2^i. \quad (10.1)$$

Each bit, i, controls a switch, S_i, that is connected to V_R when $a_i = 1$ and is connected to ground when $a_i = 0$. The reference voltage source, V_R, is assumed to have zero internal impedance. The resistors that are connected to the switches have values such as to make the current flow proportional to the binary weight of the respective input bit: the resistor in the msb position has the value R, the next has the value $2R$, etc., the resistor of the lsb having a value $2^{n-1}R$. Thus the current flowing into the summing element is

$$I = \frac{a_{n-1} V_R}{R} + \frac{a_{n-2} V_R}{2R} + \cdots + \frac{a_1 V_R}{2^{n-2}R} + \frac{a_0 V_R}{2^{n-1}R}.$$

The above equation may be rewritten as

$$I = \frac{V_R}{2^{n-1}R} \left\{ a_{n-1} 2^{n-1} + a_{n-2} 2^{n-2} + \cdots + a_1 2^1 + a_0 2^0 \right\}$$

$$= \frac{V_R}{2^{n-1}R} \sum_{i=0}^{n-1} a_i 2^i, \quad (10.2)$$

i.e., the output voltage of the DAC is proportional to a number represented by those switches that are connected to V_R, i.e., $a_i = 1$.

Maximum output current will flow when all a_i coefficients are 1

$$I_{max} = \frac{V_R}{R} \frac{2^n - 1}{2^{n-1}}. \quad (10.3)$$

EXAMPLE 10.2. Given an 8-bit DAC with a reference voltage $V_R = 10.00$ V, (a) what is the minimum value of R such that I_{max} shall not exceed 10 mA, and (b) what is the smallest quantized value of I?

(a) From eq. (10.3), $R > \dfrac{V_R}{I_{max}} \dfrac{2^n - 1}{2^{n-1}} = \dfrac{10}{10^{-2}} \cdot \dfrac{255}{128} \cong 2 \text{ k}\Omega$

(b) $I_{min} = \dfrac{V_R}{2^{n-1}R} = \dfrac{10}{128 \cdot 2} \text{mA} = 39.1 \ \mu\text{A}$

In practice, the switches represent finite impedances that are in series with the weighted resistors, and their magnitudes and variations have to be taken into account in a DAC design. Since the contribution to the total output current diminishes by a factor of 2 for each ·lower bit, the tolerances in accuracy and stability of the resistors in the lower order bits are not as exacting as in the high order bits.

EXAMPLE 10.3. Given a 10-bit DAC with a reference voltage, V_R, of 10.000 V and $R = 2$ kΩ, calculate the tolerances of the weighted resistors in (a) the lsb position, and (b) the msb position. The error due to resistance tolerances shall not exceed $\pm\frac{1}{2}$ lsb of full scale.

The smallest current step corresponding to S_0 (Figure 10.2) connected to V_R and all other switches connected to ground is, from eq. (10.2),

$$I(\text{lsb}) = \frac{10 \text{ V}}{512 \times 2 \text{ k}\Omega} \cdot 2^0 \text{ mA} \approx 9.8 \text{ }\mu\text{A}.$$

From eq. (10.3), $I(\text{full scale}) = \dfrac{10 \text{ V}}{2 \text{ k}\Omega} \cdot \dfrac{1023}{512} \approx 10 \text{ mA}.$

(a) A $\pm\frac{1}{2}$ lsb current change equals approximately ± 5 μA, corresponding to a resistance tolerance of $\pm 50\%$ in the lsb position. In practice we are mainly concerned with tolerances of the *ratios* of the weighted resistors, since in monolithic fabrication it is much easier to achieve a close tolerance of resistance ratios rather than a high precision in their absolute values. A $\pm 10\%$ tolerance of the lsb resistor is not uncommon.
(b) The resistance tolerance or the ratio tolerance of the msb is

$$\pm\tfrac{1}{2}I(\text{lsb})/I(\text{full scale}) = \pm 4.9 \text{ }\mu\text{A}/10 \text{ mA} \approx \pm 5 \times 10^{-4} = \pm 0.05\%.$$

10.3 Operational Amplifiers

Equation (10.2) shows an analog output in terms of current. This current can be converted to a voltage signal by use of an operational amplifier, OA. The OA is widely applied in electronic instrumentation. A brief description of the operational amplifier and its parameters is given below. The interested reader may find additional material in references (1) and (2).

An OA with its input and feedback resistors, R_1 and R_f, is shown in Figure 10.3. Without a feedback resistor, R_f, its *open loop gain A* is high. We will show below that the OA *with* feedback has a well defined gain, A_f, that is mainly dependent on the precision and stability of the external components and has negligibly small dependence on the variations of the open loop gain A.

Assume an ideal OA with a feedback network, infinite input impedance Z_s,

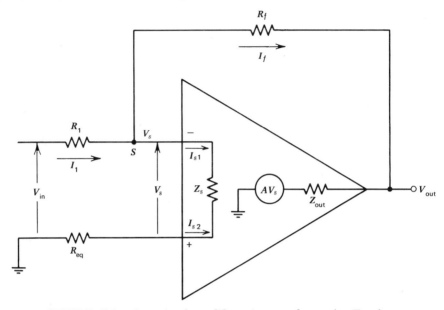

FIGURE 10.3. Operational amplifier: A = **open loop gain,** Z_s = **input impedance,** Z_{out} = **output impedance,** $R_{\text{eq}} = R_1 R_f/(R_1 + R_f)$.

zero input bias currents $I_{s1} = I_{s2}$, and zero output impedance Z_{out}. Referring to the node S in Figure 10.3, also called the *summing junction*, we have

$$V_s = V_{\text{out}}/A. \tag{10.4}$$

If we make the assumption that A approaches infinity, then it follows from eq. (10.4) that V_s approaches 0 volts. For this reason the point S is considered a *virtual ground*.

The gain of such an ideal amplifier is easy to calculate. Consider the current due to the input signal V_{in} flowing through R_1. Since the assumptions were made that (*a*) the OA at junction S demands no input current, and (*b*) the voltage at the summing junction is 0, it follows that

$$I_1 = V_{\text{in}}/R_1. \tag{10.5}$$

The sum of the currents at S is zero by Kirchhoff's law; hence all the input current flows through the feedback resistor R_f, i.e.,

$$I_f = I_1. \tag{10.6a}$$

With the direction of current flow shown in Figure 10.3, we obtain an output voltage V_{out}

$$V_{\text{out}} = - I_f R_f \tag{10.6b}$$

and a *feedback gain* $A_f = V_{out}/V_{in}$. Combining eqs. (10.5) and (10.6) we obtain

$$A_f = -R_f/R_1 \qquad (10.7)$$

independent of the open loop gain A.

When several current sources are connected to S, eqs. (10.6a) and (10.6b) are modified

$$I_f = \sum I_i \qquad (10.8a)$$

$$V_{out} = -R_f \sum I_i. \qquad (10.8b)$$

EXAMPLE 10.4. Consider a 6-bit DAC with a resistance in the lsb position $2^{n-1}R = 320$ kΩ. The reference voltage $V_R = 10.00$ V. The binary weighted resistors are connected to the summing point of an operational amplifier with $A \to \infty$; $R_f = 5$ kΩ. What is the output voltage for a binary input 101010?

From eq. (10.2) the total current flowing into S is

$$I = \frac{V_R}{2^{n-1}R} (1 \cdot 2^5 + 0 \cdot 2^4 + 1 \cdot 2^3 + 0 \cdot 2^2 + 1 \cdot 2^1 + 0)$$

$$= \frac{10 \text{ V}}{320 \text{ kΩ}} (32 + 8 + 2) \text{ A} = 1.312 \text{ mA}.$$

The output voltage $V_{out} = -IR_f = -1.312 \text{ mA} \times 5 \text{ kΩ} = -6.56$ V.

The open loop gain A in modern amplifiers varies from 10^3 to 10^7, and in most OA applications it is in the range of 10^4 to 10^6. Thus the assumption of infinite gain cannot strictly be made in very high resolution DAC's. It can be easily shown, see reference (3), that for a finite value of A the feedback gain A_f is

$$A_f = \frac{-R_f}{R_1[(1 + A)/A] + R_f/A}. \qquad (10.9)$$

Equation (10.9) reduces to eq. (10.7) when $A \gg 1$ and $AR_1/R_f \gg 1$.

Several additional OA parameters have to be considered in the design of precision converters.

Input bias currents, I_{S_1} and I_{S_2}: These are the currents flowing into the input terminals of the OA as shown in Figure 10.3 and range in values from $\approx 10^{-12}$ A to 10^{-4} A. When $I_{S_1} \neq I_{S_2}$ the input bias current may be defined as $(I_{S_1} + I_{S_2})/2$. It can be shown that the output error voltage $V_{\text{out}}(I_S)$ contributed by the bias currents is

$$V_{\text{out}}(I_S) \approx A_f \frac{R_1 R_f}{R_1 + R_f} (I_{S_2} - I_{S_1}) - A_f \left(\frac{R_1 R_f}{R_1 + R_f} - R_{\text{eq}} \right) \frac{I_{S_1} + I_{S_2}}{2} \qquad (10.10)$$

where A_f is the feedback gain as given in (10.9). It follows from (10.10) that the output error due to input bias currents will be zero if R_{eq} is chosen to equal the parallel combination of $R_1 \| R_f$ *and* if the input bias currents are equal. The quantity $I_{S_2} - I_{S_1}$, denoted the *input offset current*, is small compared to the bias current in high quality OA's.

Input offset voltage, V_{off}: This is the voltage change required at the input of the OA to bring its output voltage to 0 volts. This characteristic is inherent in all OA's and can be adjusted through an external control. Usually OA's with high input impedance utilize in their input stage field-effect transistors that are inferior to bipolar transistors in voltage stability as a function of temperature. In practice, low leakage current and low offset voltage drift are contradicting requirements. The error at the output due to offset voltage is $V_{\text{out}}(\text{offset}) = -A_f V_{\text{off}}$.

Additional characteristics of OA's to be considered for high resolution applications are: input offset voltage and bias current drifts as a function of temperature, gain and component value drifts versus time, effect of supply voltage variations, input noise, and *common mode rejection*, i.e., the property of cancelling voltages that are common to both input terminals.

In some applications the output voltage, the output load current ratings, or the *settling time*, t_s, of the OA may be of importance. Settling time is defined as the time elapsed from the application of a step function to the time when

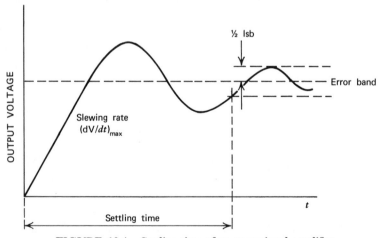

FIGURE 10.4.   Settling time of an operational amplifier.

the OA has entered and remained within a specified error band symmetrical about the final value, commonly specified as $\pm\frac{1}{2}$ lsb. The response of an OA to a step function has usually an overshoot followed by an oscillation, as shown in Figure 10.4. The rate $(dV/dt)_{max}$ at which the voltage can rise toward the maximum output following a step input is referred to as *slewing rate*. Specifications of t_s and dV/dt are of major importance in high speed converters.

10.4 Digital-to-Analog Converter Using a Ladder Network

The weighted resistor DAC of Figure 10.2 requires a wide range of resistance values and matched semiconductor switches for each bit position if high accuracy conversion is required. A DAC with an R-$2R$ ladder network, as shown in Figure 10.5, eliminates these complications at the

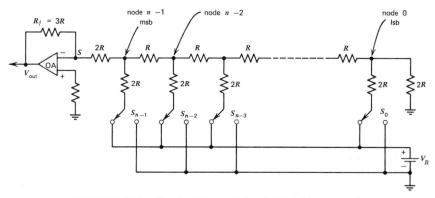

FIGURE 10.5.   DAC with a resistive R-$2R$ ladder network.

expense of an additional resistor for each bit. We will show below that in an R-$2R$ ladder network each bit position contributes to the output in proportion to its binary weight.

Since the ladder network is linear, its operation can be analyzed by superposition; that is, the contribution to the output voltage from each source may be considered independent of the other sources. Finally, all contributions are summed to yield the resultant output V_{out}.

Consider the $(n-1)$th switch (msb) at V_R and all other switches at 0. The resistance at the $(n-1)$th node is $2R$ looking to the left since the input terminal of the OA is at virtual ground. The reader can easily verify that the resistance at any node looking into either direction of the ladder network is $2R$ provided the end nodes of the network are terminated by $2R$. Thus the current due to the voltage source V_R divides equally, as shown

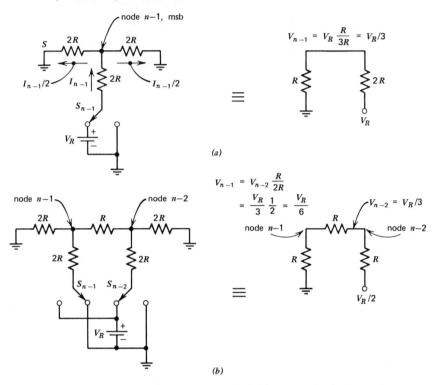

FIGURE 10.6. Equivalent circuits of the ladder network of Figure 10.5: (a) currents and voltage output at the msb node with only switch S_{n-1} connected to the reference voltage, (b) voltages at nodes $n-2$ and $n-1$ with only switch S_{n-2} connected to the reference voltage.

in Figure 10.6a, yielding an output from the contribution of the $(n-1)$th voltage only

$$V_{n-1} = V_R \frac{R}{3R} = V_R/3,$$

where V_R is the reference voltage. Since we have chosen the gain of the OA with respect to node $n-1$ to be $-3/2$ ($R_f = 3R$ in Figure 10.5), the output voltage due to the msb alone is

$$V_{\text{out}}(\text{msb}) = (V_R/3) \cdot (-3/2) = -V_R/2.$$

Considering next, switch S_{n-2} at V_R with all other switches at 0, see Figure 10.6b, we have a similar situation as before. That is, the currents at node $n-2$ divide equally to the right and to the left resulting in a voltage at node $n-2$

$$V_{n-2} = -V_R/3.$$

This voltage is attenuated by a factor of 2 at node $n - 1$ as is shown in the equivalent circuit of Figure 10.6b. Thus its contribution to the output at the operational amplifier is

$$V_{out} (\text{due to node } n - 2) = (V_R/3) \cdot (1/2) \cdot (-3/2) = V_R/4.$$

The same procedure applied to the remainder nodes yields

$$V_{out} = -V_R(a_{n-1}2^{-1} + a_{n-2}2^{-2} + \cdots + a_1 2^{-(n-1)} + a_0 2^{-n})$$
$$= -V_R 2^{-n}(a_{n-1}2^{n-1} + a_{n-2}2^{n-2} + \cdots + a_1 2^{-1} + a_0 2^0)$$
$$= -V_R 2^{-n} \sum_{i=0}^{n-1} a_i 2^i, \tag{10.11}$$

where $a_i = 0$ or 1 depending on whether the ith switch is at 0 or at V_R, respectively. Thus the output of the DAC is proportional to the sum of the weights represented by those switches that are connected to V_R.

10.5 BCD Digital-to-Analog Converters

A DAC can be designed for any weighted code using suitably weighted resistors. In the case of a biased weighted code such as the excess-3 BCD, an additional current source is introduced to cancel the offset produced by the code (see Problem 10.3).

EXAMPLE 10.5.   Given the 2-digit 8-4-2-1 BCD digital-to-analog converter with resistive summing, as shown in Figure 10.7a, calculate the output voltage corresponding to BCD input = 1001 1001 = 99_{10}.

Since the output is obtained across the finite load resistor R_L, the output voltage V_{out} is

$$V_{out} = V_R \cdot k \frac{R_L}{R_{out} + R_L},$$

where k is a function of the digital input, and R_{out} is the output impedance of the weighted resistor network

$$\frac{1}{R_{out}} = \frac{1}{R} + \frac{1}{2R} + \frac{1}{4R} + \frac{1}{8R} + \frac{1}{10}\left(\frac{1}{R} + \frac{1}{2R} + \frac{1}{4R} + \frac{1}{8R}\right);$$

$$R_{out} = \frac{80}{165} R.$$

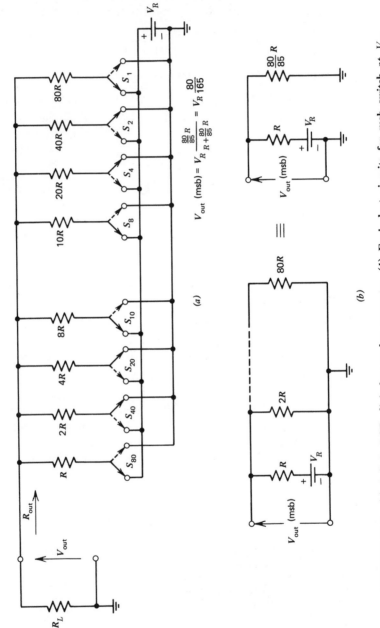

$$V_{out} \text{ (msb)} = V_R \frac{\frac{80}{85}R}{R + \frac{80}{85}R} = V_R \frac{80}{165}$$

(a)

(b)

FIGURE 10.7. (a) 2-digit BCD digital-to-analog converter. (b) Equivalent circuits for msb switch at V_R, others at 0.

The voltage due to the msb $= 1$, all other inputs being 0 and the load resistor disconnected, is

$$V'_{out}(msb) = V_R \frac{80}{165}$$

as can be seen from the equivalent circuit of Figure 10.7b. With the load resistor connected,

$$V_{out}(msb) = V_R \frac{80}{165} \frac{R_L}{R_{out} + R_L}.$$

The contributions of the other weights for which the input is 1 are obtained using the same analysis technique. The result for a BCD input of 99_{10} is, by superposition,

$$V_{out} = V_R \left(\frac{80}{165} + \frac{10}{165} + \frac{8}{165} + \frac{1}{165} \right) \frac{R_L}{R_{out} + R_L}$$

$$= V_R \frac{99}{165} \frac{R_L}{R_{out} + R_L} = 99 \times \text{constant, i.e., proportional to the BCD input.}$$

A high resolution DAC using weighted BCD resistors, such as shown in the example above, requires a large range of resistance values, which presents a serious matching problem. The BCD ladder network discussed below utilizes an identical structure for each decade and achieves superior performance. As shown in Figure 10.8a each decade excepting the msd consists of four BCD weighted resistors and a series resistance R_s. The sum of the outputs of the decade networks will be proportional to the BCD input if R_s is made to attenuate each decade output by a factor of 10 as the signal propagates towards the output (msd) end of the converter.

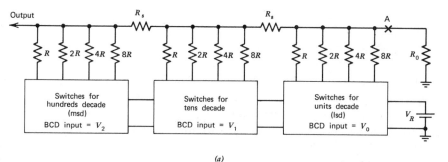

(a)

FIGURE 10.8.   (a) 3-digit BCD digital-to-analog converter; each BCD decade presents a parallel resistance $R_p = \frac{8}{15}R$. [*See next page for* (**b**).]

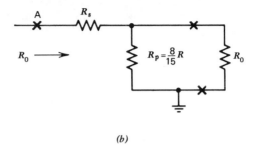

(b)

FIGURE 10.8. (*b*) Equivalent circuit showing an L section consisting of R_s and R_p, and the characteristic resistance R_0.

To calculate the value of R_s we first consider the properties of the ladder network of Figure 10.8*a* which consists of two identical L sections. Each L section has a series resistance R_s and four BCD weighted resistors that are connected either to a voltage source of ideally zero internal impedance, or to ground. Hence their parallel resistance, R_p, is constant: $R_p = \frac{8}{15}R$. The value of the terminating or *characteristic resistance*, R_0, is chosen to obtain a constant resistance (as viewed into the network from its output end), independent of the number of L sections.

EXAMPLE 10.6.   Calculate the characteristic resistance, R_0, of the ladder network of Figure 10.8*a*.

Break the network at point A and insert an L section as shown in Figure 10.8*b*. The input resistance as looked at towards the terminating end must remain unchanged since the characteristic resistance is independent of the number of L sections, by definition. Thus,

$$R_s + \frac{R_0 R_p}{R_0 + R_p} = R_0, \tag{10.12}$$

$$R_0 = \frac{R_s}{2} + \sqrt{\frac{R_s^{\,2}}{4} + R_s R_p}. \tag{10.13}$$

To calculate the output, V_{out}, due to the msd stage only, we apply the superposition principle, removing all other voltage sources and replacing them with short circuits. The equivalent circuit is shown in Figure 10.9 where the ladder network due to the lower order digits was replaced by its characteristic resistance, R_0. Thus

$$V_{out}(\text{hundreds}) = V_2 \frac{R_0}{R_0 + R_p}. \tag{10.14}$$

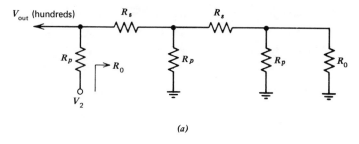

(a)

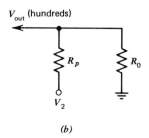

(b)

FIGURE 10.9. (a) BCD ladder network with an input to the most significant digit only. (b) Equivalent circuit.

Next we consider the tens BCD digit, see Figure 10.10a. Looking from point B to the right the circuit presents R_0; the resistance looking to the left is $R_s + R_p$. From the equivalent circuit of Figure 10.10b and substituting for $R_s + \dfrac{R_0 R_p}{R_0 + R_p}$ from eq. (10.12), we obtain

$$V_{out}(\text{tens}) = V_1 \left(\frac{R_0}{R_0 + R_p} \right) \cdot \left(\frac{R_p}{R_0 + R_p} \right). \tag{10.15}$$

Assuming that the BCD input combinations to the two decades are identical and that $V_2 = V_1 = V_{ref}$, we require that the output of the most significant digit BCD network be exactly 10 times that of the next lower order BCD. From this condition we obtain the value of the attenuating resistor, R_s.

From eqs. (10.14) and (10.15),

$$V_{ref} \frac{R_0}{R_0 + R_p} = 10 \, V_{ref} \frac{R_0}{R_0 + R_p} \frac{R_p}{R_0 + R_p},$$

$$R_0 = 9 \, R_p. \tag{10.16}$$

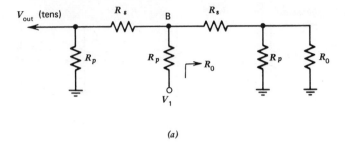

(a)

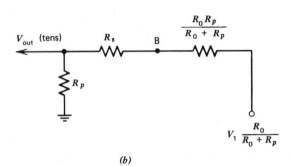

(b)

FIGURE 10.10. (*a*) **BCD ladder network with an input to the tens digit only.** (*b*) **Equivalent circuit.**

Substituting for R_0 in eq. (10.12), we obtain R_s:

$$R_s = 9 R_p - \frac{9 R_p^2}{10 R_p} = 8.1 R_p;$$

and since $R_p = (8/15) R$ we have

$$R_s = 4.32 R. \tag{10.17}$$

The BCD ladder network developed above is easily implemented with monolithic technology. Each BCD circuit requires five resistors and is identical for all decades except the msd in which R_s is omitted. The ratio of the largest to the smallest resistance is 8.

10.6 DAC's. Miscellaneous Topics

Multiplying DAC. The output voltage of a digital-to-analog converter as a function of the input may be expressed as

$$V_{\text{out}} = V_R \cdot k_1, \tag{10.18}$$

where k_1 is proportional to the digital input of the DAC. Assume that V_R is substituted by a signal, V'_{out}, derived from another DAC with an input k_2:

$$V'_{out} = V'_R \cdot k_2. \tag{10.19}$$

Substituting in eq. (10.18) we obtain

$$V_{out} = V'_{out} \cdot k_1 = V'_R \cdot k_1 \cdot k_2, \tag{10.20}$$

i.e., an analog voltage that is proportional to the *product* of two digital numbers.

Bipolar DAC. The output of the DAC's discussed thus far depended on the polarity of the reference voltage, V_R, and on the details of current summing. For resistive summing the output has the same polarity as V_R. When an operational amplifier is used, the polarity depends on whether the signal is applied to the non-inverting or to the inverting input, the latter configuration being more common.

Bipolar operation may be obtained through switching the polarity of the reference voltage, V_R, depending on the sign of the input signal. The digital input to the DAC may be in sign-and-magnitude form and the resistance of the polarity switch must be negligibly small. Alternatively, the digital input may be in 1's-complement or 2's-complement form.

Bipolar operation may also be attained by applying an offset current to either one of the inputs of the operational amplifier. The magnitude of this offset current equals one msb for a *half-scale offset code*, and its polarity is determined by whether this current is applied to the inverting or non-inverting input terminal of the operational amplifier. Comparing the 2's-complement and the offset codes shown in Figure 10.11, we note that they differ in the msb only; thus the binary value of this bit determines whether the offset current is turned on.

Signal	Digital input		Analog output
	2's-complement	Half-scale offset	
− Full scale	1 0 0 0 0 0 0 1	0 0 0 0 0 0 0 1	$-V_0$
− lsb	1 1 1 1 1 1 1 1	0 1 1 1 1 1 1 1	$-V_0/(2^7 - 1)$
zero	0 0 0 0 0 0 0 0	1 0 0 0 0 0 0 0	0
+ lsb	0 0 0 0 0 0 0 1	1 0 0 0 0 0 0 1	$+V_0/(2^7 - 1)$
+ Full scale	0 1 1 1 1 1 1 1	1 1 1 1 1 1 1 1	$+V_0$

FIGURE 10.11. The 2's complement and half-scale offset codes and the resulting bipolar analog outputs for a 7-bit DAC.

Sources of Error in Digital-to-Analog Conversion.   The conversion accuracy of a DAC is determined by the matching and temperature tracking of the resistor network, the stability of the reference voltage, and the properties of the summing amplifier.

The resistance of the switches may be considered part of the weighted resistors, $2^i R$, or of the shunt ladder resistance $2R$. A high degree of accuracy can be maintained if the temperature coefficients of the switches and of the resistor network are low and matched, hence it is desirable to make the resistance of the switches low to minimize the effect of their drifts. The resistors, including the operational amplifier feedback resistor, may be manufactured on the same substrate for close temperature tracking (see Problem 10.6).

The transistor switches are well matched when they are fabricated on the same monolithic chip. In the case of weighted DAC's the current density in the transistors is kept uniform by varying the emitter area such that it is proportional to the current switched by a given transistor.

The response time is dependent on the slew rate, on the settling time of the operational amplifier to the required resolution, and on the time constant of the resistors with the associated stray capacitances of the ladder network. The properties of the operational amplifier are usually the dominant factors in response time.

10.7   Analog-to-Digital Conversion. General Considerations

A great variety of analog-to-digital converters, ADC's, has been developed to satisfy a broad spectrum of requirements. In some applications the dominant parameters are the precision and stability of conversion; in others, conversion speed is of greatest importance. Economic considerations also enter into the choice of the conversion circuitry. Simplicity of design, however, is usually attained at the expense of lower conversion rates.

In some systems a single ADC is utilized for conversion into digital form of several analog inputs. In such a configuration two additional circuit components are usually required: (a) a *multiplexer* that performs the function of an analog switch connecting the various analog sources, one at a time, to a common ADC; and (b) a *sample-and-hold* circuit that measures the analog quantities in a brief time interval and holds these values constant (within the allowed error) until they can be applied via a multiplexer to the ADC, for conversion.

In this section we discuss ADC performance characteristics, factors affecting their conversion accuracy, and we describe components that are most commonly utilized in ADC's.

Error Sources. The *quantizing* error, or *resolution*, of an ADC is determined by the number of bits chosen to represent the analog quantity and has a maximum value of $\pm\frac{1}{2}$ lsb. *Electronic equipment error* is the sum of the errors contributed by the circuitry through which the analog signal has to pass before and during conversion. This class of errors may be divided into *random* and *systematic* errors. Some systematic errors can be eliminated through careful adjustments. Others, such as temperature dependent errors and the finite value of power supply regulation, have to be added linearly for the worst case condition, i.e., over the total temperature range, ΔT, that the ADC is specified to operate, and over the maximum power supply variations, ΔV.

The random errors are due to a statistical distribution of component values, such as resistor networks, ΔR, switch resistance, Δr, of the DAC, often used as a feedback element in an ADC, and gain stability of the amplifier, ΔA_f. Note that the components enumerated above also have systematic errors that have to be considered in an ADC of high overall accuracy.

Random errors follow a Gaussian distribution; thus their total contributed *expected error*, ε_r, is

$$\varepsilon_r = \sqrt{\sum \varepsilon_i^2}.$$

where ε_i are the individual error components.* The quantizing error, ε_Q, is a random phenomenon and its value in an n-bit ADC is

$$\varepsilon_Q(\%) = \frac{\frac{1}{2} \text{ lsb}}{\text{full scale}} \times 100 = \frac{0.5}{2^n - 1} \times 100.$$

EXAMPLE 10.7. Calculate the total error, ε, in a 10-bit ADC having the following error sources: composite errors due to temperature effects, $\varepsilon_T = 0.1\%$; composite errors due to power supply variations, $\varepsilon_V = 0.05\%$; analog multiplexer error, $\varepsilon_M = 0.1\%$; comparator error, $\varepsilon_c = 0.15\%$; and DAC error, $\varepsilon_D = 0.13\%$.

The first two error sources are systematic; the remaining errors are random. Thus, including the quantizing error,

$$\varepsilon = \varepsilon_T + \varepsilon_V + \left(\varepsilon_M^2 + \varepsilon_c^2 + \varepsilon_D^2 + \varepsilon_Q^2\right)^{1/2}$$

$$= 0.1 + 0.05 + \sqrt{10^{-2} + 2.25 \times 10^{-2} + 1.69 \times 10^{-2} + 0.25 \times 10^{-2}} \%$$

$$= 0.378 \%$$

---

* Note that the expected error, ε_r, may be exceeded in any given ADC because the *worst case* error is $\sum |\varepsilon_i|$.

As in the above example, the quantizing error commonly constitutes a negligible fraction of the total conversion error.

Conversion Rate.   The conversion word rate is the number of digital words of the required resolution encoded per unit time. The finite conversion time is a source of error when the analog signal is varying with time and sampling (to be discussed on page 354) is not employed.

EXAMPLE 10.8.   Determine the maximum allowable conversion time, t_c, for an 8-bit ADC with a sinusoidal input signal $V_{in} = \dfrac{V_{p\text{-}p}}{2} \sin 10^3 t$, as shown in Figure 10.12. The error due to conversion time shall not exceed $\frac{1}{2}$ lsb.

The maximum rate of change of a sinewave is at zero phase angle. For small angles, $\sin \omega t \cong \omega t$ radians. Equating the change of the signal during conversion time with $\frac{1}{2}$ lsb, we have

$$\frac{V_{p\text{-}p}}{2} \omega t_c = V_{p\text{-}p} \cdot \frac{\frac{1}{2}}{2^n - 1}$$

$$t_c = \frac{1}{\omega(2^n - 1)} = 3.92 \ \mu s.$$

Analog Voltage Comparators.   The comparator shown in Figure 10.13a is a dc coupled high gain amplifier with a differential input and a binary output: logical 0 or 1. Its *sensitivity*, i.e., the minimum analog voltage difference between the input terminals required for the output to cross

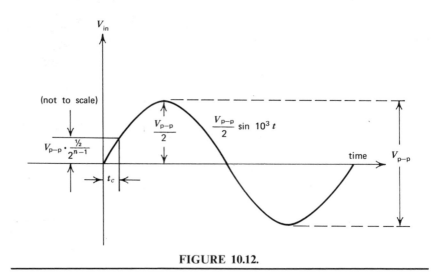

FIGURE 10.12.

the logic threshold (see Figure 10.13*b*), and its *stability* are determining factors in the overall accuracy of the conversion. Several parameters of the voltage comparator are defined in the same way as the corresponding parameters of the operational amplifier discussed in Section 10.3. Some of these are: open loop gain, input bias current, common-mode rejection, and input and output impedances.

Parameters that are of specific importance in voltage comparators are:

1. *Response time.* This is the time between the application of an input step function and the time when the output crosses the logic threshold voltage. Response time is a function of overdrive, as shown in Figure 10.13*c*.

2. *Offset voltage* is the change in input voltage required to bring the output to a predetermined potential, commonly a logic threshold voltage.

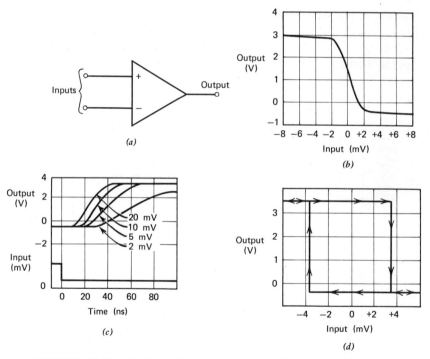

FIGURE 10.13. Analog voltage comparator: (*a*) symbol, (*b*) output voltage versus input voltage, (*c*) response time for input overdrives of 2 mV to 20 mV, (*d*) hysteresis in output voltage versus input voltage when positive feedback is employed.

3. *Hysteresis.* Some comparators employ positive feedback which results in hysteresis (backlash). In Figure 10.13d, for example, the comparator will change states from logic 1 to logic 0 at approximately $+3.7$ mV. To change its state from logic 0 to logic 1 requires an input of -3.7 mV. The difference in the "turn-on" and "turn-off" input voltages is the *hysteresis* of the comparator. An analog voltage comparator with hysteresis has usually a better noise immunity, a faster response time, and a worse input sensitivity than comparators without hysteresis.

4. *Recovery time.* The input signal to the comparator is in many applications of sufficient amplitude to drive the input stage into cut-off or saturation. The response of the circuit is slowed down when it operates in these non-linear regions. A fast recovery from overdrive is a desirable property of analog voltage comparators.

10.8   Analog-to-Digital Converters (ADC's)

Counter Ramp ADC. The analog-to-digital converter shown in Figure 10.14a incorporates a binary counter and a DAC in a feedback loop. The analog signal is applied to one input and the feedback signal to the other input of a voltage comparator. The counter is initially reset to zero; at the

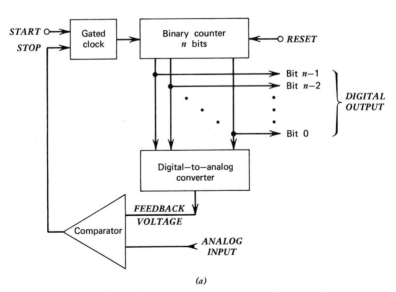

(a)

FIGURE 10.14.   Counter ramp ADC: (a) schematic.

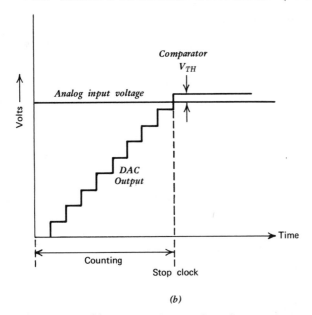

FIGURE 10.14.   (*b*) input signals to analog voltage comparator.

beginning of a conversion cycle the gated clock is actuated and will increment the counter until the output of the DAC has exceeded the analog input signal by the threshold voltage, V_{TH}, of the comparator, as shown in Figure 10.14*b*. The resultant change in the state of the comparator stops the clock. A linear ramp generator may replace the DAC in low precision ADC's. The clock period, T, of a counter ramp converter must satisfy the condition

$$T \geq t_1 + t_2 + t_3,$$

where t_1 = maximum propagation delay time of the counter
t_2 = maximum response time of the DAC
t_3 = propagation delay and slewing time of the comparator.

The maximum conversion time, t_c, equals $2^n - 1$ clock periods, where n is the number of bits of the counter.

EXAMPLE 10.9.   In a 10-bit counter ramp ADC a ripple counter is used with $t_1 = 200$ ns, a ladder network with $t_2 = 500$ ns, and a comparator with $t_3 = 40$ ns. Determine (*a*) the conversion time t_c for maximum input, and (*b*) the worst case conversion rate.

To determine the maximum allowable clock frequency, $f_{max} = 1/T$:

$$T = 200 + 500 + 40 = 740 \text{ ns}; f_{max} = 1.36 \text{ MHz}.$$

(*a*) Conversion time for maximum input is

$$t_c = T \cdot (2^n - 1) = 740 \text{ ns} \times 1023 = 756 \text{ } \mu s$$

(*b*) Worst case conversion rate: $1/t_c = 1.32$ kHz

Tracking ADC. Continuous digital monitoring of an analog voltage is possible when the unidirectional counter of Figure 10.14*a* is replaced with a bidirectional counter. The output state of the comparator determines the direction of the count (up or down), depending on whether the feedback voltage is smaller or larger than the input voltage. Tracking speed presents no problem as long as the input voltage variations are relatively slow. The initial state of the counter may be set to half scale, i.e., 01 ... 1, to reduce the average time required for the *first* conversion.

Successive Approximation ADC. The operation of this converter is based on *n* successive comparisons between the analog input, V_{in}, and the feedback voltage, V_f. It is analogous to the weighing process in which the unknown quantity is compared with a reference quantity. The first comparison determines whether V_{in} is greater or smaller than $\frac{1}{2}V_{max}$, where V_{max} is the maximum possible input to the ADC. The next step determines in which quarter of the range V_{in} is found; each successive step narrows the range of the possible result by a factor of 2. The operation of a 3-bit successive approximation ADC is illustrated in the transition diagram of Figure 10.15. The heavy lines show the transitions of an ADC for a conversion to a binary number 101.

n clock periods are required to complete a conversion cycle with a resolution of one part in $2^n - 1$. The conversion word rate is thus considerably higher than in the counter ramp ADC, at the expense of additional circuitry.

The major components of a successive approximation ADC are shown in Figure 10.16. It is essentially a feedback circuit similar to the counter ramp ADC in which the digital output is converted to an analog quantity, V_f, in the feedback loop, and is compared with the unknown, V_{in}. The addition of a shift-register and a *logic programmer* enable the circuit to decide on the next "weighing" step to be taken. Whereas the counter ramp ADC increments monotonically until the comparator threshold has just been exceeded (see Figure 10.14*b*), the successive approximation ADC increments or decrements as a result of the decision made by the comparator and executed by the logic programmer after each "weighing."

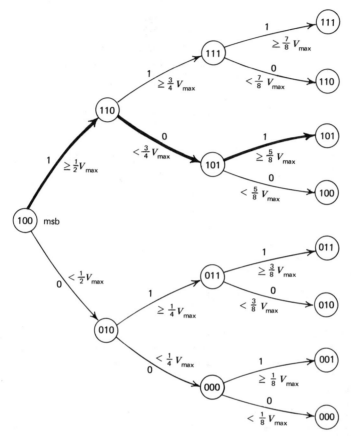

FIGURE 10.15.   Successive approximation ADC; transition diagram for a 3-bit converter.

A conversion cycle is initiated by a START pulse which sets the msb's of the two registers to "1," resetting all other flip-flops. The output of the storage register in state 100...0 is converted by the DAC to an analog quantity, V_f, which is proportional to $2^{n-1}/2^n - 1$ of full scale, i.e., it is $\frac{1}{2}$ lsb above half scale. For $V_f < V_{in}$ the msb of the storage register is left in state "1" and the shift-register shifts its "1" to the next bit $(n-2)$, loading a "1" also in the $(n-2)$ bit of the storage register. If, however, $V_f > V_{in}$ the msb in the storage register is reset. In either case, bit $(n-2)$ representing $\frac{1}{4}$ of the full scale is examined next. The conversion process is completed after the lsb has been examined. Conversion time, t_c, is thus constant, $t_c = n/f$, where n is the number of bits and f is the clock frequency. In comparison with the counter ramp ADC, the successive

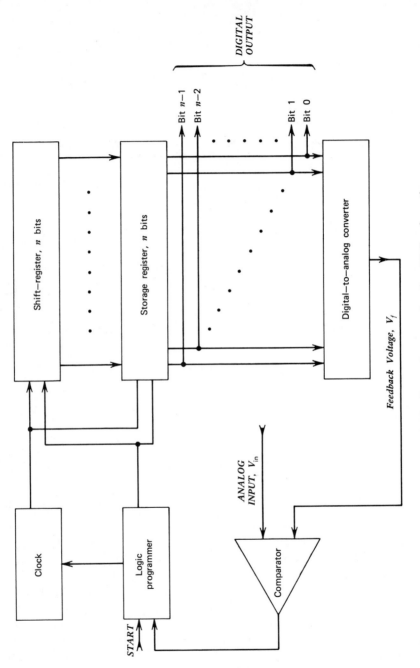

FIGURE 10.16. Successive approximation ADC; schematic.

approximation method is faster but exhibits a poorer differential linearity; the latter may be defined as the deviation from the average analog input increment required to change the digital output by one lsb.

10.9 ADC's. Miscellaneous Topics

Double Ramp ADC. Very good conversion accuracies may be obtained by the double ramp method which is often used in digital voltmeters where the conversion rate may be low. The block diagram of the ADC in Figure 10.17a has four major components: a voltage integrator, an analog voltage comparator, logic circuits, and a reference voltage V_R.

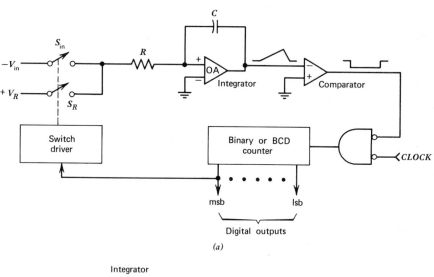

(a)

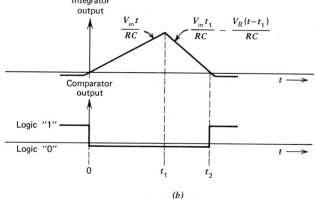

(b)

FIGURE 10.17.   Double ramp ADC: (a) schematic, (b) waveforms.

Basically the circuit is a voltage-to-time converter in which the time required for converting the analog voltage V_{in} is compared with the conversion time of a precision reference voltage V_R. The circuit operates as follows. In the initial state, C of the integrator is discharged and the counter is set to 00...0. A start pulse connects $-V_{in}$, via S_{in}, to the input of the integrator which is composed of an operational amplifier and an RC time constant network. It can be shown that the output voltage at time t_1, V_{o1}, of this configuration is the time integral

$$V_{o1} = -\frac{1}{RC} \int_0^{t_1} (-V_{in}) \, dt = \frac{V_{in} t_1}{RC}. \tag{10.21}$$

The linearly rising output voltage changes the state of the comparator as shown in Figure 10.17b. This change of logic state enables the clock gate and the counter increments until it reaches 10...00. The 1 in the counter msb changes the state of the switches, S_{in} and S_R, thus connecting the reference voltage, V_R, via S_R, to the input of the integrator. V_R and V_{in} are of opposite polarity; hence C is discharged during the time interval $t_2 - t_1$, yielding an output at time t_2

$$V_{o2} = V_{o1} - \left(\frac{1}{RC} \int_{t_1}^{t_2} V_R \, dt\right) = V_{o1} - \frac{V_R(t_2 - t_1)}{RC} \tag{10.22}$$

At $t = t_2$, C is fully discharged and $V_{o2} = 0$. The comparator senses the zero output from the integrator and changes states, which inhibits the clock and stops the counter.

Equating V_{o2} to zero in eq. (10.22) and substituting for V_{o1} from eq. (10.21), we obtain

$$\frac{V_{in} t_1}{RC} = \frac{V_R(t_2 - t_1)}{RC}. \tag{10.23}$$

t_1 is constant since it is the time required for the counter to reach 10...0; thus for an n-bit counter $t_1 = 2^{n-1}/f$. The time interval $t_2 - t_1 = N/f$, where N is the number of counts accumulated during $t_2 - t_1$. Thus, from eq. (10.23)

$$V_{in} = V_R \frac{t_2 - t_1}{t_1} = N \frac{V_R}{2^{n-1}}. \tag{10.24}$$

The dual ramp method eliminates errors due to propagation delay times in the electronic circuitry and is self-compensating for variations in the values of C, R, and in the clock frequency, which are common to both ramps. The comparator offset voltage and current are also automatically compensated since the input signal crosses the zero twice. However, drifts in V_R and in the integrator offset voltage and input currents are not compensated by the dual ramp method.

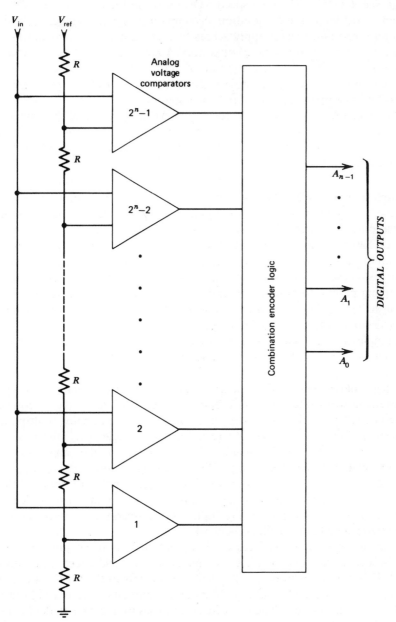

FIGURE 10.18. High speed ADC.

High Speed ADC. The high speed ADC shown in Figure 10.18 uses $2^n - 1$ comparators for an n-bit resolution. A common reference voltage, V_R, and a precision resistance chain apply a bias to each comparator that differs by one lsb from its neighboring comparators. A combinational circuit encodes the states of the $2^n - 1$ comparators into n bits. The conversion is asynchronous, its speed being determined by the sum of the propagation delays of a comparator and of the encoding logic.

The method is limited to low resolution ADC's because of the extensive amount of circuitry required. A combination of this technique and the successive approximation method yields a relatively fast ADC of high resolution (see Problem 10.19).

Sample-and-Hold Circuits. A sample-and-hold circuit is shown in Figure 10.19a. It is required in analog-to-digital conversion when the input voltage variations ΔV_{in} during the conversion time, t_c, exceed the resolution of the ADC. Sampling is effected by closing S_1 for the *sampling time* with a duration $\Delta t \ll t_c$ to charge capacitor C to V_{in}. Upon completion of a conversion cycle, S_2 is closed to discharge C, preparing it for the successive sampling period. In practice, S_1 and S_2 are semiconductor switches with a finite switching time (*aperture time*) which also has its uncertainty, denoted *aperture time jitter*.

Sample-and-hold circuits may also be used to time-share the analog output from one DAC, as shown in Figure 10.19b.

Analog Multiplexers.[4] Multiplexer circuits may be used when the required conversion word rate is not high, and thus one ADC can serve to convert several analog signals into digital form. The circuit of an analog multiplexer shown in Figure 10.20 may be regarded as a single-pole multiposition switch in which the "common" is connected to the input of the operational amplifier. The voltage follower configuration of the operational amplifier shown in Figure 10.20 exhibits a high input impedance with a gain that approaches unity. A higher gain may be obtained by connecting the "common" to the inverting input of the operational amplifier and employing a suitable feedback network.

The switches shown in Figures 10.19 and 10.20 may be junction type field-effect transistors, JFET's, or metal oxide field-effect transistors, MOSFET's. Both exhibit very low leakage currents in the OFF state, zero offset voltage in the ON state (unlike bipolar transistors), and a high ratio of OFF/ON impedance. The ON resistance varies from device to device and should be negligibly small in comparison with the input impedance of the load circuit. A small ON resistance is desirable in sample-and-hold circuits to keep the charging and discharging time constants to a low value.

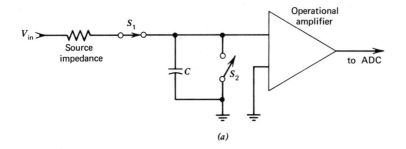

(a)

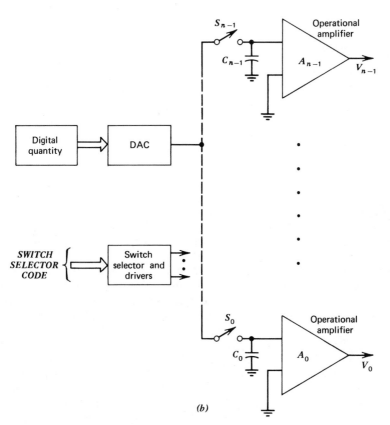

(b)

FIGURE 10.19. Sample-and-hold circuit: (*a*) application in analog-to-digital conversion, (*b*) time sharing of a DAC in digital-to-analog conversion (discharge switches not shown).

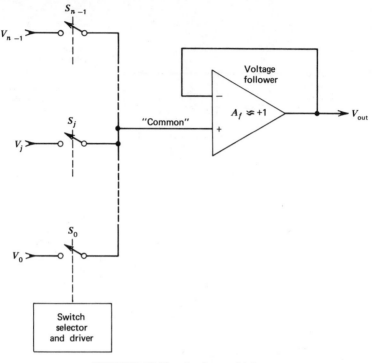

FIGURE 10.20. Analog multiplexer.

PROBLEMS

10.1 Derive eq. (10.9) of Section 10.3.

10.2 Given an operational amplifier with an open loop gain $A = 5 \times 10^4$, calculate the resolution of a DAC using this amplifier if the error due to the finite gain must not exceed a half lsb. Neglect all other error sources.

10.3 Design a weighted resistance DAC for one decade of an excess-3 BCD input when (a) the DAC output is derived across a load resistor $R_L = R_{out}$, where R_{out} is the output resistance of the decade network; and (b) the DAC output is applied to an ideal operational amplifier with a feedback resistor $R_f = R_{out}$.

10.4 Determine the minimum open loop gain of an operational amplifier used in a 10-bit DAC if the error due to its finite gain must not exceed 0.01 %.

10.5 Derive the value of the characteristic resistance of an R-$2R$ ladder network.

10.6 A 6-bit binary weighted DAC as shown in Figure 10.2 has the following temperature tracking coefficients, TTC's, over its operating temperature range: for R_0 (lsb) and R_1, TTC $= \pm 0.05\%$; for R_2 and R_3, TTC $= \pm 0.02\%$; and for R_4, TTC $= \pm 0.01\%$. (R_5 is the msb resistor to which the variations due to temperature of R_0 through R_4 are referenced; hence the absolute value of the temperature coefficient of R_5 is of no consequence.) Calculate the percentage conversion error contributed by the finite TTC's.

10.7 Calculate the resistance ratio tolerances in an 8-bit binary weighted DAC if the contribution to the overall error is not to exceed 0.04%.

10.8 A BCD digital-to-analog converter can be implemented using identical binary weighted resistors with a different reference voltage in each decade. Determine the ratio of the reference voltages for a 2-decade converter.

10.9 Using a 10-bit counter, a 10-bit DAC, and an operational amplifier, design a "staircase" generator. Assuming zero time for reset after the maximum voltage has been reached, derive the frequency of a repetitive waveform in terms of the counter clock frequency. Discuss the merits and limitations of such a system.

10.10 A *resolver* is an electromechanical device with a wound rotor and two perpendicularly oriented stator windings. The rotor is positioned at an angle x by voltages that are applied to the stator windings, one being proportional to sin x the other to cos x. Design a resolver positioning system using a 400 Hz reference signal and two DAC's. How many bits are required for a resolution of $1°$?

10.11 A BCD digital-to-analog converter can be implemented using R-$2R$ ladder networks separated by R_s as shown in Figure 10.21. Derive R_0 and R_s in terms of R.

10.12 Consider the ladder network of Figure 10.22 in which R_L is the load resistance and the voltage sources, V_i, are assumed to have zero internal impedance. (a) Calculate the resistance, R_0, as seen to the right of R_L. (b) Derive an expression for the output voltage as a function of the inputs. Assume that any voltage source may have a value of V_R or 0 volts.

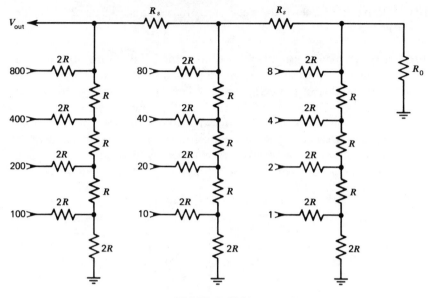

FIGURE 10.21.

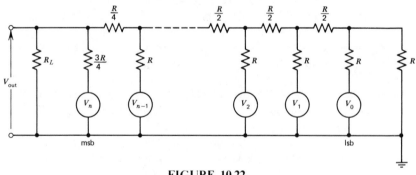

FIGURE 10.22.

10.13 The DAC shown in Figure 10.23 consists of two identical 4-bit weighted resistor networks separated by a resistive current divider R_p and R_s. Calculate the ratios R_s/R_p for (a) an 8-bit binary-to-analog converter, and (b) a 2-decade BCD digital-to-analog converter.

10.14 The DAC of Figure 10.23 uses non-saturating current switches. Discuss current and voltage switching and their effects on the reference supply, and on the conversion speed.

10.15 Design an 8-bit tracking type ADC using a DAC generated ramp and a bidirectional counter. Show the control logic for the counter.

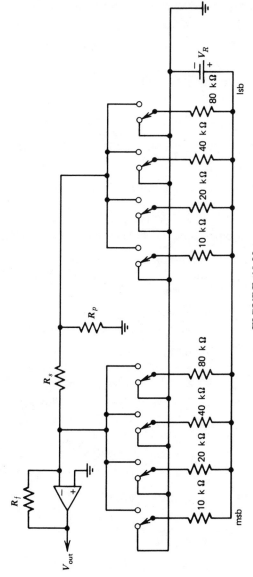

FIGURE 10.23.

10.16 A tracking type 8-bit ADC is used to measure a symmetrical triangular waveform of peak amplitude V_{max} and a period $T = 1$ ms. Calculate the minimum clock frequency if the error due to tracking is not to exceed 0.1%.

10.17 A 10 kHz sine wave having a peak-to-peak amplitude of V_{p-p} volts is applied to a tracking type ADC of eight bits. What is the minimum clock frequency required to keep the conversion error to $\frac{1}{2}$ lsb?

10.18 Consider a counter ramp type 7-bit ADC with a clock frequency of 10 MHz. Find the maximum frequency of a sine wave input signal such that the conversion error shall not exceed $\frac{1}{2}$ lsb of peak-to-peak signal.

10.19 The high speed ADC using $2^n - 1$ comparators for a n-bit resolution shown in Figure 10.18 can be combined with a successive approximation ADC to yield a p-bit conversion resolution in p/n steps. Draw a block diagram of such an ADC for $p = 12$ and $n = 3$.

10.20 Refering to the schematic of Figure 10.16 and the description of the successive approximation ADC in Section 10.8, produce a detailed design of the "logic programmer" for a 3-bit ADC.

10.21 Derive the logic equations for a combinational encoder used in a high speed ADC of the type shown in Figure 10.18. Assume an ADC of 3-bit resolution.

CHAPTER 11

Large Scale
Integrated Circuits (LSI)

11.1 Introduction

Large scale integration, LSI, is a term describing the level of complexity of gates on a single semiconductor chip. Three basic technologies are used in manufacturing LSI circuits: bipolar transistors, metal-oxide field-effect transistors, (MOSFET'S), and complementary metal-oxide field-effect transistors. In bipolar technology a circuit on a single chip is commonly considered an LSI if it contains a complexity equivalent to 100 gates or more. The definition of MOS LSI is usually made in terms of the number of FET's on a single chip: generally a MOS circuit with a complexity of 300 FET's or more is considered an LSI circuit. The limits of complexity in these LSI circuits are dictated by the maximum chip size that still allows reasonable fabrication yields and by the maximum power that can be dissipated in the small size of the circuit. 8000 FET'S on one chip are presently available: smaller transistor elements resulting in higher densities as well as lower stand-by power techniques are under development, promising an even higher level of complexity.

The advent of LSI circuits resulted in performance capabilities that were not considered possible only two decades ago: (1) A drastic reduction in the volume of equipment such as digital computers, electronic instruments, and communication circuits. (2) A considerable decrease in power required to implement a given function and the attendant decrease in the speed $\times$ power product. (3) Improved reliability, partially as a result of better semiconductor manufacturing techniques and also due to the fact that many interconnections are internal to the LSI chip. (4) Higher speeds because the shorter interconnections between the individual transistors reduce stray reactances.

The discussion in this chapter emphasizes MOS devices since their low power consumption per gate has stimulated the development of a greater

361

range of LSI circuits than has bipolar technology. Technologies alone cannot be presented without delving into device physics, but to do both would be beyond the intended scope of this book. The interested reader can find a full treatment of MOS device physics in reference (1). A discussion of MOS technology and fabrication is presented in references (2) and (3), and basic MOS concepts and operating characteristics are summarized in the Appendix.

This chapter describes static and dynamic operation of MOS devices, bipolar and MOS random-access memories (RAM's), read-only memories (ROM's), and other memories performing special functions; it concludes with a discussion of MOS programmable logic arrays. Because of the rapid pace of progress, a complete description is not feasible and only selected examples are illustrated.

11.2   Static MOS Inverters and Gates

The majority of LSI circuits are fabricated from *insulated gate field-effect transistors*, IGFET's, more commonly referred to as *metal-oxide semiconductors*, or MOSFET's (MOS). Modes of operation of digital MOS devices may be divided into two categories, namely, *static* and *dynamic*.

Three *static MOS inverters* are shown in Figure 11.1 in which Q_1 is the driver transistor while Q_2 performs the function of a load. Figure 11.1a represents a p-channel enhancement MOS inverter (see Appendix) in which terminal V_{DD} is negative with respect to ground potential, while Figure 11.1b represents an n-channel enhancement MOS inverter with V_{DD} positive with respect to ground. These MOS polarities have no significance in discussing logic properties of circuits, and thus the symbol shown in Figure 11.1c will be used throughout this chapter.

Operation of the inverter will be explained referring to Figure 11.1c. The input impedance of a MOS can be represented by a resistance $\approx 10^{14}$ Ω in parallel with a capacitance of $\approx 10^{-14}$ to 10^{-12} F at the gate terminal. The load transistor, Q_2, is normally conducting. The driver transistor, Q_1, is either conducting or off depending on the input signal V_{in}. The geometric dimensions and electrical properties of the transistors are chosen such that

$$r_{d_1}(\text{OFF}) \gg r_{d_2}(\text{ON}) \gg r_{d_1}(\text{ON}), \tag{11.1}$$

where r_d is the dynamic impedance of the transistor defined as

$$r_d = \frac{dV_{DS}}{dI_D}\bigg|_{V_{GS}} \tag{11.2}$$

with V_{DS} = drain-to-source voltage, I_D = drain current, and V_{GS} = gate-to-source voltage.

A performance in accordance with eq. (11.1) can be obtained through

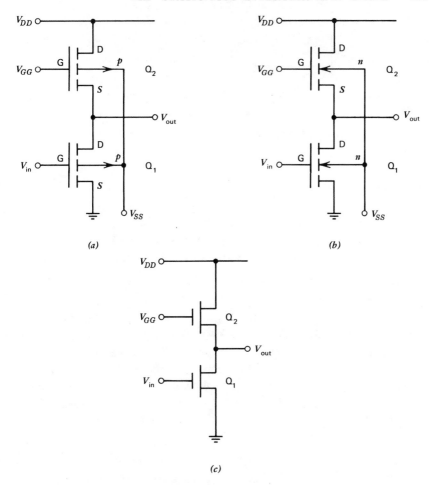

FIGURE 11.1. MOS inverter: (*a*) *p*-channel, (*b*) *n*-channel, (*c*) MOS inverter symbol without polarity (*p* or *n*) used in this chapter. D = drain, S = source, G = gate, V_{DD} = drain supply voltage, V_{SS} = substrate supply voltage.

suitable choice of *channel* dimensions of the driver and load transistors (see Appendix). Denoting by *W* the width of a channel, and by *L* its length, eq. (11.1) will be satisfied if the ratio

$$\beta_r = \frac{W_1/L_1}{W_2/L_2} \gg 1, \tag{11.3}$$

where subscripts 1 and 2 refer to Q_1 and Q_2, respectively. This inverter is referred to as a *ratio type* inverter since its performance depends on

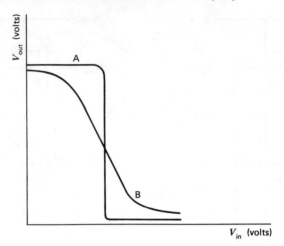

FIGURE 11.2.   Transfer characteristics of the inverter of Figure 11.1.

the ratio of the transistor geometries. Its transfer function is shown in Figure 11.2 in which curve A represents an ideal case while curve B represents a practical device. The output is the complement of the input and its state is time independent; that is, as long as the input is in one binary state the output of the inverter will be in the other binary state. This inverter is, therefore, referred to as being of the *static* type, as distinct from the dynamic inverter to be described later.

Static MOS logic functions can be implemented by suitably interconnecting

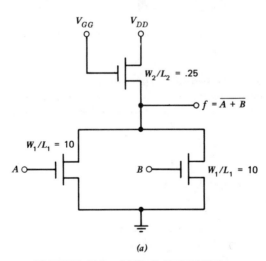

(a)

FIGURE 11.3.   (a) Static MOS NOR gate.

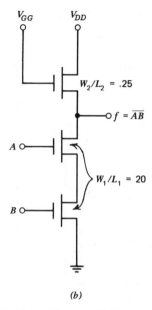

(b)

FIGURE 11.3. (b) **Static MOS NAND gate.**

several driver transistors to a common load transistor to obtain the desired logic operation. To keep the discussion general we shall omit any reference to channel polarity (p or n), and to the modes (enhancement or depletion) and regions (triode or saturation) of operation. To that effect we shall use the symbol of Figure 11.1c for our illustrations.

A dual input NOR gate is shown in Figure 11.3a: a logic 1 applied to either input, or to both, results in a logic 0 at the output node. A dual input NAND gate is shown in Figure 11.3b. Since the two NAND FET's are in series, the ratio W_1/L_1 has been doubled as compared with the NOR gate to attain a high value of β_r and hence good noise margins. More complex functions can be also readily synthesized in single stage gating.

EXAMPLE 11.1. Implement the EXCLUSIVE–OR function using static ratio-type MOSFET gates. Both the variables and their complements are given.

One solution, shown in Figure 11.4a, generates $\bar{f}$ at the common load point, which has to be followed by an inverter to obtain the desired function. Thus the result is obtained after 2-stage delays. A one-stage implementation is shown in Figure 11.4b where we used the relationship between the EQUALITY and the EXCLUSIVE–OR functions: $\overline{AB + \bar{A}\bar{B}} = A\bar{B} + \bar{A}B = A \oplus B$.

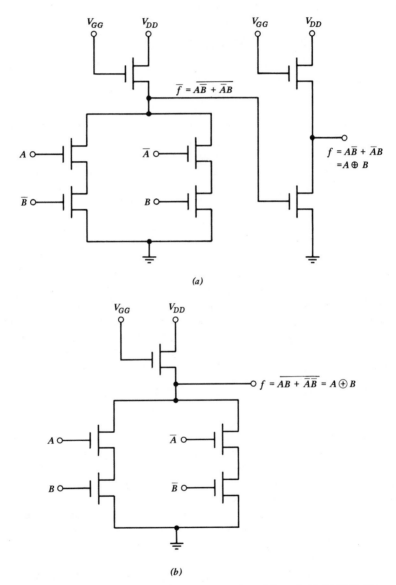

FIGURE 11.4. (*a*) Two-stage implementation of the EXCLUSIVE–OR function. (*b*) One-stage implementation of the EXCLUSIVE–OR function.

Static bistable elements can also be realized with MOSFET's by suitably interconnecting NAND and/or NOR gates to obtain flip-flops having the desired excitation functions. These are discussed later in this chapter in connection with static shift-registers and random-access memory cells.

11.3   Dynamic Ratioless MOS Inverters and Gates.

Dynamic Ratioless MOS Inverters.   These devices operate at lower power dissipations and occupy smaller chip areas than the static types. Both these properties are highly desirable in LSI circuits which incorporate several thousand devices on a chip. Operation of a dynamic inverter requires two or more clock pulses and is independent of the ratio of transistor geometries. One type of ratioless ($\beta_r = 1$) inverter, shown in Figure 11.5a, consists of two parts: the sampling circuit Q_1, Q_2, and its associated stray capacitance C_1, and the output circuit Q_3 and C_2. Clock pulse ϕ_1 samples the signal at V_{in} which is inverted in Q_1 and deposited as a charge on C_1. A subsequent clock pulse, ϕ_2, brings Q_3 into conduction and transfers the charge to C_2.

A detailed operation of the circuit will be given referring to the timing sequence shown in Figure 11.5b. Assume that at $t = t_0$, V_{in} is at logic 0, and ϕ_1 becomes 1. Q_1 is thus turned off while Q_2 is in conduction charging C_1. At time t_1, clock ϕ_2 assumes a logic 1 turning on Q_3 which allows charge to transfer from C_1 to C_2. Both capacitances are parasitic; however, $C_1 \gg C_2$; thus the voltage drop across C_1 is negligible during the transfer and the output changes to a logic 1 at time t_1. Next assume that in the

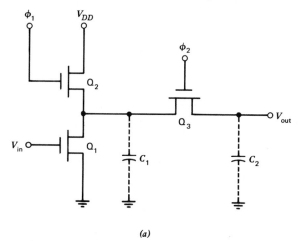

(a)

FIGURE 11.5.   Dynamic 2-phase ratioless inverter: (a) circuit. (*See next page for* (b))

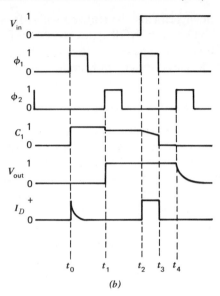

(b)

FIGURE 11.5. (*b*), 2-phase timing sequence.

interval t_2 to t_3, ϕ_1 and V_{in} are at logic 1; C_1 discharges partially but the output remains at logic 1 since Q_3 is off during this interval. When ϕ_1 is turned to logic 0 at t_3, Q_1 is fully discharging C_1. Thus at t_4, when ϕ_2 is at logic 1 causing Q_3 to conduct, C_2 discharges via Q_3 and Q_1. The output is the complement of the input with a time delay equal to $t_1 - t_0$ (or $t_4 - t_2$). Current I_D flows from V_{DD} in the circuit at t_0 to charge C_1 and also at interval t_2 to t_3 when Q_1 and Q_2 are conducting.

Power dissipation is low in comparison with the static inverter discussed in Section 11.2. This saving is achieved at the expense of adding Q_3 and a 2-phase clock. The clock signals provide 3 functions: (*i*) ϕ_1 samples the input voltage, (*ii*) ϕ_2 transfers the complement of V_{in} to the output, and (*iii*) both clocks "refresh" the information held in C_1 and C_2 through periodic recharging. A minimum clock frequency is therefore specified for dynamic circuits to prevent loss of data due to discharge of the capacitors by leakage currents.

In a ratio type inverter the geometry is set to satisfy eq. (11.3). This constraint does not apply to ratioless dynamic inverters which, therefore, can use identical transistors occupying a smaller area on the chip. Since the load transistor need not have a very high dynamic impedance, charging and discharging of the stray capacitances have shorter time constants resulting in higher operating speeds.

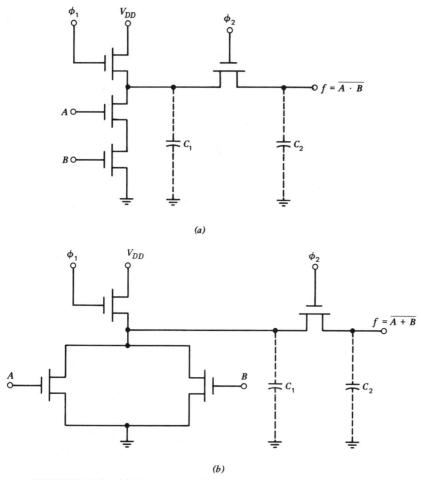

FIGURE 11.6. (*a*) Dynamic ratioless 2-phase NAND gate. (*b*) Dynamic ratioless 2-phase NOR gate.

Dynamic MOS Logic Functions. The dynamic 2-phase inverter can provide the building block for dynamic gate and memory circuits. Dual input NAND and NOR dynamic gates thus constructed are shown in Figure 11.6*a* and *b*, respectively.

Dynamic Ratioless Inverter with Low Power Dissipation. Power dissipation of the dynamic inverter of Figure 11.5 can be further reduced at the cost of added complexity as shown in Figure 11.7*a*. No V_{DD} line is connected to any of the inverter terminals; and the power required to charge stray capacitances C_2 and C_3 is supplied by clock ϕ_1. This type of inverter

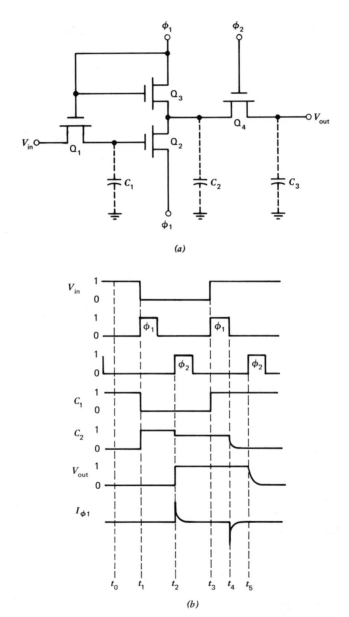

FIGURE 11.7. Dynamic 2-phase ratioless inverter having low power dissipation: (*a*) circuit, (*b*) 2-phase timing sequence.

is utilized as a building block of shift-registers having low power dissipation. In the discussion of its operation we make the assumption that the input transitions are synchronous with ϕ_1.

Assume that in the initial state at t_0, V_{in} is logic 1 and ϕ_1, ϕ_2 are logic 0, as shown in Figure 11.7b. At time t_1, $V_{in} = 0$ and $\phi_1 = 1$, discharging C_1 through Q_1. During $\phi_1 = 1$, C_2 charges through Q_3; the subsequent clock ϕ_2 transfers this charge to C_3. At t_3, V_{in} assumes logic 1 while $\phi_1 = 1$ turns on Q_1 allowing C_1 to charge. In the interval t_3 to t_4, C_2 cannot discharge since the terminals of Q_2 and Q_3 are at the same voltage Thus when clock ϕ_1 returns to 0 at time t_4, the state of the circuit is as follows: Q_1 and Q_3 are off, C_1 is charged to logic 1, and C_2 starts discharging via Q_2. At time t_5, ϕ_2 makes the transition to logic 1 and, since Q_2 is still on, C_3 discharges via Q_4 and Q_2. The output is thus seen to be the complement of the input delayed by $t_2 - t_1$.

The power for the inverter is provided primarily by ϕ_1 to charge C_2 via Q_3. As in all dynamic circuits, the information is volatile since charge states may be changed due to charge leakage to and from the capacitors. When V_{in} is constant, application of the clock pulses serves to "refresh" the information contained in the capacitors. Thus the clocks must run continuously, the period being determined by temperature dependent leakage rates of the dynamic MOS circuits.

11.4 MOS Shift-Registers

The high component density and low power dissipation make the MOSFET an ideal element for LSI shift-registers. LSI serial-in serial-out shift-registers are widely used for digital delay, manipulation of digital data, digital storage, computer peripheral circuitry, and temporary storage such as required in oscilloscope displays. In this section we show static and dynamic shift-registers of the ratio and ratioless types. Also, operation of 2-phase and 4-phase shift-registers and their properties are discussed.

Static Shift-Registers. One bit of a static MOSFET shift-register is shown in Figure 11.8. Q_2 and Q_6 are the driver and load, respectively, of one ratio-type static inverter, the other inverter consists of Q_3 and Q_7. The inverters are cross-coupled via transistors Q_4 and Q_5, and the data input is applied at Q_1. The operation of this static shift-register requires three clocks. It is common for ϕ_3 to be generated internally from ϕ_2; in some LSI shift-register circuits, ϕ_2 as well as ϕ_3 are generated internally and both are referenced in time to ϕ_1. The shift-register operates in a dynamic mode when the clock repetition rate is high; otherwise ϕ_3 is generated some 10 μs after ϕ_2 and stays ON as long as ϕ_1 is OFF. Latching is attained

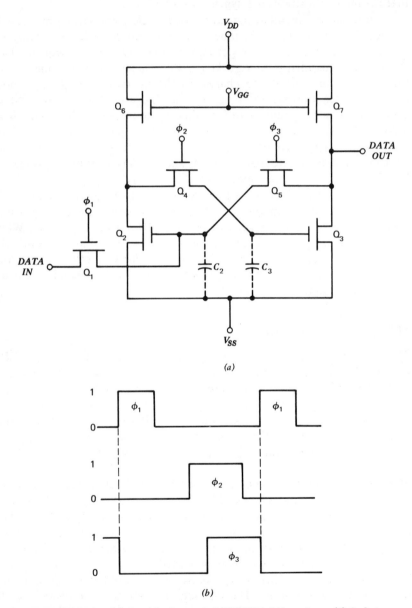

FIGURE 11.8. (a) One bit of a static MOSFET shift-register. (b) 3-phase timing sequence.

since *DATA OUT* at Q_3 has the same logic level as the gate of Q_2, and Q_5 remains ON due to $\phi_3 = 1$.

When ϕ_1 is at logic 1, the state of *DATA IN* is transferred through Q_1 to the gate capacitance C_2 of Q_2. Thus the state of Q_2 (on or off) is determined by the charge on C_2, i.e., by *DATA IN*. Next, Q_4 is turned on via ϕ_2 and the information at the drain of Q_2 is transferred to C_3, thus determining the state of Q_3. When ϕ_3 goes to logic 1, Q_5 is turned on, completing the feedback path from *DATA OUT* to the gate of Q_2. The logic information that was present at *DATA IN* during ϕ_1 is thus latched; and the shift-register will remain in this state until the state of the input has changed *and* the clocking sequence has been repeated.

2-Phase Ratio-Type Dynamic Shift-Registers. The standby power dissipation of the dynamic shift-register of Figure 11.9 is reduced, in comparison with static devices, because load transistors Q_2 and Q_5 are turned on only for the duration of ϕ_1 or ϕ_2, respectively. The circuit may be viewed as a master-slave combination, as indicated by the dashed line. Input data are stored across C_1; during ϕ_1 the master portion of the register *samples* the input data by bringing Q_2 and Q_3 into conduction. The resulting voltage that appears across C_2 is the logic complement of the input data. During ϕ_2 the slave portion of the register inverts the logic level present at the gate

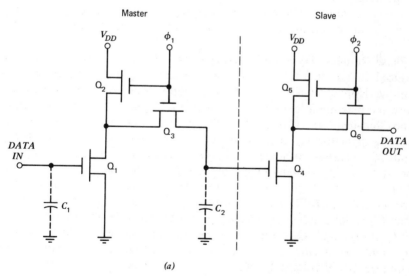

(a)

FIGURE 11.9. (*a*) **One bit of a 2-phase dynamic ratio type shift-register.**
(*See next page for* (*b*))

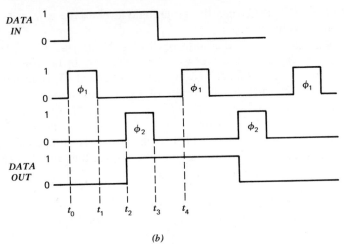

(b)

FIGURE 11.9. (b) **2-phase timing sequence.**

of Q_4 and shifts the information to the *DATA OUT* terminal. Thus one bit of information has undergone two inversions in two successive clock pulses and has been shifted from *DATA IN* to *DATA OUT*. As shown in Figure 11.9b, data appear at the output at the leading edge of ϕ_2 and are retained until the successive ϕ_2 clock pulse. Since no feedback path exists for latching, periodic clocking is required to "refresh" the information stored in the capacitances. The ratio-type shift-register dissipates dc power only while the drive and the load transistors of the inverters are on.

2-Phase Ratioless Dynamic Shift-Registers. Power dissipation is further reduced in the shift-register shown in Figure 11.10 which utilizes two non-overlapping clocks and has no V_{DD} supply. It uses identical transistors; thus its maximum operating frequency is higher than that of ratio-type shift-registers. Clocks ϕ_1 and ϕ_2 are not overlapping; during $\phi_1 = 1$, capacitor C_1 charges via Q_1 to the logic level of *DATA IN*, while C_2 charges to logic 1 via Q_3. When ϕ_1 returns to logic 0, the state of Q_2 is determined by the *DATA IN during* $\phi_1 = 1$. For *DATA IN* = 1, C_1 is charged; Q_2 conducts and thus discharges C_2. Conversely, for *DATA IN* = 0 during $\phi_1 = 1$, the charge on C_2 remains since Q_2 is not conducting. During $\phi_2 = 1$ the dynamic inverter consisting of Q_4, Q_5, and Q_6 performs in a similar manner to that of Q_1, Q_2, and Q_3 during $\phi_1 = 1$. C_2 is made much greater than C_3 to minimize loss of voltage during charge transfer when Q_4 is on. Any losses of charge in C_2 are compensated by introducing a small charge via C_5 at the leading edge of ϕ_2.

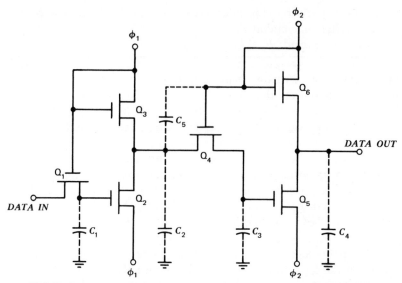

FIGURE 11.10.   One bit of a 2-phase ratioless dynamic shift-register.

Power dissipation in a 2-phase ratioless dynamic shift-register is independent of clock width but is proportional to the operating frequency because of charging of nodal (stray) capacitances via FET's when they conduct during a logic 1 of each clock phase.

EXAMPLE 11.2. Calculate the efficiency of a 2-phase ratioless dynamic shift-register circuit and the power dissipation as a function of frequency. Assume that stray capacitances C_2 and C_4 of Figure 11.10 are charged via dynamic resistances, r_d, of FET's Q_3 and Q_6 and that the charging clock duration $t_2 - t_1 \gg r_d C$.

The output, $V_c(t)$, across each capacitor is

$$V_c(t) = V_{\text{clock}}\left(1 - e^{-t/r_d C}\right)$$

and the voltage drop $V_{r_d}(t)$ across r_d is

$$V_{r_d}(t) = V_{\text{clock}}\, e^{-t/r_d C}.$$

The power dissipated in each FET during one clock cycle is

$$\frac{1}{r_d} \int_{t_1}^{t_2} V_{r_d}^{2}(t)\, dt = \frac{1}{r_d} \int_{t_1}^{t_2} V_{\text{clock}}^2\, e^{-2t/r_d C}\, dt \approx \tfrac{1}{2} V_{\text{clock}}^2 C \qquad (11.4)$$

when $t_2 - t_1 \gg r_d C$. The energy stored in the capacitor during one clock cycle is $\tfrac{1}{2} V_{\text{clock}}^2 C$; hence the energy lost in the FET is equal to the energy stored in the capacitor and is independent of the value of r_d. The efficiency of a 2-phase dynamic circuit thus approaches 50%. Power dissipation, P_d, is frequency dependent: $P_d = \frac{1}{2}V_{\text{clock}}^2 \, C \cdot f$.

4-Phase Ratioless Dynamic Shift-Registers. This type of shift-register exhibits lowest power dissipation and cell area combined with highest operating frequencies. One configuration, shown in Figure 11.11, requires six transistors and two sets of overlapping clocks. Referring to Figure 11.11a, we see that the *master* section of the register consists of Q_1 through Q_3, the *slave* section of Q_4 through Q_6. During $\phi_1 = 1$, capacitance C_1 is precharged independent of the logic state at the input. During $\phi_2 = 1$, C_1 is discharged to ground if $DATA\ IN = 1$. If $DATA\ IN$ is at logic 0, no discharge current flows since Q_3 is off. At the end of the ϕ_2 clock pulse, C_1 holds the logical complement of $DATA\ IN$.

The slave part of the shift-register operates in a similar manner. Unconditional precharging of C_2 takes place during ϕ_3 while its discharge depends on the logic input to the slave, i.e., on the potential of C_1. The final state of the slave represents the logical complement of the master; hence it is equal to the state of $DATA\ IN$ and has a delay of $t_3 - t_1$, as shown in Figure 11.11b.

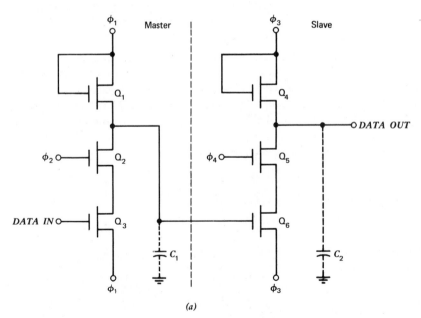

(a)

FIGURE 11.11.   One bit of a 4-phase ratioless dynamic shift register. (a) Circuit.

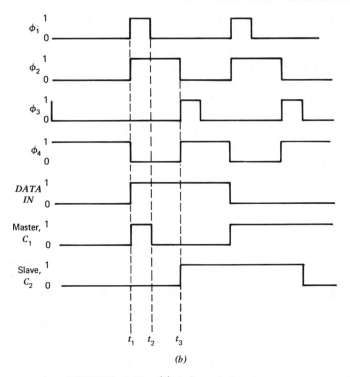

(b)

FIGURE 11.11. *(b)* **4-phase timing sequence.**

Lower and Upper Frequency Limits of Shift-Registers. In low frequency operation of dynamic shift-registers the charge stored at the node capacitances must remain within specified limits. Considering, for example, the 2-phase ratioless-type shift-register of Figure 11.10, the logic state of C_2 must be preserved for a duration $t = t_{master}$, i.e., from the trailing edge of ϕ_1 to the leading edge of ϕ_2. Similarly, the logic state of C_4 must not change appreciably until the leading edge of the subsequent ϕ_1 clock. Denoting this latter time interval by t_{slave}, we obtain the lowest operating frequency

$$f_{min} = \frac{1}{t_{master} + t_{slave}}. \tag{11.5}$$

Equation (11.5) is minimum when the clock phases are equally spaced in time, i.e., $t_{master} = t_{slave}$.

Leakage currents discharging the capacitors are temperature dependent, and the minimum allowed clock frequency decreases by about a factor of 200 as the temperature decreases from $+125°C$ to $-55°C$. In the temperature range of 0 to 70°C the minimum clock frequency changes by almost an order of magnitude.

High frequency operation is limited primarily by the charging and discharging time constants of the node capacitances, which impose a minimum value on the duration of ϕ_1 and ϕ_2 (in a 2-phase shift-register). Clock signals must not overlap, thus requiring a *clock skew* which is defined as the time interval between the trailing edge of one clock phase and the leading edge of the subsequent clock phase. In the absence of a clock skew, data could propagate through several shift-register bits during clock overlap. The maximum operating frequency is thus

$$f_{max} = \frac{1}{2t_\phi + 2t_t + 2t_s},\tag{11.6}$$

where t_ϕ = minimum duration of ϕ_1 and ϕ_2, assuming $t_{\phi_1} = t_{\phi_2}$, t_t = transition time, assuming equal rise and fall times (10% to 90%), t_s = minimum skew time between ϕ_1 and ϕ_2 pulses. Note that switching times and propagation delays in MOSFET's increase as the temperature increases.

Upper frequency limits may be extended through use of several parallel shift-registers. Input serial data are demultiplexed and latched using fast (e.g., MSI/TTL) circuits, and the outputs from the shift-registers are then synchronously reconstituted.

EXAMPLE 11.3. 1024-bit MOS static shift-registers with maximum operating frequencies of 3 MHz are utilized in a serial memory. Design a circuit that will process a 20 MHz serial data string.

The solution is shown in Figure 11.12. Eight 1024-bit registers are used in parallel, each operating at 2.5 MHz. An 8-bit addressed latch driven by a 3-bit counter presents the data sequentially to the inputs of the eight shift registers. The counter also produces synchronously the address codes for the 8-bit data selector (multiplexer) at the output of the shift-registers to reconstitute the data string at 20 MHz. The clock generator delivers an 8-phase clock, each clock pulse having sufficient duration to operate a shift-register.

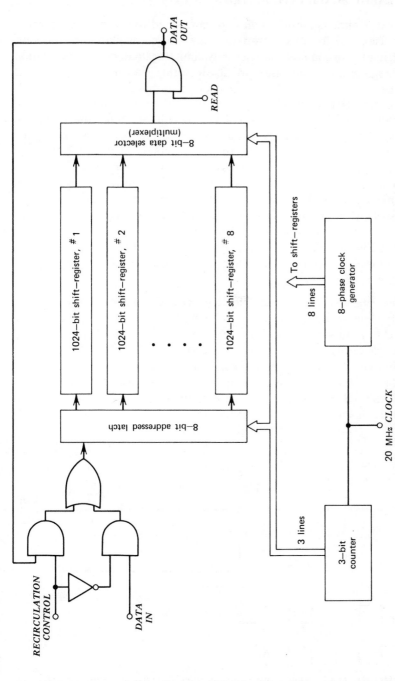

FIGURE 11.12. Extending the apparent operating frequency of MOS serial shift registers.

Ratio-type circuits require high impedance load transistors, hence wider clock pulses, and for this reason they are slower than ratioless circuits. Output load capacitances may severely limit the frequency of operation of shift-registers; in such cases circuits operating with low voltage swings are, in general, superior in high frequency performance.

Clock signals and supply voltages of shift-registers must be kept to specified tolerances. Insufficient voltage magnitudes limit the operating frequency range at both ends of the scale; too high voltages may create undesirable conduction paths due to field inversion in the FET. Voltage spikes of opposite polarity to the clock signal are especially objectionable since they forward bias junctions to the substrate, resulting in unreliable operation. These spikes should be confined to within $+0.3$ volts with respect to the substrate in a p-channel enhancement mode shift-register. In n-channel devices diffused into p-type substrates, the spikes should not be more negative than 0.3 volts with respect to the substrate. This problem of forward biasing does not exist when the substrate is not made of semiconductor material.

Complementary MOS Dynamic Shift-Registers.   One bit of a complementary MOS dynamic shift-register, shown in Figure 11.13, consists of two inverters and two transmission gates, TG's, of the type shown in Figure 5.22. The TG's are driven by complementary clock signals CL and $\overline{CL}$. When TG_1 is ON during $CL = 1$, the input signal, $DATA\ IN$, is transferred to C_1 where it is stored. This signal is isolated from the second inverter since TG_2 is OFF during $CL = 1$. When the phase of the clock reverses, i.e., $\overline{CL} = 1$, TG_1 turns off and TG_2 transmits the complement of the previous $DATA\ IN$

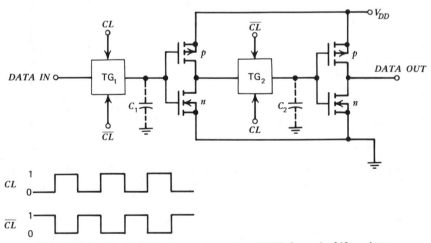

FIGURE 11.13.   One bit of a complementary MOS dynamic shift-register.

to charge C_2. The signal undergoes two inversions; thus *DATA IN* and *DATA OUT* are of the same logic level.

LSI Shift-Register Memories.   Shift-registers may be used as serial memories similar to a magnetic drum or accoustic delay-line memory. The contents of the memory are recirculated via a feedback loop from the output to the input for the three modes of operation: READ, WRITE, and RE-CIRCULATE, as shown in Figure 11.14*a* on the following page. Recirculation also provides a means of refreshing information in a dynamic storage device. The desired mode of operation is obtained through activating the respective gates as shown in the schematic. The D flip-flop facilitates inserting and deleting bits in the serial data string. To enter data, gates 1 and 2 are enabled and the new data are presented at *DATA IN*; to retain the length of the shift-register at a fixed value, some other bit must be deleted.

An *n*-word by *b*-bit recirculating memory may be obtained through driving *b* shift-registers in synchronism, each shift-register representing one bit of a word, as shown in Figure 11.14*b* on page 383. The number of serial words may be increased through cascading of shift-registers. However, *access time*, i.e., the time required to address the desired word, will increase proportionately. Recirculating shift-registers used as memories derive their address from an auxiliary shift counter which determines bit position and hence the time of entry or deletion of data, as well as the time of completion of a recirculating cycle in a purely "refresh" mode.

11.5 Random-Access Memories, RAM's

In Section 11.4 we described shift-register memories in which the time required to access the desired information was dependent on the *location* of that information in a serial data string. In *random-access memories*, RAM's, access time is independent of the location of data. In these memories, an array of bistable elements contains the binary information, and an addressing scheme permits immediate access to the desired element.

RAM's may be divided into three general classes: bipolar, static and dynamic MOS, and complementary MOS. The bipolar RAM can have access times less than 10 ns, but each cell requires a larger area and has a higher power dissipation than a MOS memory element. The size of a memory cell is an important parameter since fabrication yield, and hence cost, are closely related to the area of semiconductor occupied by a given structure. Much effort has, therefore, been directed toward reducing size and complexity to achieve memory cells of minimum possible size. Due to their smaller cells and lower power dissipation per cell, MOS memories can achieve higher densities. At present, memory LSI's with a capacity of 1024 words × 2 bits are readily available on a single LSI chip; and larger

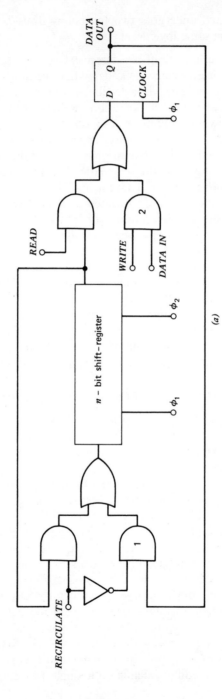

FIGURE 11.14. (*a*) **Recirculating shift-register.** [*See next page for* (**b**).]

(*a*)

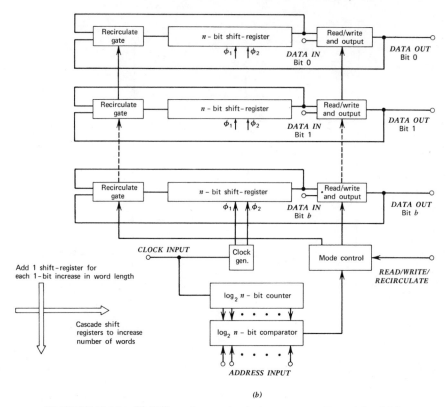

(b)

FIGURE 11.14. (b) Shift-register recirculating memory of n word $\times$ b bit capacity.

arrays are in development. The access time of a MOS RAM is typically a small fraction of 1 μs and is comparable to the fastest magnetic core memories, which they have been replacing in many applications since 1971.

A typical *memory module* is organized as an array of LSI circuits to obtain the desired *memory capacity* (number of words $\times$ number of bits/word). *Memory cells* connected to the same *address select line* constitute a *memory word*. The length of a word varies, but most often it is a multiple of *bytes* (1 byte = 8 bits). In addition to *addressing* we also require the facility to insert, i.e., *write*, binary information and to *read* it. These three requirements determine the cell configuration, with the constraint that a smaller cell area is favored provided its advantages are not offset by other considerations.

Bipolar Memory Cells. A bipolar memory cell consisting of a flip-flop with dual-emitter transistors is shown in Figure 11.15a. One of the emitters in

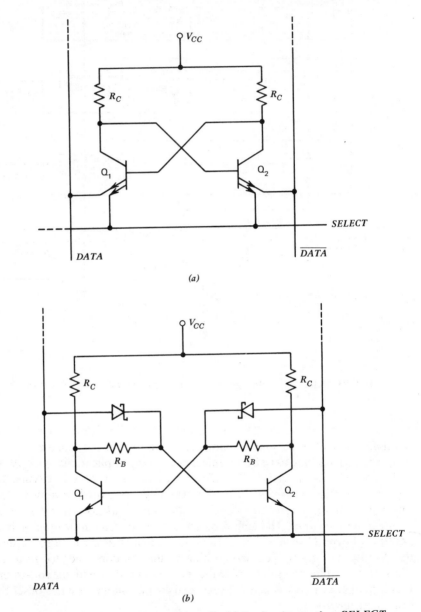

FIGURE 11.15.   Bipolar memory cells. (*a*) Dual emitter gating; *SELECT*
line is raised for READ or WRITE operations. (*b*) Single emitter gating
with Schottky diode steering.

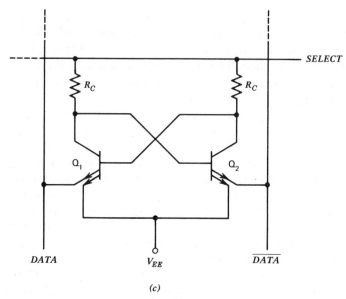

(c)

FIGURE 11.15. (c) Dual emitter gating used with ECL; *SELECT* line is raised for READ or WRITE operations.

each transistor is tied to one of the *DATA* lines; the other emitter is connected to a common word *SELECT* line. $\overline{DATA}$ is used to read or write a binary 0, *DATA* a binary 1. Both are normally held at a slightly higher voltage than the word *SELECT* line, and are thus back biased.

The cell is read by raising the *SELECT* line to steer the current (which has been flowing in the transistor in its quiescent state) to the respective *DATA* line where a current sense amplifier detects this signal. Writing is effected by raising the word *SELECT* line *and* lowering one of the *DATA* lines to force conduction in the transistor connected to it. When a cell is not selected, the *SELECT* line is "low" and sinks the memory cell current; the *DATA* lines carry no current in this condition and the sense amplifiers do not see an input signal. Similarly, when the binary states on the *DATA* lines are changed without raising the *SELECT* line to logic 1, the information in the memory cell remains unchanged. Access time depends on the current available and hence on the load resistances R_C; thus, for uniform performance, variations of R_C have to be kept within acceptable limits.

A bipolar memory cell that uses Schottky diodes for *DATA* steering is shown in Figure 11.15*b*. It dissipates less power than the cell of Figure 11.15*a* and does not require close tolerances of the load resistors, R_C. The addition of Schottky diodes does not materially increase the cell size because the diodes are fabricated within the collector region of the transistor. A memory cell is read by lowering its *SELECT* line to logic 0 and detecting, by a sense amplifier, the current unbalance transmitted via the Schottky diodes to the *DATA* and $\overline{DATA}$ lines. To *WRITE* information, the cell again is selected by lowering the *SELECT* line to logic 0. The *DATA*/$\overline{DATA}$ lines are unbalanced forcing a larger current through one of the Schottky diodes; and this current, aided by the positive feedback, rapidly changes the state of the bistable. Transition times are only slightly dependent on the value of R_C, which can be thus made large to reduce power dissipation. Further saving in power is obtained due to the fact that in the unselected cells the *SELECT* line is high resulting in a low voltage across the transistors.

A third configuration, shown in Figure 11.15c, is used with emitter coupled logic, ECL. Reading is accomplished when the *SELECT* line is raised. For writing, the *SELECT* line is raised and also the *DATA* line voltages are unbalanced to force the bistable to the desired state.

Many memory cells, such as in Figure 11.15, may be arranged in an array on a single LSI memory chip. Limitations in the number of IC pins favor an LSI organization having many words with few bits each on one chip since addresses to select the words can be coded. Common configurations range from 16 words × 4 bits, to 1024 × 1 bit. A memory module of 1024 words × 16 bits, for example, employs 16 LSI's of 1024 word × 1 bit capacity in which the sixteen bits of any word are selected by a common *SELECT* line. A chip select, *CS*, terminal is usually also included in the IC. Thus additional address lines may be decoded to provide for extension of the number of words in a memory module. *DATA* outputs are of the open collector or 3-state logic type for common busing of many IC's (see Section 5.3).

MOS Memory Cells. MOS RAM's may be divided into two categories: static and dynamic. Static storage is achieved by use of a bistable circuit similar to the memory cells shown in Figure 11.15. A MOS static memory cell with address *SELECT* lines and *DATA* lines is shown in Figure 11.16. Two cross-coupled static ratio-type inverters Q_1 and Q_2, and their loads Q_3 and Q_4, constitute the basic flip-flop. The cell employs coincident *SELECT* lines and information is read or written via the *DATA*/$\overline{DATA}$ lines. This type of memory cell requires eight transistors all of which have larger dimensions than minimum geometry because of the W/L ratios required for static operation [see eq. (11.3)]. Also, power consumption is high, limiting the number of bits that can be packaged in one IC.

Power dissipation can be reduced by clocking the V_{GG} line: load transistors Q_3 and Q_4 are turned off presenting a high impedance when V_{GG} is

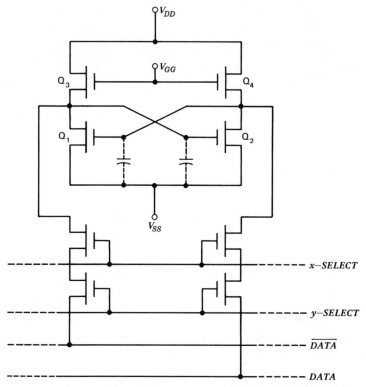

FIGURE 11.16.   MOS static memory cell utilizing eight MOS transistors per bit.

off, and correct memory information is maintained on the stray capacitances at the gates of Q_1 and Q_2. Characteristic of such operation is the time limit of charge storage imposed by leakage currents. Thus this stand-by mode of operation requires periodic *refreshing*, i.e., activating load transistors Q_3 and Q_4 by periodically turning on V_{GG}. Though the disadvantage of power dissipation has now been minimized, the size of the memory cell remains an important limitation.

Higher density can be obtained by use of a 4-transistor memory cell shown in Figure 11.17. Q_1 and Q_2 form a latch while Q_3 and Q_4 function as SELECT transistors. The cell is dynamic and operates as follows: at the start of a cycle, capacitors C_D and $C_{\bar{D}}$ at the $DATA$ and $\overline{DATA}$ lines are precharged to a voltage near V_{DD}. To READ, the $WORD$ $SELECT$ line is enabled, allowing Q_3 and Q_4 to conduct. If C_1 has been charged to a voltage above the threshold of Q_1, the FET will conduct, discharging C_D to nearly V_{SS}. Simultaneously, C_1 is recharged ("refreshed") via Q_4 since Q_2

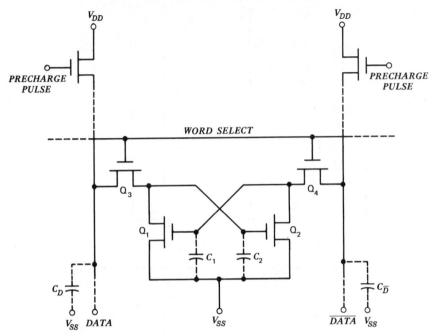

FIGURE 11.17. MOS dynamic memory cell utilizing four MOS transistors per bit.

is not conducting; C_2 is also refreshed to its logic 0 via Q_3 and via the $DATA$ line that approaches V_{SS}. An analogous description holds if Q_2 were initially in the ON state, with C_1 and C_2 interchanged as well as C_D and $C_{\bar{D}}$. The current in the DATA lines is detected by sense amplifiers. For a WRITE cycle, the desired information and its complement are placed on the $DATA$ and $\overline{DATA}$ lines while the word $SELECT$ line is brought to logic 1. Currents through Q_3 and Q_4 force the bistable to the desired state.

A 3-transistor MOS dynamic memory cell is shown in Figure 11.18. (The precharge transistor, Q_4, is common to all memory cells in a column of a cell array.) To read the memory cell content C_D is initially precharged to a voltage near V_{DD}, and the $READ$ $SELECT$ line is activated. If the voltage across C was initially above logic threshold, the NAND gate consisting of Q_2, Q_3 will conduct discharging C_D to nearly V_{SS}. Conversely, C_D will remain charged if C was initially below threshold voltage of Q_2. Thus the $READ$ $DATA$ line contains the complement of the binary information held by C, and its condition may be sensed by an on-chip sense amplifier. WRITE operation is performed by activating Q_1 via the $WRITE$ $SELECT$ line to transfer the logic level present on the $WRITE$ $DATA$ line into C. Regeneration of information, or "refresh," is required due to charge leakage at C. This

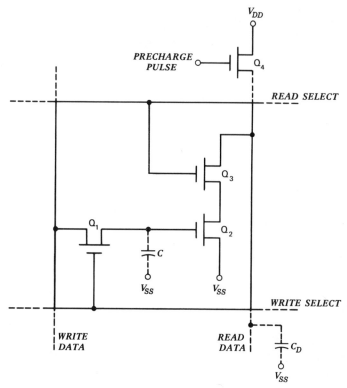

FIGURE 11.18. A 3-transistor MOS dynamic memory cell.

may be accomplished by periodically reading the contents of the cell out onto the *READ DATA* line, amplifying, inverting them, and then executing a WRITE operation.

The 3- and 4-transistor cells are widely used in LSI memory chips. The former utilizes a smaller area per cell but requires refresh amplifiers and closer tolerances on the *PRECHARGE* and *CHIP SELECT* signal timing. The 4-transistor cell is somewhat larger in area; but due to its simpler operation, less peripheral circuitry is required in the LSI. Refresh procedure and *WORD SELECT* are identical in the cell of Figure 11.17, eliminating the need for refresh amplifiers.

The single-transistor memory cell shown in Figure 11.19 represents the minimum component configuration per cell. For WRITE operation, the *WORD SELECT* line is activated; and if the *DATA* line is at logic 1, C will charge via Q_1. During READ, the charge on C modifies the voltage on the floating *DATA* line, and this voltage is sensed by an on-chip amplifier having substantial power gain. An optimum design of an LSI

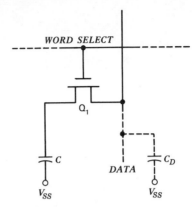

FIGURE 11.19. Single-transistor MOS dynamic memory cell.

requires C to be smaller than the stray capacitance C_D of the *DATA* line; as a result, READ operation is destructive, and the cell contents have to be restored at every READ cycle. Data restore and refresh require additional time and more peripheral circuitry on the chip than 3- or 4-transistor memory cells. The effect of the parasitic capacitance, C_D, on the *DATA* lines may be decreased through decoding so as to reduce the number of cells sharing a line; this scheme, however, requires additional peripheral circuitry.

Organization of an LSI RAM Circuit. Bipolar and MOS LSI RAM's commonly incorporate the following on-chip circuitry; (*i*) an address decoder to select the desired cell(s) within the chip; (*ii*) a chip select input signal, *CS*, which activates the LSI's addressing and/or READ/WRITE circuitry; (*iii*) a WRITE amplifier to insert *DATA* into the address-selected memory cells; (*iv*) READ or SENSE amplifiers to read information out of the selected cells; and (*v*) open-collector or 3-state output buffers. Some LSI circuits also employ a scheme whereby only stand-by power is applied during the inoperative time intervals. In this mode, voltage levels are decreased but are sufficient to retain the information in the memory, resulting in a considerable reduction of average power consumption. In addition to the above, dynamic MOS LSI's require *PRECHARGE* signals and logic to regenerate the information stored in the capacitors. "Refresh" must not interfere with the normal operation of the memory; and several timing schemes are in use, e.g., serial, cycle-steal, and burst refresh.

Address Decoding. Two address decoding schemes are common. (*i*) In small memory arrays, such as a 16 word × 4 bits, *word organization* of addressing is more prevalent: n address lines, $A_0 \cdots A_{n-1}$, are applied to a decoder which delivers 2^n unique outputs. (*ii*) *Bit organization* of addressing is commonly used in larger memories, such as a 1024 word × 1 bit array. Two decoders are employed each having an input of $\frac{n}{2}$ lines for a memory of 2^n cells, the output of each decoder delivering $\sqrt{2^n}$ lines. A cell is select when its *x*-select and *y*-select lines are in coincidence.

EXAMPLE 11.4.   Calculate the number of coded address lines for a 1024 word × 1 bit memory.

Each cell requires a coincidence of an *x*-select line with its corresponding *y*-select line: $x \cdot y = 1024$. For a square array, $x = y = \sqrt{1024} = 32$. Thus two decoders having 5 inputs and 32 outputs each are sufficient to select uniquely any cell in the memory array.

Coincidence type addressing requires two decoders that are considerably smaller than those in a single decoder scheme; also, additional gating is required to detect a coincidence. Several schemes are possible, all of which, in principle, represent two equivalent solutions. In one case, the *x-y* coincidence selection gates constitute part of the memory cell, Figure 11.20*a*. Thus the triple-emitter memory cell shown in Figure 11.20*b* is selected only if both

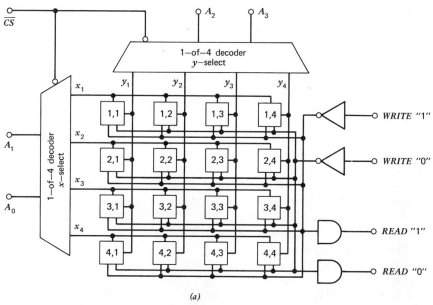

(a)

FIGURE 11.20. (*a*) A 16-bit RAM LSI with coincidence addressing. (*See next page for* (*b*))

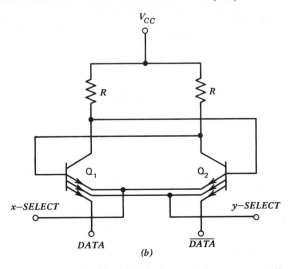

FIGURE 11.20. (*b*) **Triple-emitter bipolar memory cell.**

the x and the y lines are at logic 1. This scheme is not efficient from the point of view of chip area utilization because each individual memory cell has to be increased to accommodate the additional emitter junctions. Since memory cells are by far more numerous than any other circuit element on an LSI memory chip, better area utilization can be achieved when a gating scheme is employed external to the memory cell array to attain coincidence selection.

The internal organization of a 1024 word $\times$ 1 bit LSI memory with coincident selection and gating external to the memory cell array is shown in Figure 11.21. The 1024 cells are arranged in a 32 $\times$ 32 matrix. Five address input lines, A_0 through A_4, are applied to the x-decoder which generates 32 x-select lines; the y-select lines are generated in a similar manner. During READ operation, all cells in an addressed y-column are selected; the desired cell, however, is determined by the logic state of a unique x-select line. Amplifiers R1 through R32 sense the state of the selected cell.

The illustration shows gating for selection of memory cells number 1 $(x = 1, y = 1)$ and number 1024 $(x = 32, y = 32)$. All other memory cells are selected similarly. The number of external gates required in this selection scheme is $2n$ for an array having $N = 2^n$ cells. In comparison, the triple-emitter selection scheme would require 2^{n+1} additional emitter junctions.

Gates 1 through 32 are utilized during the WRITE operation: when $\overline{WE}$ (write enable) is at logic 0, the information at the *DATA IN* terminal is applied simultaneously to gates 1 through 32 and is written into the x-y selected cell. Note that the 3-state gating of $\overline{DATA\ OUT}$ is disabled

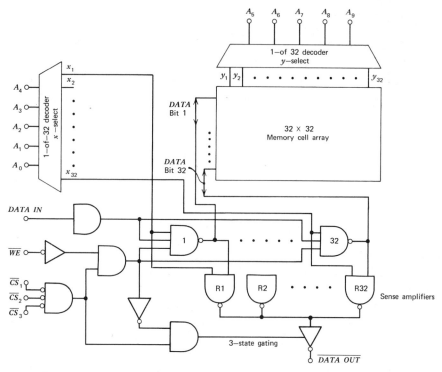

FIGURE 11.21. Organization of a bipolar 1024 word × 1 bit LSI RAM with coincidence selection.

during the WRITE operation. The use of 3-state buffers at $\overline{DATA\ OUT}$ allows paralleling many memory circuits onto one $DATA\ OUT$ bus, thus increasing the *number of words* in a large memory module containing several LSI chips. Three "chip select" inputs $\overline{CS_1}$ through $\overline{CS_3}$, shown in Figure 11.21, facilitate such an expansion. Additional address decoders, external to the LSI circuit, may be used to obtain memory modules of arbitrary word size, limited only by leakage and capacitive loading at the common $\overline{DATA\ OUT}$ bus lines and access time considerations. The *number of bits in a word* is obtained through paralleling the address and chip select lines of identical LSI circuits.

11.6 Content-Addressable Memories, CAM's

The content-addressable memory, CAM, combines logic at each bit position along with storage capacity. It has three modes of operation: READ, WRITE, and MATCH. The first two modes are identical to the READ and WRITE operations of a RAM. In the MATCH mode, $DATA$ inputs are compared with selected bits in the memory; and such comparisons may be performed simultaneously on several words. The CAM element is more complex than the RAM cell discussed in the previous section; however, the combination of memory with logic in one element provides a powerful tool for operations such as search through data files and comparisons of numbers. In addition, the CAM forms the basis for associative data processing, which deals with interrelations among data and among sets of data. It is, therefore, also known as *associative memory*. Through iteration, the simple match process of the CAM may be expanded to more complex operations of set definition and set positioning, e.g., greater than, less than, next greater, and next lower.[4]

A block diagram of a 6 word × 12 bit CAM is shown in Figure 11.22. In addition to the CAM matrix consisting of word 1 through word 6, we have shown four registers. (*i*) A *search*, or *argument, register* which stores the word that is to be compared with the words contained in the CAM. (*ii*) A *mask register* to designate which of the bits of the search word are to be included in the search; only those bits in the mask register that contain a logic 1 take part in the search operation. (*iii*) A *word select register* to select those words that are to participate in the search. (*iv*) A *search result register* to store the result of each interrogated word: 1 if a match is found, 0 otherwise.

Operation of the CAM in the search mode is explained with reference to Figure 11.22. Assume that we desire to extract all words starting with binary 1011. The search register is loaded with 1011 0101 0101, the mask register with 1111 0000 0000. A match is found in words 1, 4, and 6. The search result register, however, shows matching in words 1 and 6 only, since word 4 has not participated in the search, its word select register having a logic 0 in the position referring to word 4.

Search operations may be immediately followed by read operations in which the matched words are read out. Depending on the application, a search operation may be also followed by another search operation. This may be achieved by transferring the contents of the search result register after the first pass into the word select register in preparation for the second pass.

The logic structure of a CAM memory cell is shown in Figure 11.23, the symbol of a 4 word × 4 bit CAM in Figure 11.24. Gates 5 and 6 represent the basic memory latch, its state transitions being determined by the outputs of gates 3 and 4. In the READ mode the write enable $\overline{WE}$ signal is at logic 1 which, via the inverter, disables gates 3 and 4 and isolates the bistable latch. The word to be read is selected at the address input $\overline{A}$, and AND gate 7 yields the state of the latch.

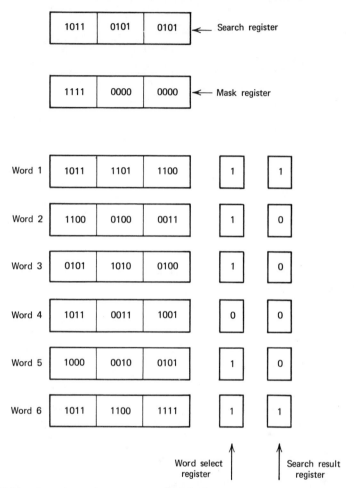

FIGURE 11.22.   Block diagram of a content-addressable memory, CAM, of 6 word × 12 bit capacity.

To WRITE new information in the memory cell, $\overline{WE}$, $\overline{A}$, and the *bit enable* inputs, $\overline{E}$, are set to logical 0 while data are applied at $\overline{D}$. A logical 0 at $\overline{D}$ will enable gate 1, and hence gate 3, which results in $Q = 1$. The other state of the latch is obtained through a logical 1 at the input.

In the MATCH mode, the circuit compares data present on the inputs, $\overline{D}$, with data stored in the memory; gates 8, 9, and 10 perform the function of EQUALITY $(= QD + \overline{Q}\overline{D})$. The output of gate 10 is "wire-AND-ed" to outputs of like binary bits of other words in the memory. Any non-matched condition will force the common line to ground.

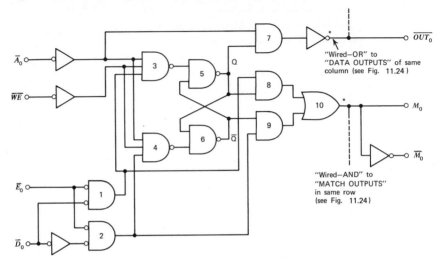

FIGURE 11.23. Logic structure of one memory cell in a content-addressable memory, CAM.

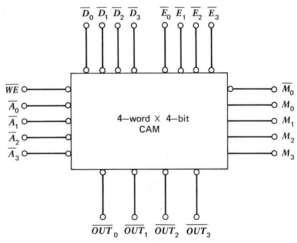

FIGURE 11.24. Logic symbol for a 4 word × 4 bits CAM.

A logic symbol of a 4 word × 4 bit CAM is shown in Figure 11.24. Each input bit $\overline{D}_i$ that is enabled by its corresponding $\overline{E}_i$ is compared with bit i of each of the four stored words in the CAM. Inputs that are not enabled are ignored in the search. If any stored word matches the data on the corresponding enabled inputs, a match will be indicated by a logical 1 at the M_i terminal.

11.7 Read-Only Memories, ROM's

A read-only memory, ROM, is an LSI circuit consisting of an array of semiconductor devices such as diodes, bipolar transistors, or FET's that are connected (programmed) to perform a particular set of switching functions. In other words, a ROM is a memory which contains a fixed set of data that can be accessed in a similar way as in a RAM. In some ROM's, data may be altered using an electrical or a mechanical process. These *programmable* and *reprogrammable* ROM's will be discussed later in this section.

A ROM is characterized by the number of words, $N = 2^n$, and the number of bits, b, in a word. Each word can be accessed (selected) by a unique address via n address bits applied to a decoder having 2^n outputs. Coincidence address decoding (discussed in Section 11.5) may also be employed. When accessed, the particular binary combination of the word appears at the ROM output terminals. The device is thus ideally suited for look-up tables in which, for example, the address is the independent variable and the word at that address contains the function, e.g., x and log x, y and e^y. However, a functional relationship between input and output is not a necessary condition for the definition of a ROM. In its most general sense, a ROM may be viewed as a listing of words or data, each word being accessed via its corresponding address.

Operation of a ROM will be explained referring to the 4×4 diode matrix of Figure 11.25a. ROM's up to 16,000-bit capacity are readily available using diodes, bipolar or MOS transistors as storage elements. The 16-cell array, Figure 11.25a, has four address inputs, $n = 4$. Inputs A and B are applied to a column decoder that generates four unique combinations; the row-decoder for C and D operates similarly. The intersections of the 4×4 decoder outputs provides a coincidence selection of sixteen cells. Current will flow through a diode only if its terminals are connected to both the column- and the row-decoded lines; otherwise the output is zero.

Referring to Figure 11.25a assume that $R_2 \gg R_1$. When a 4-bit address is placed on A through D, only one of the column lines will be at logic 1, while the remaining column lines stay at ground. Assume that row 11 and column 10 have been selected: transistor Q_3 will conduct allowing current to flow to ground via R_1. Two conducting paths exist to the collector of Q_3, one from V^+ via R_2 and D_3, and the other via the selected diode from a voltage supplied by the column decoder. Since the forward resistance of the selected diode $R_f \ll R_2$, most of the current flows through this diode. Conversely, when no diode is connected at the intersection point, current is forced to flow via R_2, Q, and R_1 to ground, where Q is the transistor of the selected row. In such a case the output will approach ground since $R_2 \gg R_1$. Diodes D_1 through

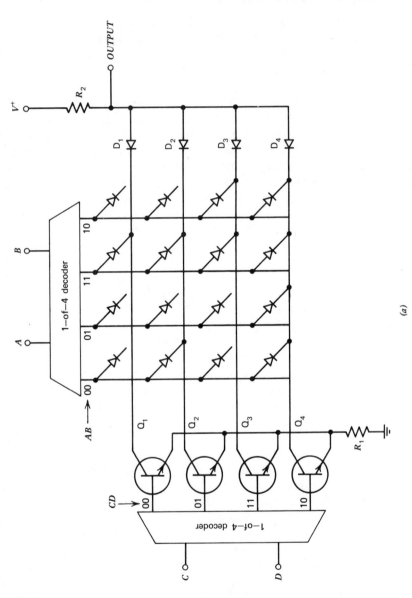

FIGURE 11.25. A 16-bit read-only memory, ROM, utilizing diode elements: (*a*) circuit diagram.

(*a*)

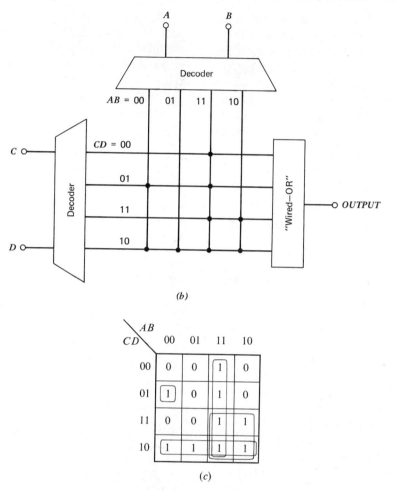

(b)

(c)

FIGURE 11.25. (*b*) simplified representation, (*c*) Karnaugh map.

D_4 block the current flow in the reverse direction and thus prevent unwanted current paths; they also provide the OR function at the common output point. A simplified representation of the ROM is shown in Figure 11.25b.

The switching function generated by the ROM of this small size may be obtained by transferring each connected intersection point to a corresponding cell of a Karnaugh map as shown in Figure 11.25c. The reader can easily verify that the ROM is Figure 11.25 yields the switching function

$$f = AB + C\bar{D} + AC + \bar{A}\bar{B}\bar{C}D.$$

A switching function that is given in a standard sum-of-products form (see Section 3.4) can be easily programmed into a ROM since each standard product term corresponds to one intersection point of the matrix (array). If the switching function is given in a minimized form, programming of the ROM is facilitated by expanding the function to a standard sum-of-products form. In general usage, when the ROM is programmed to yield an output word for each address input, discussion of minimized switching functions has little relevance.

MOS ROM's.   The physical structure of a MOS ROM is shown in Figure 11.26a. The rows are p- or n-doped semiconductors that are connected alternately to V_{SS} or to a voltage derived from a *row*, or *x-selector*. The metalized columns are connected to a *column*, or *y-selector*. At each intersection of a row and a column a transistor may be constructed by growing a

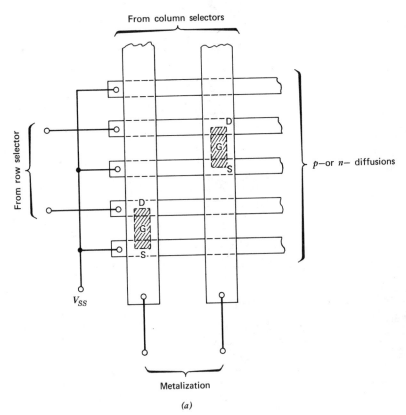

FIGURE 11.26.   MOS ROM: (*a*) physical structure with two transistors shown in detail (D = drain, G = gate, S = source).

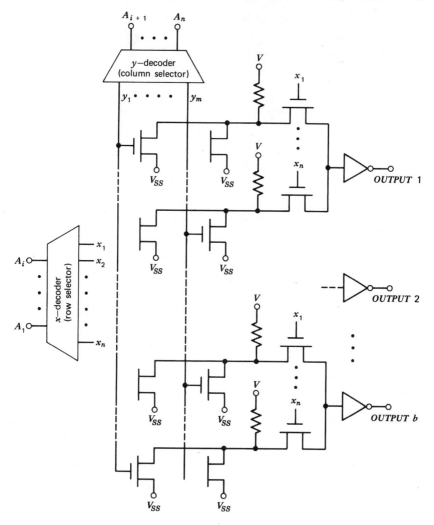

FIGURE 11.26. (*b*) organization of a MOS ROM of 2^{m+n} word $\times$ *b* bit capacity.

thin-oxide gate (see Appendix). The rows connected to V_{SS} represent the *source* terminals of the transistors, while the alternate rows represent the *drain* terminals. The organization of a MOS ROM of 2^{m+n} word $\times$ *b* bit capacity is shown in Figure 11.26*b*. Note that each of the b_i output terminals is a "wired OR" of all the cells that may be selected for the particular output bit.

The access time of a ROM is the time interval between the application of an address at the input and the appearance of a valid output. In a static ROM, the output does not change as long as the address remains stable. In dynamic ROM's, the address is clocked in and data are clocked out. These ROM's find application in synchronous circuits such as counters and sequence generators; a static ROM might also be used but it would require addition of external storage registers.

Bipolar ROM's.   Bipolar ROM's have access times of several tens of nanoseconds and are used in applications such as microprogramming (see Section 11.8) where speed is an important parameter. A bipolar transistor, often utilizing a Schottky barrier diode clamp (see Appendix), provides the basic cell. A 1024-bit bipolar ROM organized as 256 words × 4 bits is shown in Figure 11.27. Any cell may be addressed via a coincidence x-y selection scheme. Logic 1 is obtained at the output when all three electrodes of the selected transistor are connected in the matrix; otherwise the output of a selected cell is 0. Some ROM's have output latches in which new information is stored by applying a strobe pulse, as shown in Figure 11.27. The information will remain at the output until a new strobe pulse has been applied, although the input address may have changed in the meantime. Operationally, the device of Figure 11.27 is equivalent to the dynamic ROM and finds application in synchronous systems. 3-state drivers are provided at the output for ease of expansion, and similarly several "chip-select" inputs.

Electrically Programmable ROM's, PROM's.   The read-only memories described thus far consisted of an array of cells into which a fixed pattern of logical 1's or 0's has been "written" in the fabrication process. The basic matrix together with the row and column decoders, output drivers, and any optional control circuitry is fabricated first on a semiconductor chip. The particular pattern of 1's and 0's is then obtained by providing a metalization mask in the *last* fabrication step. This mask carries the desired binary information. Custom patterns are easily and economically manufactured, and the turn-around time is relatively short.

In the process of developing a digital system it is often desirable that the user be able to program a ROM in his facility. Such ROM's are referred to as programmable ROM's, or PROM's. There are two basic methods of field programming. In one method, each cell in an array incorporates a metallic link at one of its electrodes. During programming such a link may be fused by application of a high current pulse of specified duration. A broken link in a cell defines one binary state, an unbroken link represents the other state. Alternatively, a memory array may consist of cells each of

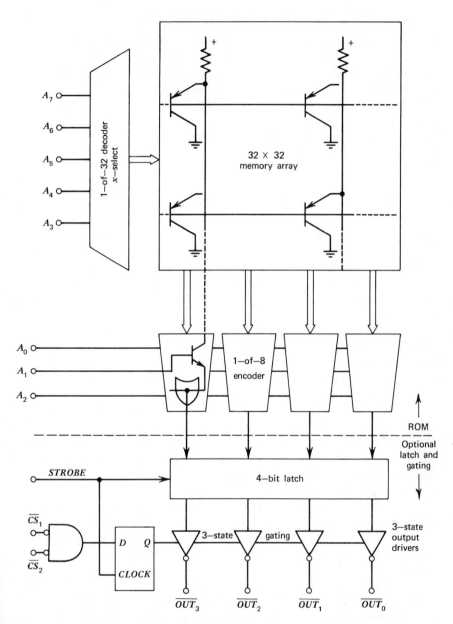

FIGURE 11.27. A 1024-bit bipolar ROM organized as 256 words × 4 bits; four ROM cells are shown in the 32 × 32 memory matrix. Transistors with emitters not connected produce a logical 0 at the output.

which has one uncommitted electrode before being "programmed". A conductive path to that electrode may be produced through a process called "avalanche induced migration" when a programming current is applied to the particular cell; this defines the logic state of the cell.

Programming procedures are usually strictly defined so that an accidental application of a pulse to the cell does not change its binary state. Programming parameters may specify rates of voltage or current change, initial and final voltage amplitudes, and duration, number and/or synchronization of programming pulses required to insert the binary information. In some instances several terminals must be activated simultaneously for any programming pulses to be effective.

Reprogrammable ROM's. Information written into a PROM cell is irreversible. Thus, although the user has the facility to program a ROM quickly to his particular pattern, he has no recourse if he needs to change even one bit, other than programming an entirely new ROM. This presents a severe restriction in an experimental situation, e.g., when a new system is being developed.

The introduction of the reprogrammable ROM allows program changes, which are often required in the developmental stage of new digital systems. One such ROM[5] consists of a MOS field-effect transistor in which no electrical contact has been made to the polysilicon gate. This gate is electrically well isolated from the substrate and from the surface of the device. Thus, once electrons have been injected into it via an avalanching process, a negative charge remains at the gate, rendering the particular FET conductive. When the avalanching voltage source is removed, no discharge path exists for the accumulated electrons. Electrical isolation is sufficient to keep this charge for several years, even when operating at elevated temperatures where the leakage is higher.

Erasure is accomplished by non-electrical methods, since the gate is electrically not accessible. When the ROM is illuminated by ultraviolet light a photocurrent discharges the gate. To that end, the ROM is sealed with a quartz window that transmits ultraviolet light. After erasure, all memory cells are left in the 0 state.

Programming is effected by applying a large voltage to selected output data terminals while addressing the requisite word by use of an x-y coincidence selection scheme. In READ operation a word is selected using the same address circuits; and the bit pattern appears at the output data terminals, as in conventional ROM's. Dynamic mode of operation is optional in some devices and refers to the decoding circuitry only and not to the memory cell. Somewhat lower power dissipation and shorter access times result from this mode of operation. The memory may be reprogrammed many times without affecting its performance. Once the final bit pattern has been established, conventional ROM's may be fabricated.

11.8   Applications of Read-Only Memories

The most common applications of ROM's are in code conversion, microprogramming, character generation, process control, and mathematical function look-up tables. For general computation it is convenient to have in a computer memory tables of frequently used functions, such as trigonometric, exponential, square root, logarithmic, etc. These may be stored either in a mass-storage medium such as a magnetic disc, or in a RAM. The former has the disadvantage of very long access time, while the latter is relatively expensive. Neither is utilized to its full operational capacity since information is written once only. Moreover, special functions that are not too often used in a computer system occupy memory space in an inefficient way. In such instances one may synthesize the function, whenever needed, through appropriate expansion algorithms which, however, put a severe time demand on the central processor unit of a computer.

ROM's have the necessary attributes to resolve these shortcomings. Their access times are short, and the cost and power dissipation are much lower than that of RAM's of comparable speed. Several applications are shown in this section.

Product of Two Numbers.   When multiplying two binary numbers the maximum number of output bits is equal to the sum of the bits in the multiplier and multiplicand. A commonly used algorithm consists of a series of shift-add operations which, in one method of implementation, requires a number of clock periods equal to the multiplier bits. When ROM look-up tables are used, the time interval between the input and the output is merely the access time of the ROM plus propagation delays through any buffer stages.

EXAMPLE 11.5.   Find the ROM organization required to obtain a look-up table of the product of two 4-bit numbers.

The number of input lines is 8, four for the multiplier and four for the multiplicand; hence the number of possible combinations (products) is 2^8. Thus the number of address bits to the ROM is 8, and the number of ROM words is $2^8 = 256$. The maximum number of bits at the output cannot exceed the sum of the multiplier and multiplicand bits. Therefore the ROM for a look-up table for a 4×4 bit product is organized as 256 words $\times$ 8 bits, a total of 2048 bits.

M_1	M_2	$(M_1 + M_2)2^{M_1 + M_2}$
4	4	2,048
5	5	10,240
6	6	49,152
7	7	229,376
8	8	1,048,576

FIGURE 11.28.   Number of ROM bits required, $(M_1 + M_2)2^{M_1 + M_2}$ for a multiplication look-up table. M_1 = number of multiplier bits; M_2 = number of multiplicand bits.

The size of the ROM grows rapidly as the number of bits in the mathematical operation is increased, as shown in Figure 11.28. Denoting by M_1 the number of multiplier bits, and by M_2 the number of multiplicand bits, it can be easily shown that the total number of ROM bits required for a $M_1 \times M_2$ look-up table is

$$2^{M_1 + M_2} \text{words} \times (M_1 + M_2) \text{bits.} \tag{11.7}$$

It is evident that ROM multiplication tables for $M_1 + M_2 > 12$ are not very practical economically even if these were technically feasible. This problem can be circumvented by representing M_1 and M_2 each as a sum of two binary numbers

$$M_1 = m_1 + \Delta m_1 \tag{11.8}$$

$$M_2 = m_2 + \Delta m_2 \tag{11.9}$$

where m_1 and m_2 are the most significant bits, and Δm_1 and Δm_2 the least significant bits. The product $M_1 \times M_2$ can thus be expanded

$$M_1 \times M_2 = (m_1 + \Delta m_1)(m_2 + \Delta m_2)$$
$$= m_1 m_2 + m_1 \Delta m_2 + m_2 \Delta m_1 + \Delta m_1 \Delta m_2. \tag{11.10}$$

Four ROM's and several adders are required for the implementation of eq. (11.10); the number of bits in each ROM, however, has decreased drastically, as shown in the example below.

EXAMPLE 11.6.   Implement a ROM look-up table for the products of two 8-bit numbers using the expansion of eq. (11.10). How many ROM bits would be required in a look-up table using one ROM only?

Let us express M_1 as

$$M_1 = \underbrace{----}_{m_1} 0000 + 0000 \underbrace{----}_{\Delta m_1}$$

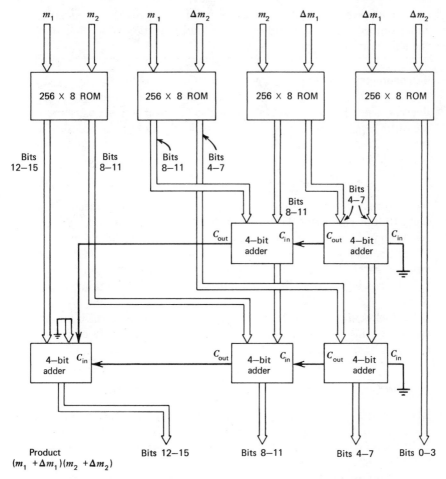

FIGURE 11.29. ROM look-up table for the product of two 8-bit binary numbers. Use of small ROM's and adders reduces the number of ROM bits required by a factor of about 125.

where each "–" represents a binary 0 or 1; M_2 may be expressed similarly. Following eq. (11.10), the result is obtained from the sum of four 4×4 bit products. Each such product requires a 256 word $\times$ 8 bit = 2048 bit ROM, as was demonstrated in Example 11.5. Thus the total number of ROM bits required for an 8×8 bit product is $4 \times 2048 = 8192$ bits, a factor of about 125 down from the single ROM realization, as may be seen from Figure 11.28. Implementation of the circuit is shown in Figure 11.29. Five 4-bit adders are required to sum the terms of eq. (11.10). The signal will appear at the output after a maximum of four propagation delays. This time compares very favorably with the multiplication procedure that uses the shift-add algorithm.

Trigonometric Functions.[6]   Consider a ROM look-up table for sin X. The oscillatory nature of the function reduces the required range to $0 \le X \le 90$ degrees, since the absolute values of the function in all quadrants cover identical ranges. For a resolution of 0.01 degree and an accuracy of 16 bits a look-up table of 9000 words $\times$ 16 bits $= 1.44 \times 10^5$ would be required if a single ROM were to be used. However, writing the function in the form

$$\sin X = \sin(x + \Delta x) = \sin x \cos \Delta x + \cos x \sin \Delta x, \qquad (11.11)$$

where x are the integer and Δx the fractional degrees, we reduce the requirement to four small ROM's. The two ROM's for sin x and cos x require a resolution of 1 part in 90, the functions involving Δx require a resolution of 1%. Equation (11.11) may be implemented with four ROMS of 128 word $\times$ 16 bit capacity, i.e., a total of 8192 bits. The product terms in eq. (11.11) can be realized with multipliers as shown in Section 8.4 or with ROM's.

Further reductions in ROM bit capacity are possible, since for $\Delta x \ll 1$, sin $\Delta x \approx x$ in radians, i.e., sin $\Delta x = x/57.2958$ in degrees. Equation (11.11) can thus be rewritten

$$\sin X = \sin x \cos \Delta x + (\cos x)(\Delta x/57.2958)$$
$$= \sin x \cos \Delta x + [(\cos x)/57.2958]\Delta x. \qquad (11.12)$$

The $(\cos x)/57.2958$ term requires eleven bits only since the leading five bits are always 0; thus a 128 word $\times$ 12 bit ROM will suffice. A further reduction in ROM bit requirements is recognized in the term cos Δx, since the range of cos Δx is between 1 (for $\Delta x = 0$) and 0.9998 (for $\Delta x = 0.99$). Thus, although the resolution of Δx is 1 part in 100, the *accuracy* of cos Δx is easily achieved by use of sixteen words only. Moreover, a detailed inspection of the cos Δx table, in the range under discussion, reveals that in the *binary* representation, bits 2^{-1} through 2^{-12} are identical for all sixteen words. Thus a 16 word $\times$ 8 bit ROM is more than sufficient to represent all values of cos Δx.

Implementation of the sin X look-up table is shown in Figure 11.30. It requires two multipliers, one adder, and 3712 ROM bits, a reduction of a factor of 40 in the ROM size in comparison with the single ROM realization. Output accuracy is fourteen bits, i.e., better than 0.007 degrees. (The accuracies of the ROM's and multipliers to achieve this 14-bit output accuracy of the sin X function was arrived at through detailed error analysis which has not been included in the above discussion.)

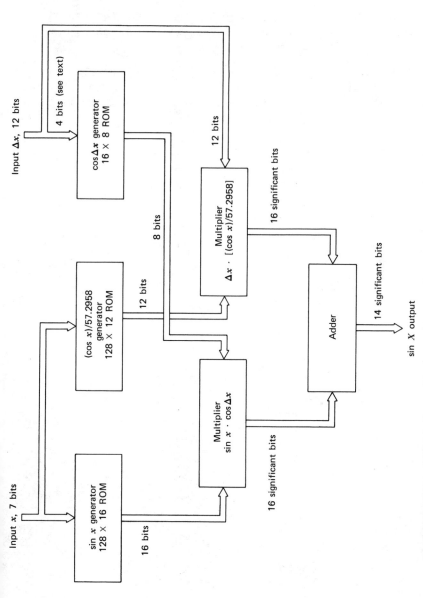

FIGURE 11.30. ROM look-up table for sin X, realized through the expansion sin $X = \sin x \cos x + [(\cos x)/57.2958] \, \Delta x$, (see text).

Alternative solutions are possible, one of which requires four 1024-bit ROM's and three 4-bit adders.[7] Such a circuit generates 2048 values of sin X with an output accuracy of 0.015%. A similar technique implemented with four 4096-bit ROM's would yield a 15-bit resolution with an output accuracy of about 0.0025%.

Code Conversion.   Several examples of binary-to-BCD and BCD-to-binary code converters using MSI or LSI circuits were shown in Chapter 9. These conversions were based on algorithms due to the well defined relationships between the converter output values and their corresponding input variables. Simple SSI circuits could be employed in some serially oriented converters.

When no mathematical relationship exists between the data to be converted, LSI's provide the most practical solution. A variety of codes is used in the computer and its peripherals to express alphanumeric characters. Standard ROM's are available for the following: Baudot teletypewriter codes; ASCII (American Standard Code for Information Interchange); BCDIC and EBCDIC (Binary Coded Decimal Interchange Code and Extended BCD Interchange Code); Hollerith Code, which is used with punch cards; IBM Selectric typewriter code; and the EIA Numerical Control Code.

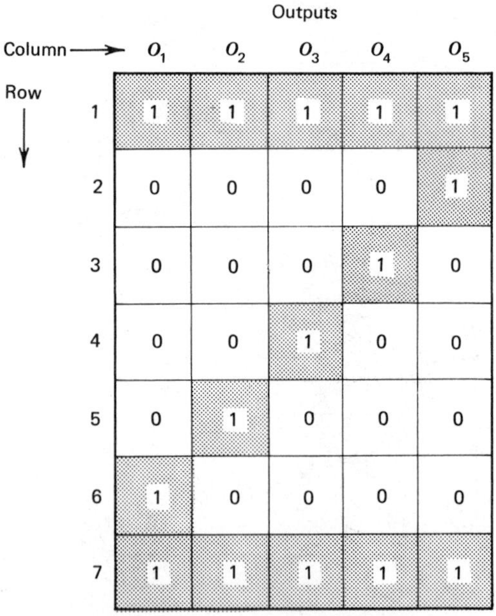

Outputs

Column ⟶	O_1	O_2	O_3	O_4	O_5
Row					
1	1	1	1	1	1
2	0	0	0	0	1
3	0	0	0	1	0
4	0	0	1	0	0
5	0	1	0	0	0
6	1	0	0	0	0
7	1	1	1	1	1

FIGURE 11.31.   A 5 × 7 dot matrix for alphanumeric display. A ROM generates the 0's and 1's required to constitute a character.

Widely used ROM character generators for oscilloscope display have a 64- or 128-character capacity. Each character is defined as a specific combination of logic 1's and 0's in a 5×7 or 5×9 dot matrix. A 5×7 matrix for the letter Z is shown in Figure 11.31. The block diagram of Figure 11.32 shows a *"row-select"* scheme: when a specific character is selected via its 6-bit binary code, seven words of 5-bits each appear sequentially at the output. In a *"column-select"* character generator, five columns are sequentially selected, providing a 5-word sequence of seven parallel bits per word at the outputs for each character selected by the coded address inputs.

The variety of codes utilized in computers and computer peripheral equipment has stimulated ROM's for conversion from one code to another. Multiple converters of 256 word $\times$ 8 bit capacity are available in which each 64 words are reserved for a different code. A parity bit (see Section 9.4) may be added when a particular code requires less than the available eight coding bits.

In Section 9.5 we have shown the generation of check bits in a Hamming single-error correcting code. These check bits were generated with SSI's and MSI's and were added to the information bits, each unique information word requiring a corresponding unique set of check bits. This correspondence between the information and check bits suggests a ROM as the check

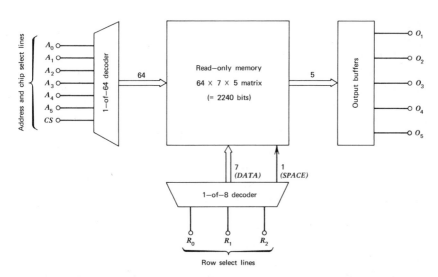

FIGURE 11.32. ROM dot-matrix character generator using a "row-select" scheme on the address lines. The input is an 6-bit encoded ASCII, the output is presented sequentially as seven rows of five bits each.

bit generator: the information bits are presented at the address input of the ROM which in turn outputs the appropriate check bits; the two sets of bits are combined to yield the coded word.

EXAMPLE 11.7.   An information word of $I = 10$ bit length is to be coded with the Hamming single-error correction code. Determine the ROM size required to generate the check bits, C.

The number of check bits, C, required for I information bits is derived from (see eq. 9.7)

$$2^C \geq C + I + 1.$$

Thus for $I = 10$, $C = 4$; a ROM of 2^{10} words $\times$ 4 bits = 4096 bits is required to generate the Hamming single-error check bits.

Microprogramming.[8]   The concept of microprogramming was originally conceived to systematize information transfers in a computer.[9] A growing interest in this technique has developed in the last few years when fast ROM's of sizeable bit capacity became more readily available since, as we shall see later, the data paths and the transfer sequences in a microprogram are controlled by a ROM. Also, the concept has much more general applicability since any sequential circuit may be viewed as a network in which information is transferred via data paths between various storage elements. Microprogramming techniques are now being applied in the design of complex instrumentation and control processes, in addition to its originally intended use in computers.

The name *microprogram* is borrowed from computer terminology, where (*i*) operations consisting of a single step are called *micro-operations*, (*ii*) a collection of micro-operations executed within one computer cycle is called a *micro-instruction*, and (*iii*) a sequence of micro-instructions is called a *micro-program*. Arithmetic operations in computers are greatly streamlined by use of microprograms which arrange the data stream to follow a prescribed sequence, i.e., the algorithm that has to be executed. Each algorithm requires a different length of micro-operations. Each such operation determines the next step in the micro-instruction. A micro-programming circuit also incorporates decision or *branch logic* to allow choices of the next micro-instruction step, depending on the results obtained from the last step. Due to the general nature of the procedure it is applicable to any sequential circuit, and in practice it is utilized when the level of complexity is sufficient to warrant the use of a multibit ROM in place of SSI's and MSI's.

A microprogrammed device is shown in Figure 11.33. The outputs from

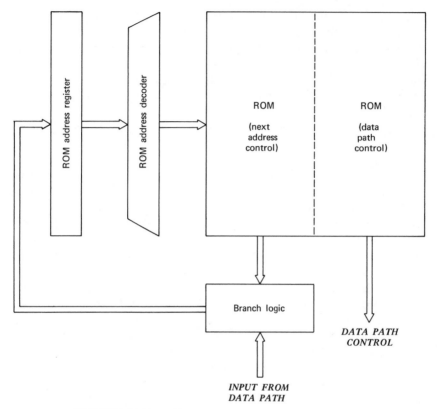

FIGURE 11.33. Microprogrammed device using a ROM.

the ROM are divided into two sets: (*i*) *data path control* and (*ii*) *next address control*. The latter is fed into a *branch logic network* where it is combined with status information from the data paths and determines the address of the next micro-operation. The data path control directs information flow between various registers by activating the appropriate gates. The information may be modified in its transmission, e.g., complemented, shifted right or left one or several positions; these modifications are executed under the data path control.

11.9 Programmable Logic Arrays, PLA's[10]

A programmable logic array, PLA, consists of combinational logic matrices (ROM's) and a set of memory elements (flip-flops); it also includes input inverters and output buffers. These elements are first fabricated on a semiconductor chip without any interconnections; that is, the gates of the ROM FET's are initially not connected, nor are there any feedback paths from the memory elements to the matrices. Thus a basic building block is attained having a considerable capacity for performing combinational and sequential logic functions. The specific function is determined in the final processing step and involves programming a photographic mask to yield the desired interconnection patterns in the basic structure. PLA's provide a reasonable compromise solution in cost and fabrication time between custom designed LSI circuits and an implementation with SSI and MSI circuits.

Consider the combinational circuit of Figure 11.34a which implements the sum function of a full adder. It consists of two matrices: the upper matrix (matrix 1) is a 6 row × 4 column array of FET's, while matrix 2 is a 1 × 4 array. Each column or row has a common load to which matrix FET's have been selectively connected, yielding a static NOR circuit of the type shown in Figure 11.3a; thus matrices 1 and 2 represent a 2-stage NOR–NOR gating network. In Section 4.2 we have shown the correspondence between NOR and NAND gates; redefining the logic levels with respect to logic "true" and "false," we may view the matrices as a 2-stage NAND network. Furthermore, a NAND–NAND network is logically equivalent to a 2-stage AND–OR gating network; hence matrix 1 performs the function of AND gates, and therefore it is also referred to as the *product terms generator*. Matrix 2 performs the function of OR gates and is also called the *sums-of-product terms generator*. Design of combinational networks using AND–OR gates may be done from truth tables by inspection if minimization of terms is not required, which is often the case when designing with PLA's of constant bit capacity.

EXAMPLE 11.8.   Using a PLA, realize the sum, Σ, and the carry out, C_o, functions of a full adder having inputs A, B, and C_i ($=$ carry-in), where

$$\Sigma = \overline{A}\overline{B}C_i + \overline{A}B\overline{C}_i + A\overline{B}\overline{C}_i + ABC_i \qquad (11.13)$$

and

$$C_o = AB\overline{C}_i + A\overline{B}C_i + \overline{A}BC_i + ABC_i. \qquad (11.14)$$

The gate connections for eq. (11.13) are shown in Figure 11.34a. Note that in the 24 element AND matrix only 12 gates have been selectively connected to the input lines. Each column represents a product term: $P_1 = \overline{A}\overline{B}C_i$, $P_2 = \overline{A}B\overline{C}_i$, etc. The unconnected gates and their transistors do not contribute to the logical properties of matrix 1. The sum of the product terms is obtained at Σ of matrix 2 (OR matrix)

$$\Sigma = P_1 + P_2 + P_3 + P_4.$$

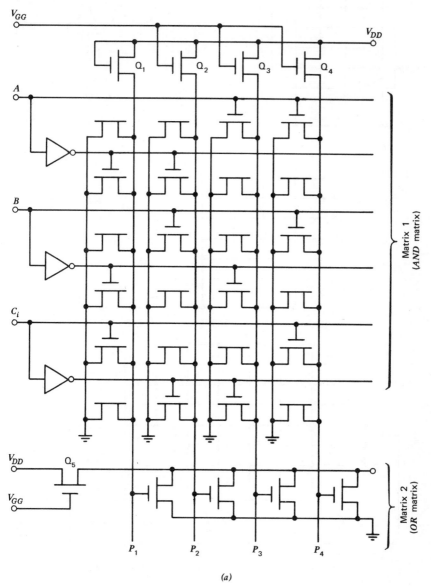

FIGURE 11.34. (a) Implementation of the *sum* function, Σ, of a full adder. (*See next page for* (*b*), (*c*), *and* (*d*))

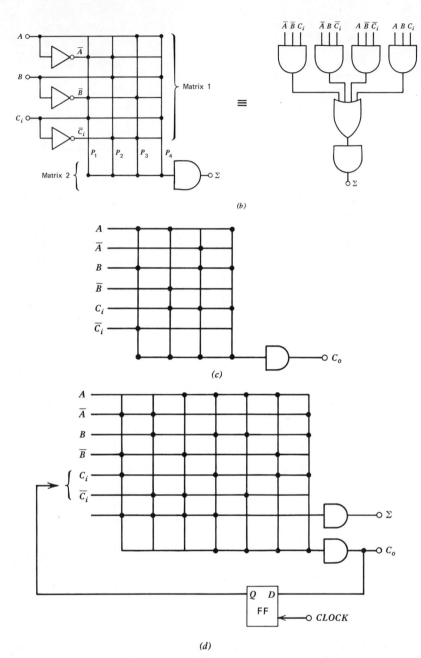

FIGURE 11.34. (*b*) Symbolic representation of the circuit of Figure 11.34*a*; dots show the gate connections, (*c*) symbolic representation of the *carry* function, C_o, of a full adder, (*d*) sum and carry functions executed in one PLA.

The representation as shown in Figure 11.34a is rather cumbersome and can be reduced to the form shown in Figure 11.34b, in which each connected FET is shown by a dot at the appropriate intersection. Note that Figures 11.34a and b are identical in their topology.

The carry-out, C_o, is designed by inspection of eq. (11.14) and is shown in Figure 11.34c. In the final step, we add a delay element from the output of C_o to provide the carry-in on the subsequent clock pulse. We also combine the Σ and C_o matrices, noting that input variables and the term ABC_i are common to both. The resultant PLA configuration is shown in Figure 11.34d and consists of a 6×7 AND matrix and a 2×7 OR matrix.

Note that about half of the transistor elements in the matrices of the previous example have not been utilized. In general, many transistors of both matrices may not be connected. Thus the savings obtained through minimization of Boolean expressions are of significance only if such minimization results in the reduction of the matrix sizes. Elimination of a literal in a Boolean product will not reduce the matrix size in most cases; however, every product term eliminated due to a minimized function reduces the ROM by one column.

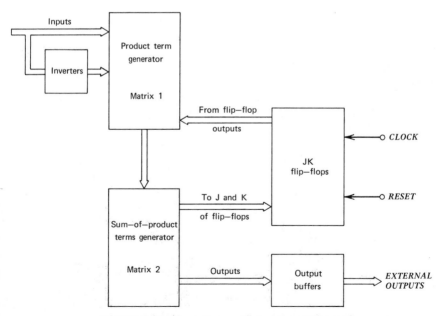

FIGURE 11.35. A 2-stage PLA sequential circuit.

The size of a PLA is determined primarily by the dimensional limitations of the semiconductor chip and may typically consist of 10^3 to 10^4 elements including inverters, gates, and memory elements (flip-flops), as shown in the block diagram of Figure 11.35. The memory elements and a feedback network enhance greatly the capabilities of a PLA: with the addition of an external clock and a reset pulse any sequential network can be realized within the size limitations of the PLA. For example, the Texas Instruments PLA type TMS 2000JC can handle up to 17 logic inputs and provides 18 logic outputs. The memory elements consist of eight J-K master-slave flip-flops.

EXAMPLE 11.9. Calculate the ROM equivalent of the PLA type TMS 2000 JC that was described above.

The inputs to the product term generator consist of 17 external signals, n_1, and 8 signals, n_2, that are fed back from the flip-flops. The number of different words that can be generated, $N = 2^{n_1 + n_2}$. The sum-of-product-terms generator has 16 outputs, n_3, that are applied to the eight J-K inputs and to 18 external outputs, n_4. The ROM equivalent is, therefore,

$$2^{n_1 + n_2}(n_3 + n_4) = 2^{25} \cdot 34 \approx 10^9 \text{ bits}$$

The level of complexity may be raised through cascading two PLA's to obtain a 4-stage PLA network; furthermore, clocked feedback may be added between the cascaded PLA's. Cascaded circuits may become very complex when feedback is employed, and thus far no systematic design procedures are available to guide the designer.

Two-stage sequential circuits may be designed using the procedures of Chapter 7.

EXAMPLE 11.10. Design a BCD counter using a PLA with four D-type flip-flops. Show the additional matrix connections required for a 7-segment display.

From the transition diagram of a decade counter and the excitation tables of D flip-flops we obtain the following minimized input functions to flip-flops Q_3(msb) through Q_0(lsb):

$$\begin{aligned}
D_3 &= Q_2 Q_1 Q_0 + Q_3 \bar{Q}_0 \\
D_2 &= Q_2 \bar{Q}_1 + Q_2 \bar{Q}_0 + \bar{Q}_2 Q_1 Q_0 \\
D_1 &= \bar{Q}_3 \bar{Q}_2 Q_0 + Q_1 \bar{Q}_0 \\
D_0 &= \bar{Q}_0.
\end{aligned} \qquad (11.15)$$

Implementation of eq. (11.15) is shown in the dashed rectangle P of Figure 11.36. Next, the AND and OR matrices are extended to generate

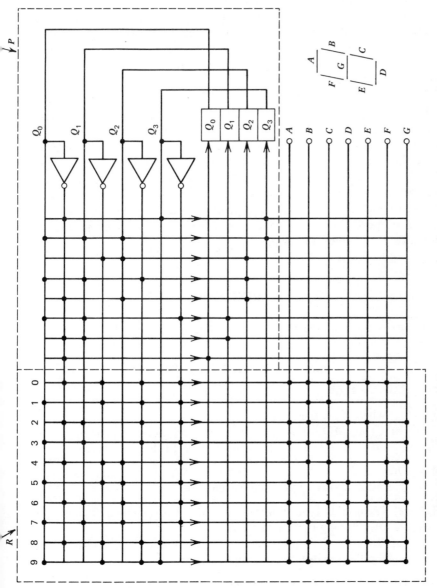

FIGURE 11.36. PLA implementation of a decade counter and 7-segment display coding.

the inputs required for coding the 7-segments; this is shown in the dashed rectangle R. Note that the dots in rows Q_0, Q_1, etc., represent the 1's in a natural BCD coding. The ten lines representing 0 through 9 are applied to an OR matrix to activate the desired segments. For example, numeral 2 requires that segments A, B, D, E, and G be activated; thus the column "2" is OR'ed in the lower matrix at the appropriate intersections with lines A, B, etc.

In summary, a single PLA has the logical capabilities to replace a great number of SSI and MSI devices. The advantages that ensue from the use of PLA's are: (*i*) A drastic reduction of input and output connections since the feedback variables are connected internally in the chip; a high function-to-pin ratio is thus attained. (*ii*) Considerable decrease in propagation and rise times of signals due to proximity of components and reduced stray reactances. (*iii*) The process is amenable to production of special purpose networks through suitable programming of the gates and flip-flops available on the chip.

PROBLEMS

11.1 (a) Show that the full adder may be represented by the switching functions

$$S(\text{sum}) = \bar{C}_o(A + B + C_i) + ABC_i$$

$$C_o(\text{carry}) = AB + (A + B)C_i,$$

where C_i is the "carry-in."
(b) Using MOSFET static inverters and gates, implement the full adder as given by the expressions in Problem 11.1(a).

11.2 Draw a dynamic 2-phase MOSFET NAND gate and explain its operation showing requisite waveforms.

11.3 (a) Using a D-type flip-flop and gates, implement a 2-phase non-overlapping clock generator.
(b) Generate the 4-phase clock pulses of Figure 11.11*b*, using D-type flip-flops, gates, and a Johnson counter.

11.4 Using MOS static inverters and NAND or NOR gates, implement the following clocked flip-flops:
(a) R-S,   (b) J-K,   and   (c) toggle, T.

11.5 Implement a 2-phase ratio-type dynamic J-K flip-flop.

11.6 The static shift-register of Figure 11.8 may be modified for use in either a static or a purely dynamic mode. Show the minimum gating to be added to the circuit of Figure 11.8 to implement the static-dynamic control.

11.7 Discuss how the effective operating frequency of a serial shift-register may be quadrupled by use of MSI TTL devices at the inputs and outputs of four parallel registers. Show the logic circuit required to generate a 4-phase skewed clock.

11.8 A serial memory of the type shown in Figure 11.14b has a 1024 word $\times$ 16 bit capacity and utilizes ratioless 1024-bit dynamic shift-registers. The clock inputs at each bit present a capacitance of 10^{-13} F for ϕ_1 and the same value for ϕ_2. The clock amplitude is 10 V, and transition times (rise and fall) are 5×10^{-8} s. (a) Calculate the peak current that each phase must be able to supply on clocking. (b) Calculate the clock power required by the memory at operating frequencies of 500 Hz and 5 MHz.

11.9 Given an LSI memory with internal decoding as in Figure 11.21, including three chip select, $\overline{CS}$, inputs, show the organization of a memory module having 16,384 words $\times$ 16 bits. Include any external buffers and address decoders that may be required.

11.10 Using the content-addressable memory, CAM, of Figure 11.23, establish an algorithm to find the word in the memory containing (a) the maximum binary value and (b) the minimum binary value.

11.11 Given a CAM of 512 word $\times$ 16 bit capacity, what is the maximum number of passes required to find the maximum valued binary number?

11.12 The LSI shown in Figure 11.24 contains sixteen CAM cells organized as 4 words $\times$ 4 bits. Draw a block diagram showing how this LSI can be expanded to process (a) 16 words $\times$ 4 bits, and (b) 4 words $\times$ 16 bits. (Note that terminals $\overline{OUT}$ and M are of the open collector type.)

11.13 (a) Using four 256 word $\times$ 8 bit ROM's and 4-bit adders, implement a ROM look-up table for the quotient $Q = N/D$, where N and D are 8-bit numbers. Use the expansion $Q = \dfrac{n}{D} + \dfrac{\Delta n}{D}$ where $n + \Delta n = N$.

(b) How many significant bits after the binary point do you obtain, taking round-off into consideration? (c) How many 4-bit adders are required to implement the circuit? (d) How many bits would be required to realize this division look-up table using one ROM only?

11.14 Using the floating point representation discussed in Section 2.7, implement the function e^x. Use two ROM's of 128 word × 8 bit capacity and assign eight output bits to the exponent and eight output bits to the mantissa. Establish a table showing the maximum output value as a function of the resolution (step size = Δx) of the input variable, for binary steps of $2^{-3} \leq \Delta x \leq 1$.

11.15 A square-summing multiplier based on the binomial theorem yields the product of two numbers, A and B,

$$AB = \tfrac{1}{2}[(A + B)^2 - A^2 - B^2].$$

The multiplier may be realized with ROM's and add/subtract circuits. Calculate the largest values of A and B that may be used in the product if the ROM capacities must not exceed 2^{14} bits, each. Draw a block diagram of the complete multiplier showing the circuit interconnections for the correct bit positions in the adders and subtractors.

11.16 Compare the number of ROM bits required in the square-summing multiplier of Problem 11.15 with the ROM multiplier look-up table of Figure 11.28.

11.17 The logic diagram of Figure 11.37 represents an open loop micro-programmed controller that generates sixteen cycles of 16 words × 4 bits. To avoid race conditions, the counters operate on the positive clock transition while the decoder is enabled when the clock is low. (a) Discuss the operation of the circuit. (b) Show how the circuit may be utilized in a variety of applications.

11.18 Repeat Example 11.10 and show the PLA connections when no minimization is applied for the flip-flop input functions. Compare the number of PLA gates required in both solutions.

11.19 Using a programmable logic array, (a) design a modulo-10 counter using J-K flip-flops in an excess-3 code. (b) Show the ROM connections required for coding of a 7-segment display. (c) How many matrix elements are required for the counter? How many for the decoder?

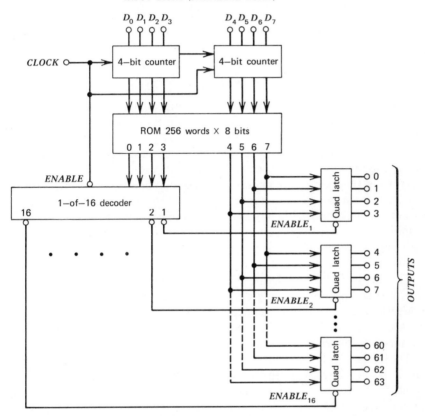

FIGURE 11.37. **Open loop microprogrammed controller of sixteen 4-bit cycles.**

CHAPTER 12

Practical Considerations

The design of a practical digital system encompasses several important factors reflecting unwanted interactions with its surroundings. Therefore, it is necessary to consider these factors as part of the design in order to assure a system that performs reliably. Some of these were already touched upon. Interconnections have capacitances that result in added propagation delays and they contribute undesirable coupling between signals, necessitating "noise margins"; interference is also contributed by the grounding and power system, especially in large systems. Voltages along ground conductors resulting from dc loads and from transients can have significant magnitudes, and capacitive coupling between two points can introduce undesired transients.

This chapter deals with the effects of interconnections and undesired interference and noise. Many details pertaining to grounding and power systems, however, will not be covered. For SSI and MSI this material can be found in handbooks written specifically for TTL, ECL, and MOS circuits; see references (1), (2), and (3).

The chapter also introduces three auxiliary circuits: the Schmitt trigger circuit is used for regenerating noisy signals that have slow transition times, single-shots are utilized to generate fixed time intervals, and oscillators provide the basic clock that is required in synchronous digital systems.

12.1 Interconnections

Although one may be tempted to ignore the effects of interconnections between integrated circuits, this could be disastrous: interconnections can be a source of signal loss, signal distortion, added propagation delays, and noise.

Signal loss on interconnections is usually easy to deal with: if too much signal is lost due to the resistance of the interconnection, heavier conductors

424

should be used. In complex situations, logic levels, tolerances, and noise margins may have to be designed to accomodate a large loss. Thus, for example, the high-threshold circuit of Problem 15 in Chapter 5 would be more tolerant to signal losses than a standard TTL circuit.

The finite length of an interconnection results in an increase of the rise time, fall time, and propagation delay. When the delay contributed by the interconnection is short compared to the rise and fall times, it can be approximated by a capacitance; the computation of the resulting signal deterioration is quite straightforward (see examples in Chapter 5), provided the capacitance is known. If the capacitance is not known, it can be measured if the circuit has already been built, or it can be estimated by methods that are covered in the next section.

12.2 Transmission Lines

In many applications the length of an interconnection cannot be neglected and it has to be treated as a *transmission line*.* Transmission lines may take many different forms such as a single wire above a ground plane, two parallel wires, a stripline, or a coaxial cable.

Characteristics. The maximum velocity of propagation attainable by any signal is that of light, which in vacuum or in air is approximately $c \cong 30$ cm/ns, or 1 foot/ns. Practical cables with little dielectric material can approach 99% of this speed; cables with solid dielectric commonly have speeds as low as $\frac{2}{3}c$. The delay, T, of a transmission line of length l and propagation velocity v can be written as **

$$T = \frac{l}{v}. \tag{12.1}$$

EXAMPLE 12.1. A coaxial cable has a length of $l = 60$ cm and a velocity of propagation v that is $\frac{2}{3}$ of that of light. Find the delay, T, of the cable in nanoseconds.

The velocity of propagation in the cable is $v = \frac{2}{3} c = \frac{2}{3} 30$ cm/ns = 20 cm/ns. By substituting into eq. (12.1), the delay is

$$T = \frac{l}{v} = \frac{60 \text{ cm}}{20 \text{ cm/ns}} = 3 \text{ ns.}$$

---

* A detailed treatment of transmission line theory is beyond the scope of this book, and only some results will be presented. The reader interested in the derivation of these results and in further details should consult references (4) and (5).

** v is related to c and to the relative dielectric constant ε_r as $v = c/\sqrt{\varepsilon_r}$.

An important property of a transmission line is its characteristic impedance, Z_0; this is the ratio of the voltage and the current of a signal
traveling along the line. Z_0 is independent of length, and is a real positive
number if the line is lossless. Characteristic impedances of various transmission lines are shown in Figure 12.1.

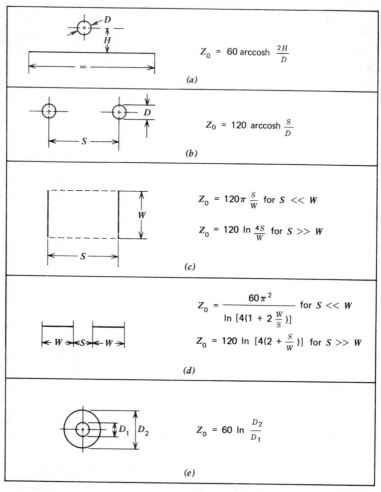

$$Z_0 = 60 \operatorname{arccosh} \frac{2H}{D}$$

(a)

$$Z_0 = 120 \operatorname{arccosh} \frac{S}{D}$$

(b)

$$Z_0 = 120\pi \frac{S}{W} \text{ for } S \ll W$$

$$Z_0 = 120 \ln \frac{4S}{W} \text{ for } S \gg W$$

(c)

$$Z_0 = \frac{60\pi^2}{\ln\left[4\left(1 + 2\frac{W}{S}\right)\right]} \text{ for } S \ll W$$

$$Z_0 = 120 \ln\left[4\left(2 + \frac{S}{W}\right)\right] \text{ for } S \gg W$$

(d)

$$Z_0 = 60 \ln \frac{D_2}{D_1}$$

(e)

FIGURE 12.1. Various transmission line configurations and their characteristic impedance Z_0 in vacuum or in air. For a relative dielectric
constant of ε_r, Z_0 has to be divided by $\sqrt{\varepsilon_r}$. (*a*) Single wire above infinite
ground plane, (*b*) two parallel wires, (*c*) parallel strips, (*d*) stripline, (*e*)
coaxial cable.

The capacitance C of the line is related to the delay T of eq. (12.1) and to characteristic impedance Z_0 as

$$C = \frac{T}{Z_0}. \tag{12.2}$$

EXAMPLE 12.2. A coaxial cable has a characteristic impedance of $Z_0 = 50\ \Omega$ and a velocity of propagation that is $\frac{2}{3}$ of that of light. Find the capacitance for a cable with a length of 1 foot ($\cong 30$ cm).

First we compute the velocity of propagation, v: $v = \frac{2}{3}c = 20$ cm/ns. By substituting into eq. (12.1), the delay, T, is

$$T = \frac{l}{v} \cong \frac{30\ \text{cm}}{20\ \text{cm/ns}} = 1.5\ \text{ns}.$$

Now, by substituting into eq. (12.2), we get for the capacitance

$$C = \frac{T}{Z_0} = \frac{1.5\ \text{ns}}{50\ \Omega} = 30 \times 10^{-12}\ \text{s}/\Omega = 30\ \text{pF}.$$

Propagation and Reflections. If a voltage step V_{in} is applied to the input of a transmission line, initially a current step of V_{in}/Z_0 will result, since the ratio of the voltage and the current traveling along the line is Z_0, a constant. If V_{in} remains constant at the input, a signal with a voltage V_{in} and with a current V_{in}/Z_0 will propagate along the line. If there are no discontinuities along the line, the signal will travel undisturbed to the other end. If that end is terminated by a resistance having a value equal to Z_0, an equilibrium is instantly established and no reflection of signal occurs. If the termination is different from Z_0, a reflection will take place; the polarity and magnitude of such a reflection can be computed from transmission line theory. When the load or the source is not linear, however, such as in a TTL circuit, computations of the reflections can become complex. In what follows, we shall discuss a graphical method that can be applied both to linear and to nonlinear signal sources and loads.

The Graphical Method.[1,6,7] Consider the circuit of Figure 12.2, consisting of two logic inverters interconnected by a transmission line. Although a coaxial line is shown in the figure, considerations are valid for other types of lines as well. The output of the first inverter is characterized by two voltage-current characteristics: one for the low and one for the high output state. The input of the second inverter is characterized by a single voltage-current characteristic.* If the inverters are of the standard TTL

---

* If the second inverter is replaced by a NAND gate, its characteristics at an input may depend on the states of the other inputs.

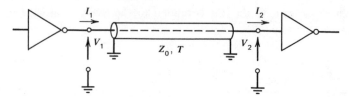

FIGURE 12.2. Two logic inverters interconnected by a coaxial trans-mission line with a characteristic impedance Z_0 and a delay T.

family (see Figure 5.13a), these characteristics can be approximated by straight line segments as shown in Figure 12.3. These may look strange at first sight; however, each segment can be correlated with Figure 5.13a by use of the voltage and current polarities specified in Figure 12.2. Thus, the three segments of V_2 vs. I_2 in Figure 12.3 approximate, from left to right, the forward characteristics of the protection diode D_1 and the active and the cutoff regions of Q_1 in Figure 5.13a. The V_1 vs. I_1 characteristics at

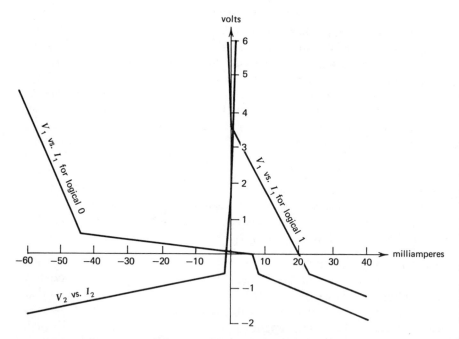

FIGURE 12.3. Straight line approximations of the input and output characteristics of the TTL logic inverter of Figure 5.13a, using the voltage and current polarities given in Figure 12.2.

the output of a TTL circuit can also be correlated with Figure 5.13*a* and Figure 5.13*c*.

The quiescent operating points for both high and low outputs in the circuit of Figure 12.2 are at $V_1 = V_2$ and $I_1 = I_2$. These are given by the intersection of the characteristics of Figure 12.4. For logical 0, the operating point is at $V_1 = V_2 \cong 0.08$ V and $I_1 = I_2 \cong -1.2$ mA; for logical 1, it is at $V_1 = V_2 \cong 3.5$ V and $I_1 = I_2 \cong 0.5$ mA (somewhat high for a standard TTL inverter).

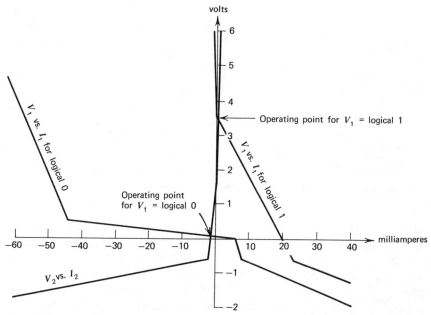

FIGURE 12.4.   Determination of the quiescent operating points for logical 0 (low) and logical 1 (high) output levels, using the characteristics given in Figure 12.3, and the voltage and current polarities given in Figure 12.2.

In addition to finding the operating points, the graphical method also enables us to determine the transitions between them when Z_0 and T of Figure 12.2 are given. The method is illustrated in Figure 12.5 for the transition from logical 0 to logical 1 with $Z_0 = 50$ Ω. The transition takes place from the initial operating point marked by ① to the final operating point marked by FINAL. The resulting transients of V_1 and V_2 shown in Figure 12.6 consist of an infinite number of steps and are obtained from Figure 12.5 as follows.

From the initial operating point ① we draw a straight line with a slope

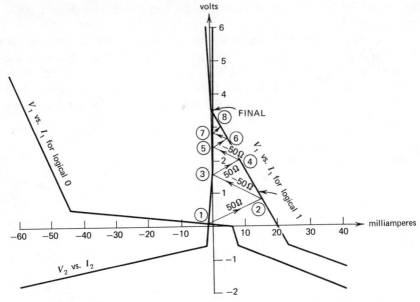

FIGURE 12.5. Determination of the transition from logical 0 (low) to logical 1 (high) in the circuit of Figure 12.2 when $Z_0 = 50$ Ω.

of $Z_0 = 50$ Ω, and intercept the "V_1 vs. I_1 for logical 1" line at point ②. This point represents the initial value of V_1 (and of I_1) following the change from state 0 to state 1, and is plotted as such in Figure 12.6a. This new voltage now propagates towards the right end of the line in Figure 12.2 and reaches it in a time T. The resulting V_2, however, also depends on the V_2 vs. I_2 characteristics and is found by drawing a straight line from point ② with a slope of -50 Ω, intercepting the V_2 vs. I_2 curve at point③.* This point will provide the voltage V_2 following time T, and is plotted as such in Figure 12.6b. The process is continued in Figure 12.5 and Figure 12.6 for three additional reflections; in practice it is stopped when V_1 and V_2 are sufficiently close to their final values.

The determination of the transient for the logical 1 to logical 0 transition is shown in Figure 12.7, the resulting V_1 and V_2 in Figure 12.8. In this case, the transient converges from the initial value of 3.5 V to the final value of 0.08 V in an oscillatory manner.

The graphical method can be also used when the source or the load, or both, are linear.

---

* The negative sign associated with this 50 Ω slope is a result of the current polarities assumed in Figure 12.2.

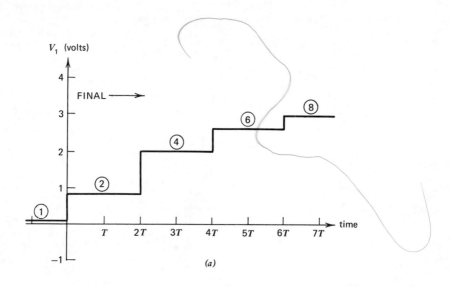

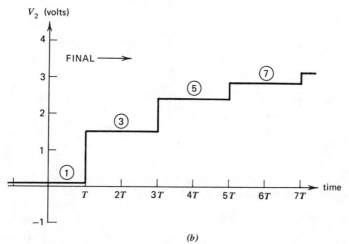

FIGURE 12.6. Transients for $Z_0 = 50 \ \Omega$ obtained from Figure 12.5. Circled numbers refer to those of Figure 12.5. (a) V_1 vs. time, (b) V_2 vs. time.

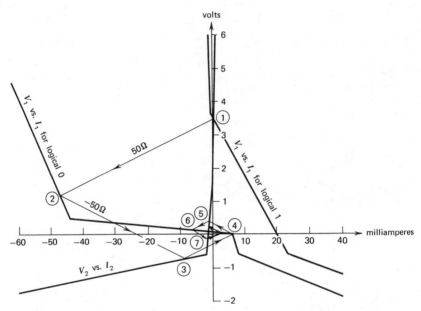

FIGURE 12.7. Determination of the transition from logical 1 (high) to logical 0 (low) in the circuit of Figure 12.2 when $Z_0 = 50 \ \Omega$.

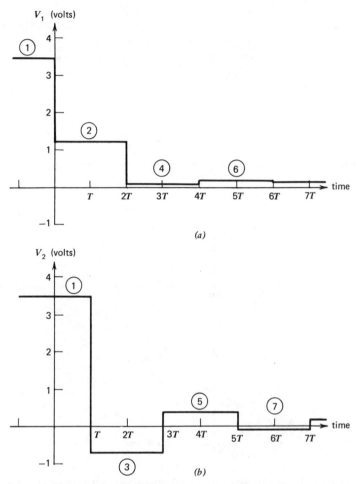

FIGURE 12.8. Transients for $Z_0 = 50 \ \Omega$ obtained from Figure 12.7. Circled numbers refer to those of Figure 12.7. (a) V_1 vs. time, (b) V_2 vs. time.

EXAMPLE 12.3. In Figure 12.9, two ECL circuits of Figure 5.18 are interconnected by a 50 Ω coaxial line. The output characteristics of the circuits are given by the V_1 vs. I_1 curves of Figure 12.10. The input currents of the circuits are neglected; hence the V_2 vs. I_2 characteristic in Figure 12.10 is given by 50 Ω to -2 V. The transition from the low to the high output is from point ① to point ② in Figure 12.10. Following the procedure of Figure 12.5, we first draw a straight line with a slope of $Z_0 = 50 \ \Omega$ from point ①; this now coincides with the V_2 vs. I_2 line. The

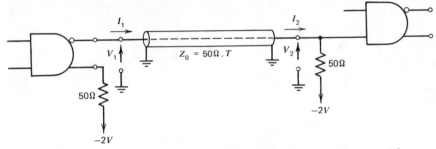

FIGURE 12.9. Emitter-coupled logic circuits interconnected by a coaxial line.

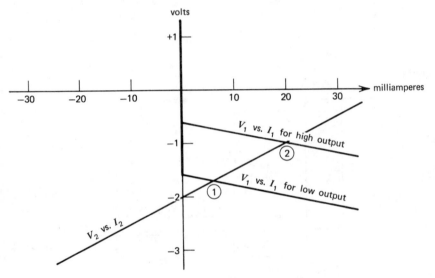

FIGURE 12.10. Determination of the transients in the circuit of Figure 12.9.

straight line intersects the "V_1 vs. I_1 for high output" line at point ②; this will be the output after the transition. Following the procedure outlined earlier, a straight line with a slope of $-50\ \Omega$ has to be drawn next, from point ② to intersect the V_2 vs. I_2 curve. However, this intersection is at point ② itself; and so will be all subsequent intersections. Thus, at zero time V_1 steps from -1.65 V to -1 V and remains there; also, at time T, V_2 steps from -1.65 V to -1 V and remains there. No reflections take place; this is due to the fact that the resistance seen by the line at the right is equal to its characteristic impedance.

12.3 Noise

In practice, signals carrying useful information are corrupted by undesired interfering "*noise*." Many effects are collected under this designation, some of these are really interferences from power supplies, ground currents, capacitive coupling from other circuits, etc.

Capacitive Coupling. Transitions in neighboring circuits can result in undesired noise when the capacitance between the circuits is not negligible. If the transition can be approximated by a step voltage with a magnitude of V_0 from a voltage source with zero source impedance (see Figure 12.11), the resulting noise voltage will be

$$V_{\text{noise}} = V_0 e^{-t/RC_c}, \tag{12.3}$$

where C_c is the coupling capacitance between the source and the pickup point, and R is the resistance seen at the pickup point.*

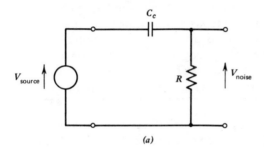

(a)

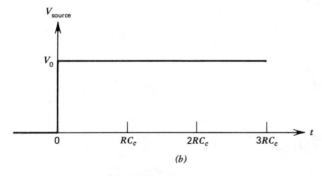

(b)

FIGURE 12.11. Pickup from capacitive coupling when the source is a step-voltage: (*a*) circuit, (*b*) source waveform. (*See next page for* (*c*))

---

* In what follows, the illustrations show positive-going transitions; the considerations are identical for negative-going transitions.

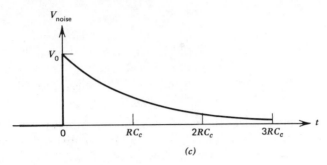

FIGURE 12.11. *(c)* **noise waveform.**

EXAMPLE 12.4. The wiring capacitance between two unrelated TTL circuits is 20 pF (rather high). The voltage swing is approximately $V_0 = 3$ V, the output resistance $R = 150\ \Omega$ (see Chapter 5, page 122). Thus the resulting noise transient for a step input can be approximated as $V_{noise} = 3$ V $e^{-t/(150\ \Omega\ 20\ pF)} = 3$ V $e^{-t/3\ ns}$. Such a transient is too fast to interfere with standard TTL circuits having transition times in the vicinity of 10 ns. If, however, the circuits are of the fast Schottky-TTL type with ≈ 3 ns transition times, V_{noise} may result in false signals. In such a case, an effort should be made to reduce the 20 pF capacitance between the wires, either by shortening or separating them, or by appropriate shielding.

In reality, V_{source} has a finite rise time. If this can be approximated by a straight line as shown in Figure 12.12, the resulting noise voltage can be written as

$$V_{noise} = V_0 \frac{RC_c}{T_s} (1 - e^{-t/RC_c}) \quad \text{for} \quad 0 \le t \le T_s, \tag{12.4}$$

and as

$$V_{noise} = V_0 \frac{RC_c}{T_s} (e^{T_s/RC_c} - 1)e^{-t/RC_c} \quad \text{for} \quad t > T_s. \tag{12.5}$$

Thus the magnitude of the noise transient V_{max} is obtained by equating $t = T_s$ in eq. (12.4) or eq. (12.5):

$$V_{max} = V_0 \frac{RC_c}{T_s} (1 - e^{-T_s/RC_c}). \tag{12.6}$$

It can also be shown that at a height of $V_{max}/2$ the width of the noise transient is T_s.

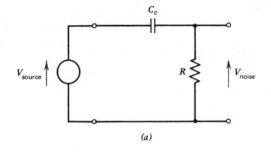

(a)

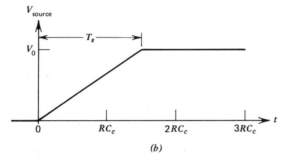

(b)

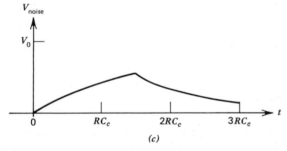

(c)

FIGURE 12.12. Pickup from capacitive coupling when the source has a finite risetime: (a) circuit, (b) source waveform, (c) noise waveform.

EXAMPLE 12.5. In the circuit described in Example 12.4, the source has a rise time of $T_s = 2.5$ ns. The resulting noise voltage will have a magnitude of

$$V_{max} = 3 \text{ V} \frac{150\ \Omega\ 20\ \text{pF}}{2.5\ \text{ns}} (1 - e^{-2.5\ \text{ns}/150\ \Omega\ 20\ \text{pF}}) = 2.3 \text{ V}.$$

Thus, the resulting noise transient has a width at half height of 2.5 ns and a magnitude of 2.3 V. Such a transient may result in a false signal in a Schottky-diode TTL circuit.

Series Noise Voltages.   Noise can also appear in series with the input, as shown in Figure 12.13a, where V_{noise} is introduced between the grounds of two logic gates. Series noise can also be introduced along the interconnection via coupling from a nearby magnetic field as shown in Figure 12.13b. We should try to minimize the magnitude of such noises by use of a suitable ground system and heavy ground connections, and by shielding against time-varying magnetic fields if their strength is significant. In some cases, however, we may have to live with the introduced series noise voltage.

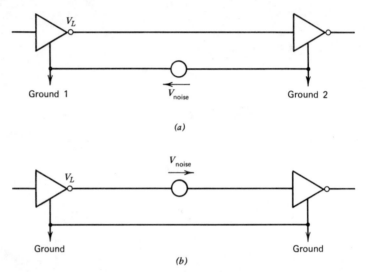

FIGURE 12.13.   Introduction of series noise by (a) ground noise, (b) induced noise.

The situation can be improved significantly if a complementary driver ("*differential line driver*") is used to drive two interconnecting lines that are observed at the receiving end by a comparator ("*line receiver*"). Such an arrangement is shown in Figure 12.14. Since there is a noise voltage, V_{noise}, present between Ground 1 and Ground 2, the (+) input of the comparator receiver with respect to Ground 2 has a voltage of $V_{noise} + V_L$, the (−) input a voltage of $V_{noise} + \overline{V_L}$. The comparator reacts to the *difference* of its two inputs; hence V_{noise} is cancelled.* This cancellation also applies for noise induced in both lines, and can be aided by equalizing them using for the interconnections a twisted pair of wires terminated by its characteristic impedance R_T (typically between 75 Ω and 200 Ω). Line drivers designed for such a purpose are capable of delivering the logic

---

* A description of comparator properties can be found in Section 10.7.

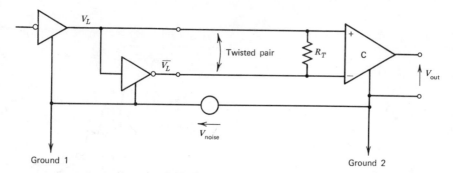

FIGURE 12.14. Complementary line driver and receiver comparator reducing the effects of common mode ground noise.

voltage into R_T, and include compensation for the delay of the inverter generating $\overline{V_L}$. Some line receivers incorporate attenuators at their inputs enabling them to accomodate noise voltages with magnitudes larger than the maximum common-mode input voltage of the comparator itself.

12.4 Schmitt Trigger Circuits

We have seen in the preceding section that in practical circuits undesired interference and noise are present in addition to the desired logic signals. Consider the logic inverter circuit of Fig. 12.15 that has a threshold of

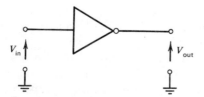

FIGURE 12.15. A logic inverter.

$V_{TH} = 2.5$ V. Under severe noise conditions its input signal may be as V_{in} shown in the upper half of Figure 12.16. If such a signal is entered at the input, the output signal may be as V_{out} shown in the lower half of Figure 12.16. We can see that as a result of the non-monotonic input, the threshold voltage V_{TH} is crossed several times both on the rising and the falling edges of the input signal. This results in several transitions of V_{out} when there should be only one transition on the leading edge and one on the trailing edge of the input signal. Also, because of the slow rise of input V_{in}, the transitions of V_{out} are slow, even if the transition times and the propagation delay of the

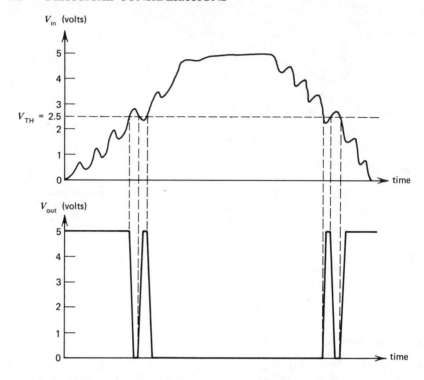

FIGURE 12.16. A noisy input signal V_{in} entered into the input of the logic inverter of Figure 12.15, and the resulting output signal V_{out}.

inverter itself are negligible. These disadvantages can be overcome by use of a *Schmitt trigger* circuit.

Consider the comparator of Figure 12.17. When $V_p - V_n < 0$, $V_{out} = 0$; when $V_p - V_n > 1$ mV, $V_{out} = 5$ V. By adding two resistors in a positive feedback loop, as shown in Figure 12.18, the properties of the comparator have been altered considerably. Such a circuit configuration is referred to as a Schmitt trigger, and its operation is described below.

In Figure 12.18, voltage V_n is identical to V_{in}; also voltage $V_p = 2.5$ V + $(V_{out} - 2.5$ V$)R_1/(R_1 + R_2) = 2$ V + $V_{out}/5$. Thus, $V_p = 2$ V when $V_{out} = 0$, $V_p = 3$ V when $V_{out} = 5$ V, and 2 V $\leq V_p \leq 3$ V at all times, since $0 \leq V_{out} \leq 5$ V. The resulting V_{out} vs. V_{in}, and V_p vs. V_{in} are shown in Figure 12.19a and b. V_p is always between 2 V and 3 V, hence for $V_{in} < 2$ V, V_{out} will be 5 V and V_p will be 3 V. If we now raise the input voltage $V_{in} = V_n$, the state of the circuit will not change until V_{in} gets within 1 mV of $V_p = 3$ V (point ⓐ). At this point, according to Figure 12.17, V_{out} starts decreasing, hence V_p decreases too, and due to the positive feedback the circuit

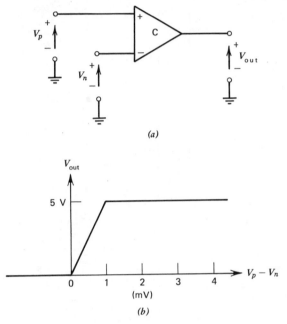

FIGURE 12.17. A comparator: (a) circuit, (b) output voltage V_{out} as a function of the difference $V_p - V_n$ of the input voltages.

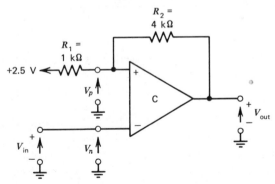

FIGURE 12.18. A Schmitt trigger circuit utilizing the comparator of Figure 12.17 and a positive feedback connection.

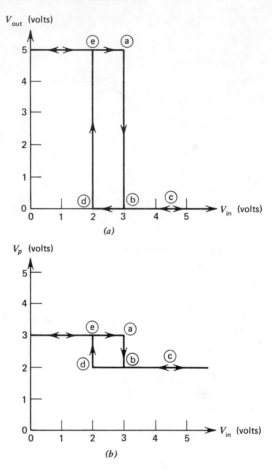

FIGURE 12.19. Characteristics of the Schmitt trigger circuit of Figure 12.18: (*a*) V_{out} as a function of V_{in}, (*b*) V_p as a function of V_{in}.

makes a sudden transition to $V_{out} = 0$ and $V_p = 2$ V (point ⓑ). The circuit will remain in this state as V_{in} is raised further (point ⓒ). However, as V_{in} is lowered down to 2 V (point ⓓ), $V_p - V_n$ becomes positive and V_{out} starts to rise from 0 V (see Figure 12.17). This will start raising V_p, and a sudden transition to $V_{out} = 5$ V and $V_p = 3$ V takes place (point ⓔ). The circuit will now remain in this state as V_{in} is lowered further.

For input voltages of 2 V $\leq V_{in} \leq$ 2.999 V, the Schmitt trigger circuit of Figure 12.18 can be in one of two states; thus there is a backlash, or *hysteresis*, of approximately 1 V in this circuit. There are two consequences of this:

(*i*) the transitions between the two states are fast even when the input signal changes slowly; and (*ii*) as long as the magnitude of the noise on the input signal is less than the hysteresis, there will be only one transition on the rising edge and one transition on the falling edge of the input signal as shown in Figure 12.20.

Schmitt trigger circuits can also be built by use of active elements other than the comparator.

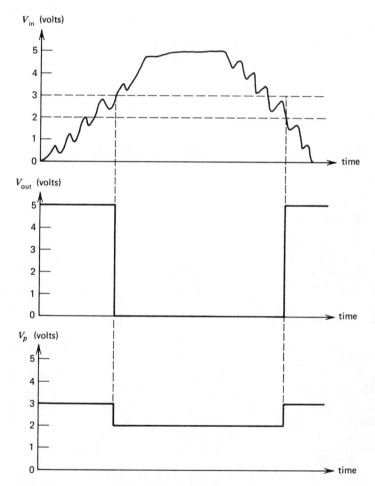

FIGURE 12.20. Operation of the Schmitt trigger of Figure 12.19 for a noisy input V_{in} that is identical to that of Figure 12.16.

EXAMPLE 12.6.   Figure 12.21 shows the emitter-coupled logic circuit of Figure 5.18 used as a Schmitt trigger. It can be shown (see Problem 12.11) that when driven from a low-impedance source at the input (V_{in}), the thresholds of the circuit are at approximately -1.15 V and -1.3 V; hence its hysteresis is 0.15 V.

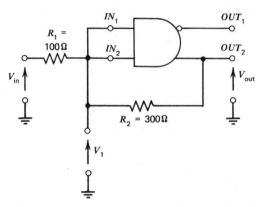

FIGURE 12.21.   The emitter-coupled logic circuit of Figure 5.18 used as a Schmitt trigger.

12.5   Single-Shots

It is usually straightforward to generate a time interval with a specified duration in a digital system that has a clock. By counting the number of clock pulses in a counter, any interval that is an integer multiple of the clock period can be generated. While this is always possible and results in accurate time intervals, the accuracy is not always required, and the resulting circuitry may be unnecessarily complex.

EXAMPLE 12.7.   At the completion of its 5-minute task, a digital system alerts the operator by ringing a bell for two seconds. The system has a 1 MHz clock; thus the two-second interval can be generated by a count-by-2×10^6 circuit. Such a circuit would consist of 21 flip-flops and additional control circuitry. The realization of the circuit by MSI is straightforward, although perhaps unnecessarily complex.

A simple circuit for generating time intervals with moderate accuracies (0.1% to 10%) is the *single-shot*, also known as *one-shot*, *univibrator*, or *monostable multivibrator*. Instead of a clock, these circuits rely for timing on passive elements, such as resistors, capacitors, and delay lines. Their active components may consist of comparators, logic gates, or circuitry designed specifically for the purpose.

A simple single-shot utilizing a comparator as an active element is shown in Figure 12.22; the propagation delay of the comparator is neglected in what follows. Initially $V_{in} = 0$, $V_{out} = 5$ V, and $V_p = 2.5$ V. For an input

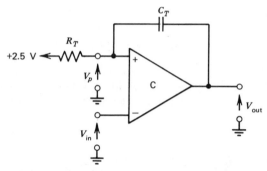

FIGURE 12.22. A single-shot circuit utilizing a comparator as an active element, R_T and C_T as timing elements.

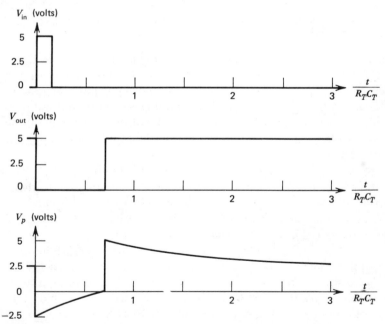

FIGURE 12.23. Signals in the single-shot circuit of Figure 12.22 for a single narrow input pulse.

pulse V_{in} with a height of 5 V, at $t = 0$ V_{out} changes to 0 and V_p to -2.5 V. For $t \geq 0$, the voltage at the $(+)$ input of the comparator is

$$V_p = -2.5 \text{ V} + 5 \text{ V} \, (1 - e^{-t/R_T C_T}). \tag{12.7}$$

This waveform, shown in Figure 12.23, will reach $V_p = 0$ at a time

$$t = R_T C_T \ln 2 \approx 0.7 \, R_T C_T. \tag{12.8}$$

If by this time V_{in} has returned to 0, as is the case in Figure 12.23, V_{out} changes to 5 V, and V_p to 5 V. Thus, a time interval of $T = R_T C_T \ln 2$ has been generated at the output.

EXAMPLE 12.8.   The single-shot circuit of Figure 12.22 with V_{in} as given above is used for generating a time interval $T = 2$ seconds. Resistor R_T is chosen as 100 kΩ, and the value of the capacitor can be obtained from eq. (12.8) as

$$C_T = \frac{T}{0.7 \, R_T} = \frac{2}{0.7 \times 10^5} = 28.6 \, \mu\text{F}.$$

The accuracy of the timing interval T is limited principally by the accuracies of C_T, R_T, the 2.5 V and 5 V supply voltages, and the input impedance of the comparator at its input (see Problem 12.12).

Simple as it is, the single-shot circuit of Figure 12.22 has several short-comings. One of these is the long time it takes V_p to recover to its original value of 2.5 V (see Figure 12.23). If a second input pulse arrives during this recovery period, the width of the resulting second output pulse can be significantly shorter than $R_T C_T \ln 2$. Also, input pulses arriving during the generated interval are ignored, which may be desirable in some applications, undesirable in others.

A retriggerable single-shot circuit utilizing a Schmitt trigger is shown in Figure 12.24. Initially input *IN* is at 0 V, the output of the logic inverter at 5 V, the $(-)$ input of the comparator at 5 V, and output *OUT* at 0 V. An input pulse will force the voltage at the $(-)$ input of the comparator to 0.7 V, from where it will rise towards 5 V as 0.7 V $+ 4.3$ V $(1 - e^{-t/R_T C_T})$. This signal will reach the 3 V threshold of the comparator (Schmitt trigger) at a time $t = T_1 = R_T C_T \ln (4.3 \text{ V}/2 \text{ V}) \approx 0.8 \, R_T C_T$ after the *trailing edge* of the input pulse. If a second pulse arrives during interval T_1, the output will be extended to last until a time 0.8 $R_T C_T$ *after* the trailing edge of the second input pulse. If the second pulse arrives after interval T_1, a second identical output pulse will be generated. This retriggerable feature of the one-shot is useful in many applications where the operation of a circuit or a process has to be inhibited for a specified period.

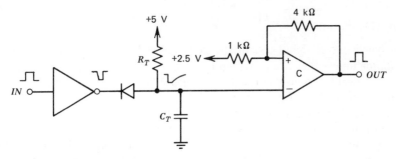

FIGURE 12.24.   A retriggerable single-shot circuit.

The preparation of a timing diagram for this circuit is left to the reader as an exercise.

In many applications it is desirable to ignore input pulses that arrive within the duration of the output. In other cases it may be required that the output pulse width be independent of the width and the arrival time of the input pulse. The single-shot circuit shown in Figure 12.25 will generate a single output pulse of width $T \approx 0.8\ R_T C_T$ at the *leading edge* of the input pulse. It will, however, ignore input pulses that arrive during T. A similar circuit utilizing a delay line as a timing element is shown in Figure 12.26; the duration, T, of the output pulse is determined by the length of the delay line.

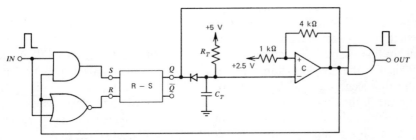

FIGURE 12.25.   A retriggerable single-shot circuit with trigger lockout.

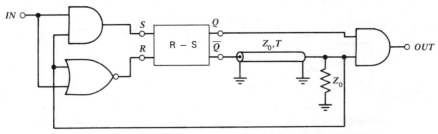

FIGURE 12.26.   A retriggerable delay-line single-shot circuit with trigger lockout.

12.6   Oscillators

In a synchronous digital system, the clock pulses have to be generated by an *oscillator* unless they are provided by an external source. One of the most accurate oscillators in use is the *crystal oscillator*, built around a crystal with a well defined frequency of mechanical oscillation that is coupled with an electrical signal. The accuracy and stability of the frequency exceed those of mechanical clocks, and are sufficient in most digital systems.

When the accuracy of a crystal oscillator is not required, simpler *RC oscillators* can be built. One of these, based on the circuit of Figure 12.25, is shown in Figure 12.27; its accuracy is a function of R_T, C_T, the $+5$ V

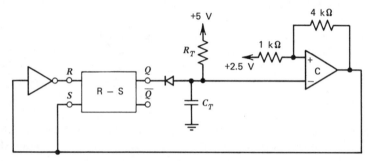

FIGURE 12.27.   An *RC* oscillator circuit.

and $+2.5$ V supplies, and the delays of the active components. Somewhat better accuracy can be attained by *delay line oscillators*. A gated delay line oscillator is shown Figure 12.28. When gate input *GATE* is at logical 1, the flip-flop cannot be set and its quiescent state is $Q = 0$; thus the quiescent state of D is also 0. When the gate input is switched to logical 0, the flip-flop is set and the circuit immediately starts generating a square wave with a frequency that is $1/2T$, provided the propagation delays of the flip-flop and the NOR gate are negligible. If gate input *GATE* is switched to logical 1, the circuit will complete the cycle under

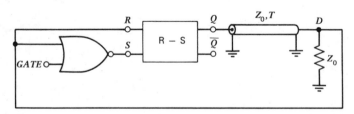

FIGURE 12.28.   A gated delay-line oscillator circuit that oscillates when gate input *GATE* is at logical 0.

way and return to $Q = 0$. Thus, unlike a crystal oscillator, the circuit of Figure 12.28 is easy to start in synchronism with the gate signal, and it always provides a full output cycle when stopped.

PROBLEMS

12.1 The capacitance of a solid dielectric coaxial cable is specified as 15 pF/foot; the velocity of propagation is given as $\frac{2}{3}$ of that of light. Find the characteristic impedance Z_0.

12.2 Find the characteristic impedance of a transmission line consisting of two 1-cm wide parallel strips (Figure 12.1c) separated by 0.2 cm, if the relative dielectric constant is $\varepsilon_r = 2$.

12.3 Find the characteristic impedance of a coaxial cable with $D_1 = 0.003$ inch (# 40 wire), $D_2 = 1$ inch, and $\varepsilon_r = 1$.

12.4 Sketch currents I_1 and I_2 as functions of time for the transitions of Figure 12.5 and Figure 12.7.

12.5 Find the transients corresponding to those of Figure 12.5 through Figure 12.8 in the circuit of Figure 12.2 when $Z_0 = 100 \ \Omega$.

12.6 Find the transients corresponding to those of Figure 12.5 through Figure 12.8 if a 200 Ω resistor is added in Figure 12.2 between $+5$ V and the input of the inverter on the right.

12.7 Figure 12.29 shows two methods of connecting the inputs of three ECL circuits to the output of an ECL circuit. Discuss the advantages and disadvantages of both approaches. What is the purpose of resistor R_B in Figure 12.29a, and what is its minimum value for

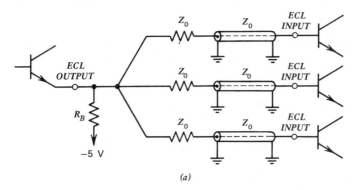

(a)

FIGURE 12.29. (*See next page for* (*b*))

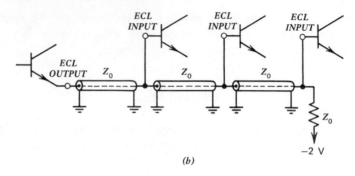

(b)

FIGURE 12.29. (b)

correct operation of the circuit if logic levels are -0.8 V and -1.6 V? Assume infinite input impedances to the ECL circuits and $Z_0 = 100\ \Omega$.

12.8 The output of an ECL circuit has a rise time $T_s = 1.5$ ns and is coupled to the input of another ECL circuit via a series stray capacitance C_c into a resistance of 50 Ω to ground. Find the value of C_c that results in a noise transient at the input that has a magnitude of 0.15 V.

12.9 Sketch the waveforms of the circuit of Figure 12.14 if V_L and $\overline{V_L}$ are 10-μs wide TTL signals (0 V to 3 V) and V_{noise} is a 1 MHz sine wave with a peak-to-peak magnitude of 20 V. Neglect the delays of the inverters and the comparator.

12.10 Sketch the hysteresis curves as in Figure 12.19 if the resistors in the circuit of Figure 12.18 are (a) $R_1 = 10\ \Omega, R_2 = 9990\ \Omega$; (b) $R_1 = 10\ \Omega$, $R_2 = 49990\ \Omega$; (c) $R_1 = 10\ \Omega, R_2 = 99990\ \Omega$. Find the hysteresis for each case.

12.11 Find the thresholds of the Schmitt trigger circuit of Example 12.6, if the logic levels of the ECL gate are -0.8 V and -1.6 V, and if V_{out} can be approximated by

$$V_{\text{out}} \cong -0.7\ \text{V} - \frac{0.9\ \text{V}}{1 + e^{(V_1 + 1.2\ \text{V})/30\ \text{mV}}}.$$

(Hint: To find the thresholds, find the incremental amplification $A \equiv dV_{\text{out}}/dV_1$, and set $AR_1/(R_1 + R_2)$ equal to unity.)

12.12 In Example 12.8, the $(+)$ input of the comparator can be represented as a 1 MΩ resistor to ground. Find the resulting time interval T.

12.13 Sketch a timing diagram for the retriggerable one-shot circuit of Figure 12.24.

12.14 Sketch a timing diagram that shows the signals in the one-shot circuit of Figure 12.25 if the width of the input pulse is (a) 0.5 $R_T C_T$ and (b) 2 $R_T C_T$.

12.15 Sketch a timing diagram that shows the signals in the one-shot circuit of Figure 12.26 if the width of the input pulse is (a) 0.5 T and (b) 2 T, where T is the delay of the delay line.

12.16 Sketch a timing diagram for the RC oscillator of Figure 12.27.

12.17 Sketch a timing diagram that shows all signals in the gated delay line oscillator circuit of Figure 12.28. Demonstrate that the circuit always completes the cycle of oscillation.

12.18 Design a gated delay-line oscillator using two ECL logic gates and a delay line.

12.19 Show that for $H/D \gg 1$, Z_0 of Figure 12.1a can be approximated as $Z_0 \approx 60 \ln 4H/D$.

APPENDIX

Summary of Semiconductor Device Characteristics

Semiconductors. Semiconductors are solid or liquid materials which have an electrical conductivity intermediate between that of *conductors* and that of *insulators*. In conductors, free electrons are available to move under the influence of an electric field; most metals are good conductors, having resistivities in the range of $\approx 10^{-6}$ to 10^{-4} ohm-cm. In contrast, the resistivity of insulators (e.g., diamond, quartz, or mica) is typically in the range of 10^9 to 10^{20} ohm-cm at room temperature. Silicon, and to a lesser extent germanium, are the semiconductors used principally in the manufacture of diodes, transistors, and integrated circuits. The resistivities of *intrinsic* (pure) silicon and germanium are 2.3×10^5 ohm-cm and 45 ohm-cm, respectively.

The properties of semiconductors are determined by crystalline imperfections that include lattice defects and by *chemical impurities*. The process of adding these impurities to change the properties of semiconductors is called *doping*, and the added material is referred to as *dopant*. A semiconductor to which an impurity has been added is referred to as an *extrinsic* semiconductor. Germanium and silicon both are in column IV of the periodic table; that is, each atom has four valence (outer-shell) electrons which may create bonds with adjacent atoms. A small portion of the periodic table is shown in Figure A1, in which the semiconductor elements appear in column IV, while columns III and V contain doping elements. Semiconductor materials useful for transistors and diodes are one of two types: *n*-type (extrinsic) semiconductors, in which there exists an excess of negative charge carriers, i.e., *electrons*; and *p*-type materials, in which the excess charge carriers are positive.

n-type semiconductors are obtained through adding impurities from column V of the periodic table. For example, if we add phosphorus (which has five valence electrons) to a silicon crystal, some of the silicon atoms will be replaced by phosphorus atoms which will be held in the crystalline structure through *covalent bonds* with four surrounding silicon atoms. The fifth phosphorus electron may be considered free at $\gtrsim 100$ °K. Thus an *n*-type semiconductor has been achieved through adding an impurity of a *donor* atom, e.g., a phosphorus atom in the above example (see Figure A2).

453

III	IV	V
3 valence electrons	4 valence electrons	5 valence electrons
Boron ^{5}B		
Aluminum ^{13}Al	Silicon ^{14}Si	Phosphorus ^{15}P
Gallium ^{31}Ga	Germanium ^{32}Ge	Arsenic ^{33}As
Indium ^{49}In		Antimony ^{51}Sb

FIGURE A1.   Small portion of the periodic table showing semiconductors in column IV and doping elements in columns III and V.

If an impurity from column III of the periodic table, such as boron, is added to the semiconductor material, a deficiency of an electron is created in the covalent bonds, since column III elements have three valence electrons only. This deficiency, referred to as a *hole*, attracts electrons readily from adjacent atoms. A semiconductor material with column III impurities is called a *p*-type (extrinsic) semiconductor, and the impurity element is called an *acceptor*. A hole can be cancelled by an electron only at the cost of creating a hole in another atom and for each random jump of a hole, there is a reciprocal electron motion. Electrons and holes are thus charge carriers moving in opposite directions under the influence of an electric field. Each impurity atom may be viewed as a charge embedded in the semiconductor crystal but the net charge of the crystal is neutral because the " free " (excess) electrons or holes are cancelled by the charges of their corresponding nuclei.

Semiconductor Diodes.   Intrinsic or extrinsic semiconductors conduct current equally in any direction; however, a junction of a *p*-type and an *n*-type material exhibits preference in one direction due to a charge carrier density distribution that is asymmetrical. A semiconductor junction diode consists of a *p-n* junction, see Figure A3; characteristics of diodes are shown in Figure A4. For example, when a silicon diode is forward biased, i.e., the *p* region is positive with respect to the *n* region, appreciable current commences to flow only for $V_{\text{forward}} \geq 0.6$ volts. A potential barrier at the junction allows only a few charge carriers to cross the boundary between the *p* and *n* doped materials when the forward bias is less than $\simeq 0.6$ volts in silicon diodes. (The corresponding value for germanium diodes is $\simeq 0.3$ volts.)

The current, *I*, as a function of the applied voltage, can be approximated by

$$I = I_0(e^{V/V_T} - 1) \qquad (A1)$$

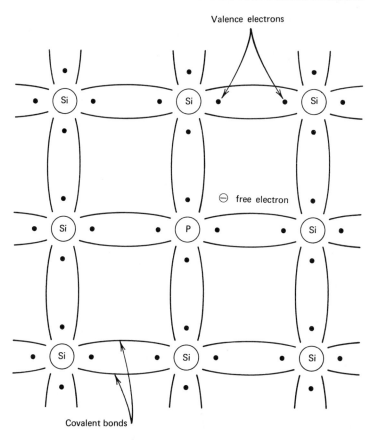

FIGURE A2. Covalent bonds in a silicon crystal with a phosphorus impurity atom. The 5-valent phosphorus donor atom leaves one free electron producing an *n*-type semiconductor.

in which I_0 = saturation current, a constant independent of the junction voltage, and

$$V_T = \frac{kT}{q} \approx 26 \text{ mV at room temperature}$$

where k = Boltzmann's constant = 1.38×10^{-23} joules/°K, T = Temperature °K($= °C + 273$), q = electronic charge = 1.6×10^{-19} coulomb. In practice the *I-V* characteristic of a diode does not exactly follow eq. (A1). First, for silicon diodes V_T has to be replaced by αV_T, where $1 \leq \alpha \leq 2$. Second, the ohmic series resistance, R_s, in the semiconductor diode becomes dominant at high values of current. The voltage drop, *V*, for a given forward current can be derived from eq. (A1), taking

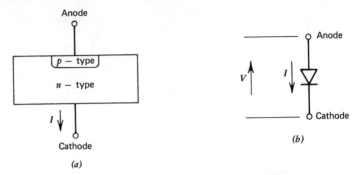

FIGURE A3. Junction diode: (*a*) structure, (*b*) symbol.

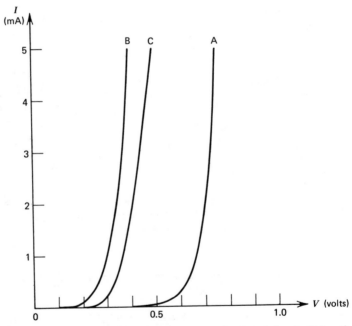

FIGURE A4. Diode voltage versus current characteristics. A: Si junction diode, B: Ge junction diode, C: Schottky diode.

into consideration R_s and assuming $e^{V/V_T} \gg 1$ (e.g., $V \geq 600$ mV, $V_T \approx 26$ mV to 50 mV)

$$V = V_T \ln \frac{I}{I_0} + IR_s.$$ (A2)

The *dynamic* or *incremental* resistance, r_f, of a forward biased diode may be found by differentiating eq. (A2)

$$r_f = \frac{dV}{dI} = \frac{V_T}{I} + R_s.$$ (A3)

The temperature coefficient of V is negative for silicon and germanium

$$\frac{dV}{dT} \cong -2.5 \text{ mV/°C}.$$ (A4)

Schottky Diodes. These devices, also known as *hot-carrier* or *Schottky barrier* diodes, consist of a metal-semiconductor contact in which the current flows by means of majority carriers. In a *p*-type material the majority carrier is a hole; in an *n*-type material it is an electron. Most Schottky diodes are made of *n*-type silicon in contact with gold or aluminum. When the metal is biased positively with respect to the semiconductor, a current flows across the junction due to transport of the electrons from the *n*-type material. This current is a majority carrier current and thus differs from an ordinary *p-n* junction in which the current is due to minority carriers, i.e., holes drifting into the *n* region and electrons into the *p*-region under the influence of an electric field. Hence, when the field across an ordinary junction is reverse biased, minority carriers are stored for some time, and conduction of current (in the direction of the field) continues until the minority carriers have been swept out or have recombined. In contrast, no minority carrier conduction exists in a Schottky diode: electrons which have crossed the junction to enter the metal are not distinguishable from the free metal electrons. When the diode is reverse biased, no phenomenon of minority carrier storage exists, and conduction ceases in a very short time after reverse bias has been applied. This *reverse recovery time* of a Schottky diode may be as short as 10 ps, and the device is therefore widely used in high-speed switching circuits.

When forward biased, the electrons injected into the metal have a higher energy than the free metal electrons and are referred to as "hot"; hence the name *hot-carrier* diode. The characteristics of a Schottky diode are shown by curve C of Figure A4, and a symbol of the device is shown in Figure A5.

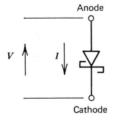

FIGURE A5. Symbol for a Schottky diode.

Zener Diodes. When the electric field in a *reverse* biased diode reaches sufficient magnitude to extract electrons out of their covalent bonds, a sudden increase in current results as shown in Figure A6. The additional electron-hole pairs thus created contribute greatly to the current flow, and the diode is referred to as operating in the *Zener* region. The dynamic impedance of the diode $\dfrac{\Delta V_{Zener}}{\Delta I_{Zener}}$ is low and is dependent on the current. The value at which breakdown sets in can be controlled by the doping density, and Zener diodes are available in the range of two volts to several hundred volts.

When the applied voltage is less than required for Zener breakdown, the diode current is $\approx -I_0$, from eq. (A1). Raising the voltage *beyond* the Zener breakdown

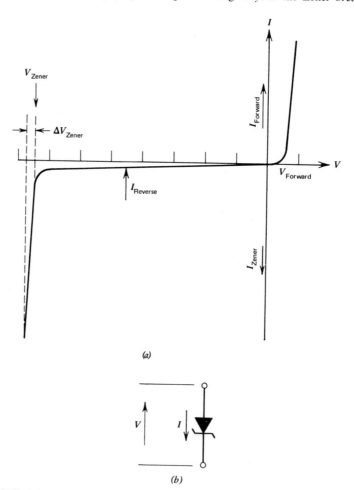

(a)

(b)

FIGURE A6.   Zener diode: (a) voltage versus current characteristics, (b) symbol.

brings the diode into the *avalanche* region in which carriers have sufficient energy to knock out electrons from covalent bonds. When forward biased, the device operates as an ordinary diode, as shown in Figure A6a.

Bipolar Transistors.   A *bipolar*, or *junction*, transistor consists of three semiconductor regions separated by two junctions. An idealized structure of an NPN transistor which consists of two *n* regions and one *p* region is shown in Figure A7a, a more realistic device cross-section in Figure A7b, and a symbol in Figure A7c. Either or both junctions may

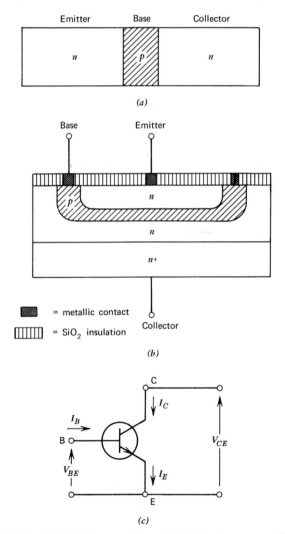

FIGURE A7.   NPN bipolar transistor: (*a*) **idealized structure,** (*b*) **device cross-section,** (*c*) **symbol.**

	Emitter-base junction forward biased	Emitter-base junction reverse biased
Collector-base junction forward biased	Saturated region	Reverse active region
Collector-base junction reverse biased	Forward active region	Cutoff region

FIGURE A8.   Operating regions of a bipolar transistor.

be operated in the forward or reverse bias conditions, resulting in four operating regions, shown in Figure A8. Typical transfer characteristics of an NPN transistor are shown in Figure A9.

In the *forward active region* of operation the emitter junction is forward biased (i.e., $V_{BE} > 0$ in Figure A7c.) and the collector junction is reverse biased, i.e., $V_{CE} > V_{BE}$. An emitter current I_E flows under these conditions that consists mainly of electrons injected from the emitter. The base is made very narrow so that most of the injected minority carriers diffuse across the base into the collector region, and only a small fraction combines with holes in the base region. The ratio of the collector and emitter currents is denoted α and typically varies from 0.9 to 0.999:

$$I_C = \alpha I_E. \tag{A5}$$

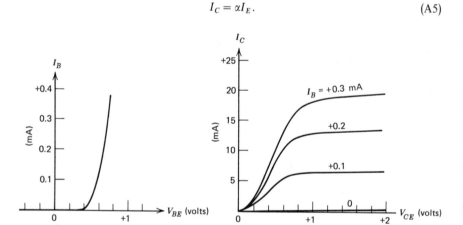

(a)   *(b)*

FIGURE A9.   Transfer characteristics of an NPN silicon bipolar transistor: *(a)* base current, I_B, *versus* base-emitter voltage, V_{BE}, *(b)* collector current, I_C, versus collector-to-emitter voltage, V_{CE}; the variable parameter is I_B.

We have previously stated that I_E constitutes the sum of collector and base currents. Thus

$$I_C = \alpha I_E = \alpha(I_C + I_B). \tag{A6}$$

When a current is applied to the base of a transistor, the resulting collector current is, from eq. (A6),

$$I_C = \frac{\alpha}{1 - \alpha} I_B = \beta I_B. \tag{A7}$$

Equation (A7) defines the *current gain* factor β, which typically varies in bipolar transistors from 10 to 1000. In the above discussion we have neglected a current across the collector junction, I_{CBO}, that is due to thermally generated minority carriers. This *cut-off* current is usually negligibly small in silicon transistors at room temperature. The forward active region of operation is used in emitter-coupled logic circuits.

Another region of operation, the *cut-off region* is used by emitter-coupled logic and by TTL logic; the latter also operates in the *saturation region*. In the cutoff region $V_{BE} \gtrsim 0$ *and* $V_{CE} - V_{BE} > 0$. In the saturation region both junctions are forward biased. For example, referring to Figure A9a, $I_B = 0.2$ mA for $V_{BE} = +0.7$ volts. If concurrently $V_{CE} = +0.1$ volts, the transistor will be in the saturation region since $V_{CE} - V_{BE} = -0.6$ volts < 0; and only few minority carriers diffuse across the base into the collector region because the electric field aiding this diffusion process is weak. When the transistor is cut off by a voltage at the base, these minority carriers keep the transistor in conduction for some time until they have recombined with majority carriers or have been swept out of the base region. The response of the saturated transistor is therefore relatively slow. For this reason saturated transistors are used principally in the slow and medium speed TTL logic. In a Schottky-TTL circuit, the saturation effect is greatly diminished by a Schottky diode that is formed in parallel to the base-collector junction. The forward bias of this junction is limited to about $+0.4$ volts, resulting in a significant reduction of minority carrier storage and hence improved switching speeds.

The *reverse active region* (see Figure A8) is characterized by low β values and very low leakage currents. It is rarely used in digital circuits.

The above discussion, based on an NPN transistor, is equally applicable to a PNP transistor if p is replaced by n, n is replaced by p, and if voltage and current polarities are reversed. A PNP transistor symbol is shown in Figure A10.

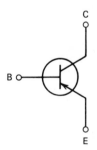

FIGURE A10.   PNP transistor symbol.

Insulated-Gate Field-Effect Transistors, IGFET's. The IGFET, also known as MOSFET (metal-oxide FET), consumes little power; therefore, it is widely used in large-scale integrated circuits, LSI's. A cross-sectional view of a *p*-channel MOSFET is shown in Figure A11. (Symbols are shown in Figure 11.1) The basic substrate consists of *n*-type silicon into which *p*-type *source* and *drain* regions have been diffused. An insulating layer of SiO_2 is deposited between the drain and the source, as shown. This insulator, which is about 1000 Å thick, is called *gate oxide*. The *gate* consists of a metal, usually aluminum, that is deposited on top of the gate oxide. Other methods of gate fabrication have been developed, and various chemical compounds have been used as gate insulators to achieve desirable device characteristics.

Operation of a MOSFET will be outlined referring to a *p*-channel device. The input impedance at the gate terminal is 10^9 to 10^{12} ohms in parallel with a capacitance <1 pF. The source-to-drain current, I_D, is in the range of 10^{-9} to 10^{-12} amperes at zero gate voltage, since source and drain are isolated by the high resistivity substrate. A negative voltage applied to the gate induces a corresponding positive charge in the substrate under the gate. When the gate voltage has reached sufficient magnitude, the *channel* between source and drain will become *p*-type, i.e., *inverted*, as shown in Figure A12, and will conduct current if a voltage is applied between the source and drain electrodes.

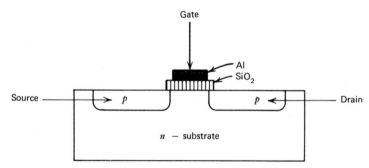

FIGURE A11.   Cross-sectional view of a *p*-channel MOSFET.

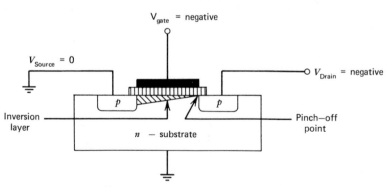

FIGURE A12.   Inversion layer in a *p*-channel MOSFET.

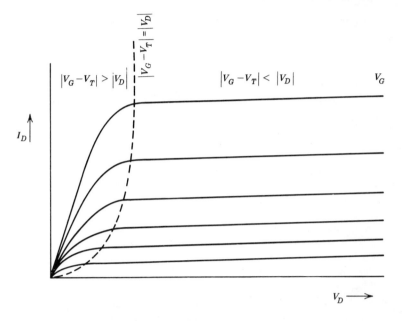

FIGURE A13.   Typical transfer characteristics of a *p*-channel MOSFET.

Typical transfer characteristics of a MOSFET, shown in Figure A13, may be divided into two regions of operation: on the left side of the dashed line the drain current, I_D, rises rapidly with the source-to-drain voltage, V_D, and the source-to-gate potential, V_G:

$$I_D = \beta \left[(V_G - V_T)V_D - \frac{V_D{}^2}{2} \right] \tag{A8}$$

in which V_T is the *threshold voltage* and β is a factor including a geometry dependence

$$\beta \propto W/L, \tag{A9}$$

where W = width of channel, L = length of channel. This is the region of *triode operation* given by

$$|V_G - V_T| > |V_D|. \tag{A10}$$

To the right of the dashed curve of Figure A13 the MOSFET operates in its *saturation region*. As V_D is increased beyond the triode region, the electric field across the gate oxide becomes insufficient to maintain the mobile charges at the drain end of the channel. The channel is *pinched-off* and no further increase in current is possible. The drain current in the saturation region is

$$I_D = \beta \frac{(V_G - V_T)^2}{2}, \tag{A11}$$

and the device *transconductance*, g_m, may be derived from eq. (A11):

$$g_m = \frac{\partial I_D}{\partial V_G}\bigg|_{V_D} = \beta(V_G - V_T). \tag{A12}$$

From eq. (A8) and eq. (A11) we note that, in both regions of operation, I_D is modulated by the gate-to-source voltage and is also dependent on the device geometry through the coefficient β [see eq. (A9)].

The above discussion of p-channel MOSFET's may be applied to n-channel devices if all p's and n's are interchanged and the polarities of voltages and currents are reversed. In general an n-channel MOSFET has better high frequency characteristics since the mobility of electrons is higher by a factor of ~ 2.5 than that of holes. p-channel MOSFET's are mostly *enhancement* mode devices, i.e., they conduct when V_G is negatively biased and are off when no bias is applied to V_G. In n-type MOSFET's the drain is positive with respect to the source and both these electrodes are n-type semiconductors. Many n-channel MOSFET's operate in the *depletion* mode; when zero bias is applied to the gate, the MOSFET conducts its maximum current I_{DSS} while a negative bias at V_G reduces the current and, if of sufficient magnitude, cuts it off to a negligible value. p-channel depletion mode and n-channel enhancement mode MOSFET's are also in use.

The above discussion dwelt on metal-oxide field-effect transistors. Other FET's have also been used in the construction of LSI circuits. One such device is the *MESFET* in which two ohmic contacts form the source and drain, and a rectifying Schottky contact forms the gate. The MESFET is characterized by relatively high switching speeds and low power dissipation.

The substrate in the field-effect transistors described above is a p-type or an n-type semiconductor. The capacitances can be significantly reduced, however, by using an insulator for the substrate, such as saphire which yields *silicon-on-saphire MOS*, or *SOSMOS*, devices. These devices can provide reduced capacitances, resulting in increased operating speeds and lower dynamic power dissipations as compared to the regular MOS technology.

References

CHAPTER 1

1 Shannon, C. E., "A Symbolic Analysis of Relay and Switching Circuits," *Trans. AIEE*, **57**, 713–723 (1938).

CHAPTER 3

1 Boole, G., *The Mathematical Analysis of Logic*, Blackwell, England (reproduced 1948).

2 Shannon, C. E., "A Symbolic Analysis of Relay and Switching Circuits," *Trans. AIEE*, **57**, 713–723 (1938).

3 Karnaugh, M., "The Map Method for Synthesis of Combinational Logic Circuits," *AIEE Comm. Electronics*, **9**, 593–598 (1953).

4 McCluskey, E. J., "Minimization of Boolean Functions," *Bell Syst. Tech. J.*, **35**, No. 5, 1417–1444 (Nov. 1956).

5 McCluskey, E. J., *Introduction to the Theory of Switching Circuits*, McGraw-Hill Book Co., New York, 1965.

CHAPTER 4

1 Maley, G. A., *Manual of Logic Circuits*, Prentice-Hall Inc., Englewood Cliffs, N.J., 1970.

2 Yau, S. S., and Tang, C. K., "Universal Logic Circuits and Their Modular Realizations," *Spring J.C.C., AFIPS Conf. Proc.*, **32**, 297–305 (1968).

3 Anderson, J. L., "Multiplexers Double as Logic Circuits," *Electronics*, **42**, No. 22, 100–105, (Oct. 27, 1969).

4 Richards, R. K., *Digital Design*, Wiley-Interscience, New York, 1971.

5 Hill, F. J., and Peterson, G. R., *Switching Theory and Logical Design*, John Wiley & Sons, New York, 1968.

CHAPTER 5

1 Millman, J., and Taub, H., *Pulse, Digital, and Switching Waveforms*, McGraw-Hill Book Co., New York, 1965.

2   Barna, A., *High-Speed Pulse Circuits*, Wiley-Interscience, New York, 1970.
3   RCA, *COS/MOS Integrated Circuits Manual*, Technical Series, CMS–270, RCA Solid State Division, Sommerville, N.J., 1971.

CHAPTER 7

1   Hill, F. J., and Peterson, G. R., *Introduction to Switching Theory and Logical Design*, John Wiley & Sons, New York, 1968.
2   Unger, S. H., *Asynchronous Sequential Switching Circuits*, Wiley-Interscience, New York, 1969.
3   Gill, A., *Introduction to the Theory of Finite-State Machines*, McGraw-Hill Book Co., New York, 1962.

CHAPTER 8

1   Richards, R. K., *Digital Design*, Wiley-Interscience, New York, 1971.
2   Chu, Y., *Digital Computer Design Fundamentals*, McGraw-Hill Book Co., New York, 1962.
3   Hill, F. J., and Peterson, G. R., *Digital Systems: Hardware Organization and Design*, John Wiley & Sons, New York, 1973.

CHAPTER 9

1   Susskind, A. K., *Notes on Analog-Digital Conversion Techniques*, M.I.T. Press, Cambridge, Mass., 1957.
2   Rhyne, V. T., "Serial Binary-to-Decimal and Decimal-to-Binary Conversion," *IEEE Trans.*, **C-19**, No. 9, 808–812 (Sept. 1970).

CHAPTER 10

1   Barna, A., *Operational Amplifiers*, Wiley-Interscience, New York, 1971.
2   Graeme, J. G., Tobey, G. E., and Huelsman L. P., *Operational Amplifiers*, McGraw-Hill Book Co., New York, 1971.
3   See Reference 1, page 20.
4   Hoeschele, D. F. Jr., *Analog-to-Digital/Digital-to-Analog Conversion Techniques*, John Wiley & Sons, New York, 1968.
5   Sifferlen, T. P. and Vartanian, V., *Digital Electronics with Engineering Applications*, Prentice-Hall Inc., Englewood Cliffs, N.J., 1970.
6   Schmid, H., *Electronic Analog/Digital Conversions*, Van Nostrand, New York, 1970.
7   Sheingold, D. H., Editor, *Analog-Digital Conversion Handbook*, Analog Devices Inc., Norwood, Mass., 1972.

CHAPTER 11

1 Gosling, W., Townsend, W. G., and Watson, J., *Field-Effect Electronics*, Wiley-Interscience, New York, 1971.

2 Carr, W. N., and Mize, J. P., *MOS/LSI Design and Application*, McGraw-Hill, New York, 1972.

3 Eimbinder, J., Editor, *Semiconductor Memories*, Wiley-Interscience, New York, 1971.

4 *ibid.*, chapter 8.

5 Frohman-Bentchkowsky, D., "A Fully Decoded 2048 Bit Electrically Programmable FAMOS Read-Only Memory," *IEEE J. Solid-State Circuits*, **SC-6**, No. 5, 301–306 (Oct. 1971).

6 Hemel, A., *Electronics*, **43**, No. 10, 104–111 (May 11, 1970).

7 Percival, R., *Electronic Design*, **12**, No. 12, 66–71 (June 8, 1972).

8 Davidow, W. H., "General Purpose Multicontrollers. Part II: Design and Applications," *Computer Design*, **11**, No. 8, 69–76 (Aug. 1972).

9 Wilkes, M. V., "The Growth of Interest in Microprogramming," *Computing Surveys*, **1**, No. 3, 139 (Sept. 1969).

10 See reference 2, chapter 9.

11 Special Issue on Semiconductor Memories and Digital Circuits, *IEEE J. Solid-State Circuits*, **SC-6,** No. 5 (Oct. 1971); **SC-7**, No. 5 (Oct. 1972).

12 Hodges, D. A., *Semiconductor Memories*, IEEE Press, New York, 1972.

CHAPTER 12

1 Morris, R. L., and Miller, J. R., *Designing with TTL Integrated Circuits*, McGraw-Hill Book Co., New York, 1971.

2 Blood, W. R. Jr., *MECL Systems Design Handbook*, Motorola Semiconductor Products Inc., 2nd Edition, Phoenix, Ariz. 1971.

3 RCA, *COS/MOS Integrated Circuit Manual*, Technical Series, CMS–270, RCA Solid State Division, Sommerville, N.J., 1971.

4 Matick, R. E., *Transmission Lines for Digital and Communication Networks*, McGraw-Hill Book Co., New York, 1969.

5 Barna, A., *High-Speed Pulse Circuits*, Wiley-Interscience, New York, 1970.

6 Singleton, R. S., *Electronics*, **41**, No. 22, 93–99 (Oct. 28, 1968).

7 DeFalco, J. A., "Reflection and crosstalk in logic circuit interconnections," *IEEE Spectrum*, **7**, No. 7, 44–50 (July 1970).

APPENDIX

1 Grove, A. S., *Physics and Technology of Semiconductor Devices*, John Wiley & Sons, New York, 1967.

2 Gosling, W., Townsend, W. G., and Watson, J., *Field-Effect Electronics*, Wiley-Interscience, New York, 1971.

Answers to
Selected Problems

CHAPTER 2

2.1 100110111

2.2 665_8

2.4 45445_6

2.5 (a) $.42334_6$ (b) $.739_{10}$

2.6 (a) 1B   (b) 1D3   (c) 1FF

2.10 (a) 515   (b) 898   (c) 1791   (d) 4095
(e) .123046   (f) .013671   (g) .410156   (h) .001949

2.14 (c) 714179_{10}

2.17 31.2314_8

CHAPTER 3

(*Answers to problems involving simplification may not be unique*).

3.1 $A \cdot B + A \cdot C + B \cdot C$

3.4 $f = X + Y$

3.5 $\bar{X} + Y \cdot [(\bar{Z} + W)(V + \bar{S})]$

3.6 $X + Y$

3.8 On solution is $\bar{A}B + \bar{B}C + A\bar{C}$

3.10 $f = 1$

3.15 (a) $f(V,W,X,Y,Z) = \sum m(3,5,6,9,10,12,17,18,20,24)$

3.17 (a) Complement the sum of those minterms that do not appear in the standard S-of-P function.

3.19 The least number of planes to supply all plants is two. $f(A,B,C,D,E,F) = AE + BC + AC$. Of these three possibilities AC is the least expensive.

3.20 (a) $f = \bar{A}\bar{D}$ (both, S-of-P and P-of-S)
   (b) $f = \bar{A}\bar{C} + B\bar{C}$, (S-of-P); $f = (\bar{A} + B)\bar{C}$
   (c) $f = \bar{A}\bar{B}\bar{D}\bar{E} + AB\bar{D}E + A\bar{B}D$;
   (d) $f = \bar{B}\bar{C}EF + \bar{B}CDE + AB\bar{C}\bar{E} + AB\bar{D}\bar{E} + \bar{C}DEF + \bar{A}DEF + \bar{A}BEF$
       $+ AB\bar{C}\bar{D}F + \bar{A}BCDE + A\bar{B}CDE + BC\bar{D}EF$,
       $f = (A + C + E)(B + C + E)(A + D + E)(A + C + F)(B + C + F)$
       $\quad \cdot (B + \bar{C} + D)(A + B + E + \bar{F})(C + \bar{D} + \bar{E} + F)(\bar{B} + \bar{C} + \bar{E} + F)$
       $\quad \cdot (\bar{A} + \bar{B} + \bar{C} + \bar{D})(\bar{A} + B + \bar{C} + \bar{E} + \bar{F})(\bar{A} + \bar{B} + C + D + \bar{E} + \bar{F})$.

CHAPTER 4

4.3 Minterm $\overline{A\bar{B}CD\bar{E}}$; maxterm $(\bar{A} + B + \bar{C} + D + E)$

4.6 (a) $XY(W + \bar{Z}) + \bar{X}\bar{Y}$

4.8 $f_1 = g + W\bar{Z} + Y\bar{Z}$, $f_2 = g + \bar{W}X$, $f_3 = g + \bar{W}\bar{Z} + \bar{Y}\bar{Z}$; where $g = \bar{W}\bar{X}\bar{Y}Z + WXY\bar{Z}$.

4.12 (a) Five NAND gates   (b) Sum: three gate delays. Carry: two gate delays.

4.16 One possible solution is $f_1(A,B,C,D) = h + \bar{A}\bar{B}CD + ABD$, $f_2(A,B,C,D) = h + \bar{A}B\bar{C} + ABD$, $f_3(A,B,C,D) = h + A\bar{B}\bar{C}$, where $h = A\bar{B}C\bar{D}$.

CHAPTER 5

5.9 140 ns, ≈ 0

5.20 130

CHAPTER 7

7.27 80 ns

CHAPTER 8

8.5 60 ns

8.9 80 ns

8.10 50 ns

8.11 200 ns

8.12 80 ns

CHAPTER 9

9.13 $B_3 = E_3 E_1 E_0 + E_3 E_2$
$B_2 = \bar{E}_2 \bar{E}_0 + E_2 E_1 E_0 + E_3 \bar{E}_1 E_0$
$B_1 = E_1 \oplus E_0$
$B_0 = \bar{E}_0$

9.17 $Y_3 = X_3 + X_2 X_1 + X_2 X_0$
$Y_2 = X_3 + X_2 X_1 + X_2 \bar{X}_0$
$Y_1 = X_3 + \bar{X}_2 X_1 + X_2 \bar{X}_1 \bar{X}_0$
$Y_0 = X_0$

9.19 $Y_4 = B_3 + B_2 B_0 + B_1 B_0$
$Y_3 = B_3 B_0 + B_2 \bar{B}_1 \bar{B}_0$
$Y_2 = B_1 B_0 + \bar{B}_2 B_1 + B_3 \bar{B}_0$
$Y_1 = B_3 \bar{B}_0 + \bar{B}_3 \bar{B}_2 B_0 + B_2 B_1 \bar{B}_0$
$Y_0 = 0$

9.21 (a) $J_3 = Q_2 \bar{Q}_1 \bar{Q}_0$
$K_3 = \bar{Q}_0$
$J_2 = Q_1 + Q_3 \bar{Q}_0$
$K_2 = \bar{Q}_1 + \bar{Q}_0 = \overline{Q_1 Q_0}$
$J_1 = \bar{Q}_3 \bar{Q}_2 Q_0 + Q_3 \bar{Q}_0$
$K_1 = \bar{Q}_2 \bar{Q}_0 + Q_2 Q_0$
$J_0 = C_i, K_0 = \bar{C}_i$
(b) $C_{i+1} = Q_3 + Q_2 Q_1 + Q_2 Q_0$

9.24 $J_3 = Q_2 \bar{Q}_0 + Q_2 C_i$
$K_3 = \bar{Q}_2 Q_0 + \bar{Q}_0 \bar{C}_i$
$J_2 = Q_1 (\bar{Q}_3 + Q_0 + \bar{C}_i)$
$K_2 = \bar{Q}_1 (Q_3 + \bar{Q}_0 + C_i)$
$J_1 = Q_2 \bar{Q}_0 + Q_3 Q_0 + Q_2 \bar{C}_i + \bar{Q}_2 C_i$
$K_1 = \bar{Q}_2 C_i + \bar{Q}_3 C_i + Q_0 \bar{C}_i$
$J_0 = Q_2 Q_1 + Q_2 \bar{C}_i + Q_3 C_i + \bar{Q}_3 Q_1 \bar{C}_i$
$K_0 = \bar{Q}_2 \bar{Q}_1 + \bar{Q}_2 \bar{C}_i + \bar{Q}_3 \bar{C}_i + Q_3 \bar{Q}_1 C_i$
$C_{i+1} = Q_3$

9.26 (a) Inputs Y_3 to X_0, $\bar{Y}_2$ to X_3, $\bar{Y}_1$ to X_1, Y_0 to X_2
(b) Inputs Y_3 to X_3, $\bar{Y}_2$ to X_1, Y_1 to X_2, Y_0 to X_0
(c) Inputs Y_3 to X_3, $\bar{Y}_2$ to X_1, $\bar{Y}_1$ to X_2, Y_1 to X_0
(d) Inputs Y_3 to X_3, Y_2 to X_2, Y_1 to X_1, Y_0 to X_0

Outputs: MSI output	Connect to display numeral			
	XS3-Gray	Gray	2-4-2-1	XS3-BCD
A_0	1	7	6	7
A_1	8	6	7	8
A_2	4	0	2	9
A_3	5	1	3	0
A_4	2	4	4	1
A_5	7	5	5	2
A_6	3	3	0	3
A_7	6	2	1	4
A_8	0	8	8	5
A_9	9	9	9	6

Note: Alternative solutions are possible in each case for input and output connections

9.28 $V_3 = X_3 \overline{X}_2 \overline{X}_1$
$V_2 = \overline{X}_3 X_2 + X_2 X_1$
$V_1 = \overline{X}_3 X_1 + X_3 X_2 \overline{X}_1$
$V_0 = X_0$

CHAPTER 10

10.5 $2R$

10.12 (b) $V_{out} = \dfrac{R_L}{2^n(R_L + R_0)} \displaystyle\sum_{i=0}^{n} V_i 2^{i-1}$

10.17 16 MHz

CHAPTER 12

12.1 100 Ω

12.3 350 Ω

12.12 2.05 seconds

Index